D0397668

Fitting the Human

INTRODUCTION TO ERGONOMICS

Sixth Edition

KARL H. E. KROEMER

Fitting the Human

INTRODUCTION TO ERGONOMICS

Sixth Edition

CRC Press
Taylor & Francis Group
Boca Raton   London   New York

CRC Press is an imprint of the
Taylor & Francis Group, an **informa** business

The front cover illustration uses **Marzenna Beata Musial's** poster *Perfect Match* in *Placat Ergonomia 2000* (ISBN 83-87354-24-4) from the periodic Occupational Safety Poster Competition organized by the Central Institute for Labour Protection National Research Institute — as part of the dissemination of the results of the National Programme "Adaptation of Working Conditions in Poland to European Union Standards" 4th edition "Ergonomics". On 27 April 2007, Bozena Popowicz, Head of the Publishing Department CIOP-PIB, Centrainy Instytut Ochrony Pracy, Warshaw, Poland, sent the permission to use the poster design.

CRC Press
Taylor & Francis Group
6000 Broken Sound Parkway NW, Suite 300
Boca Raton, FL 33487-2742

© 2008 by Taylor & Francis Group, LLC
CRC Press is an imprint of Taylor & Francis Group, an Informa business

No claim to original U.S. Government works

Printed in the United States of America on acid-free paper
Version Date: 20110715

International Standard Book Number: 978-1-4200-5539-9 (Hardback)

Library of Congress Cataloging-in-Publication Data

Kroemer, K. H. E., 1933-
   [Physiologische Arbeitsgestaltung. English]
   Fitting the human : introduction to ergonomics / Karl H.E. Kroemer. -- 6th ed.
     p. ; cm.
   Rev. ed. of: Fitting the task to the human / by K.H.E. Kroemer and E. Grandjean. 5th ed. c1997.
   Includes bibliographical references and index.
   ISBN 978-1-4200-5539-9 (hardback : alk. paper)
   1. Work--Physiological aspects. 2. Human mechanics. 3. Human engineering. 4. Industrial hygiene. I. Kroemer, K. H. E., 1933- Fitting the task to the human. II. Title.
   [DNLM: 1. Human Engineering. 2. Occupational Health. WA 400 K93p 2008]

QP309.G7313 2008
612'.042--dc22
                                         2008020478

Visit the Taylor & Francis Web site at
http://www.taylorandfrancis.com

and the CRC Press Web site at
http://www.crcpress.com

For Hiltrud

Half a century of caring, helping, love.

Contents

Foreword: the sixth edition   xiii
About the author   xvii
The first page   xix

SECTION **I**

The human body

1 Body sizes   **3**

    1.1   Our Earth's populations   3
    1.2   Measurements   3
    1.3   Designing to fit the body   24

2 Mobility   **31**

    2.1   Work in motion   31
    2.2   Body joints   33
    2.3   Designing for mobility   43
    2.4   Workspaces   45

3 Muscular work   **53**

    3.1   Physiological principles   53

3.2	Dynamic and static efforts, strength tests	58
3.3	Fatigue and recovery	61
3.4	Use of muscle strength data in design	63

4 Body strength and load handling   **69**

4.1	Static and dynamic strength exertions	70
4.2	Maximal or minimal strength exertion	71
4.3	Hand strength	72
4.4	Whole body strength	75
4.5	Designing for easy load handling	89

SECTION II
The human mind

5 How we see   **103**

5.1	Our eyes	103
5.2	Seeing the environment	106
5.3	Dim and bright viewing conditions	113

6 How we hear   **121**

6.1	Our ears	121
6.2	Hearing sounds	123
6.3	Noise and its effects	129

7 How we sense objects and energy   **141**

7.1	Sensing body movement	141
7.2	The feel of objects, energy, and pain	143
7.4	Designing for tactile perception	146

8 How we experience indoor and outside climates   **153**

8.1	Human thermoregulation	153

8.2   Climate factors: temperatures, humidity,
       drafts   158
8.3   Our personal climate   161
8.4   Working in hot environments   165
8.5   Working in cold environments   167
8.6   Designing comfortable climates   170

SECTION III
Body and mind working together

9 Mental activities   177

9.1   The brain–nerve network   177
9.2   Taking up information   185
9.3   Making decisions   189
9.4   Actions and reactions   192

10 Hard physical work   199

10.1   Physiological principles   199
10.2   Energy consumption   200
10.3   Heart rate as a measure of work demands   205
10.4   Limits of human labor capacity   207
10.5   Designing heavy human work   210

11 Light and moderate work   215

11.1   Physiological and psychological principles   216
11.2   Tiredness, boredom, and alertness at work   219
11.3   Suitable postures at work   222
11.4   Accurate, fast, skillful activities   226

12 Workload and stress   235

12.1   Stress at work   235
12.2   Mental workload   238
12.3   Physical workload   241
12.4   Underload and overload   241
12.5   Psychophysical assessments of workloads   242

SECTION IV
Organizing and managing work

13 Working with others   **251**

13.1   Getting along with others   252
13.2   Motivation and behavior   254
13.3   Task demands, job rewards   258

14 The organization and you   **263**

14.1   Structure, policies, procedures, culture   263
14.2   Design for motivation and performance   268
14.3   A good place to work   271

15 Working hours   **277**

15.1   Circadian body rhythms   277
15.2   Rest pauses, time off work   281
15.3   Daily and weekly working time   283

16 Night and shift work   **291**

16.1   Need for sleep to be awake   291
16.2   Performance and health considerations   295
16.3   Organizing shift work   297

SECTION V
Human engineering

17 Designing the home   **307**

17.1   Designing for mother and child   307
17.2   Designing for impaired and elderly
         persons   308
17.3   Access, walkways, steps, and stairs   309
17.4   Kitchen   310

17.5   Bedroom, bath, and toilet   311
17.6   Light, heating, and cooling   312
17.7   Home office   313

18 Office design   317

18.1   Office rooms   317
18.2   Office climate   321
18.3   Office lighting   325
18.4   Office furniture and equipment   329
18.5   Ergonomic design of the office workstation   335
18.6   Designing the home office   342

19 Computer design   349

19.1   Sholes' "type-writing machine" with its
        QWERTY keyboard   350
19.2   From typewriter to computer keyboard   352
19.3   Human factor considerations for
        keyboarding   353
19.4   Input-related anthromechanical issues   358
19.5   Possible design solutions   359
19.6   Design alternatives for keyboards   363
19.7   Designing for new syntax and diction   364
19.8   Designing "smart" software   365
19.9   Designs combining solutions   365

20 Workplace design   369

20.1   Sizing the workplace to fit the body   369
20.2   On the feet or sitting down?   372
20.3   Manipulating, reaching, grasping   376
20.4   Handling loads   376
20.5   Displays and controls   382

21 Making work pleasant and efficient   391

21.1   Using our skills and interests; getting along
        with others at work   391
21.2   Setting up our own work, workplace, and
        work environment   396

The last page 405

References 407

Index 421

Foreword: the sixth edition

This text, *Fitting the Human*, is the sixth edition in a series of books. The first printing appeared in 1963, in German, under the title *Physiologische Arbeitsgestaltung*. The author was Etienne Grandjean. He edited his text in 1969 and 1979. Ott Verlag Thun published all three editions. In 1980, Taylor & Francis put out the third edition, translated into English, with the title *Fitting the Task to the Man*, and printed a revised version in 1988 as the fourth edition.

Etienne Grandjean was a pioneering researcher, a person who had great rapport with industry, an outstanding professor at the renowned ETH in Switzerland, and an author who could write easily and plainly about topics of great complexity. His death in 1991 was a great loss to the ergonomic community and his friends.

The fifth edition, which I co-authored, appeared in 1997 under the title *Fitting the Task to the Human*. Richard Steele of Taylor & Francis and I had decided to keep this text close to the 1988 edition, as our homage to Etienne Grandjean.

This sixth edition follows the motto of the previous editions: "Sound science; easy to read, easy to understand, easy to apply." This edition's aim is still the same: "human engineering" workplaces, tools, machinery, computers, lighting, shift work, work demands, the environment, offices, vehicles, the home—and everything else that we can design to fit the human. We want to avoid needless stress and effort; instead, we should be able to perform our tasks safely and efficiently, even with enjoyment.

Since the 1960s, work requirements and conditions have been changing substantially. Hard physical labor in agriculture, forestry, industry, and commerce still exists but has lost much of its importance in many economies. The changes are not only in the nature of jobs, but also in the common mindset. In Europe and North America, during the 1950s, the clichéd work for a woman outside her family and household was to be a secretary or teacher. The title of the third edition in 1980, *Fitting the Task to the Man*, reflects this frame of mind. Just a few decades later, women had attained equal status to men in most professions. Women now were carpenters and truckdrivers, firefighters and soldiers, engineers and computer programmers, airline pilots and cosmonauts, and personnel supervisors and heads of corporations—hence the change in the title of the 1997 fifth edition to *Fitting the Task to the Human*.

Mechanization, robots, electronics, and new work systems are continually changing the demands on the human: not only in task content, but in utensils and equipment, and in the physical and social environments as well. Therefore, the title of this sixth edition is simply *Fitting the Human*.

This brief title points out the two basic tasks in ergonomics: one is matching individuals with work requirements by personnel selection and training; this is mainly in the domain of industrial psychology, an area of study that widely overlaps with ergonomics. However, the basic ergonomic task is to adapt work requirements and procedures, equipment and tools, physical and social conditions, and working hours and shift arrangements to accommodate the human. This is the fundamental and most successful approach—the topic of this book—make the details and the overall work system fit the human.

The sixth edition is entirely new. I regrouped major topics from the fifth edition, added and updated, and reworked and designed new figures, tables, and graphs to convey the latest information in contemporary style and appearance.

In this new edition, I strive to continue Etienne Grandjean's direct, lucid, down-to-earth style of providing essential knowledge about design that fits the human body and mind. I hope the book will appeal to students and professionals in design, engineering, safety, and management, and to everybody else interested in making work safe, efficient, satisfying, and even enjoyable.

About references in this edition

Basic human anatomical, biomechanical, physiological, and psychological characteristics did not change in recent years. They are well described in the previous editions, which contain exhaustive listings of pre-1997 publications. I don't want to repeat those listings of references here, so I suggest that the interested reader check the earlier books. In this sixth edition, I refer mostly to recent publications, along with a few selected classic ones.

Traditional practice is to support statements in the text by listing the names of the authors and their co-authors, who wrote previously on that topic. That wordy custom disrupts the flow of reading, especially when there are strings of names and dates in parentheses. To avoid that problem, I am following Jacques Barzun's practice (as in his 2000 book, *From Dawn to Decadence: 500 years of Western Cultural Life: 1500 to the Present*. New York: HarperCollins): he simply places a small marker, °, in the text when references or explanations are in order. These appear in a separate "Notes" section, which the reader may skip or consult.

Do you want more information?

Like its predecessors, this edition is an introductory text. Other books are—and will be—on the international market that contain in-depth discussions of ergonomics issues. If you want to stay with me as author/co-author, you may check out our other books: the 1997 *Engineering Physiology* (3rd Edn, Wiley), our 2001 *Office Ergonomics* (CRC Press), and the 2003 *Ergonomics: How to Design for Ease and Efficiency* (2nd Edn, Prentice-Hall/Pearson), or my 2006 *Extra-Ordinary Ergonomics: How to Accommodate Small and Big Persons, The Disabled and Elderly, Expectant Mothers and Children* (CRC Press).

Thank you!

I wish to thank Taylor & Francis/CRC's outstanding professionals for their dedication and many years of support. I particularly enjoyed working with Richard Steele in the 1990s and recently with Cindy Carelli and James Miller.

Sincere thanks to you, the reader, for your interest. I will be glad to receive your comments and suggestions for future improvements in *Fitting the Human*!

Karl Kroemer
kroemer@vt.edu

About the author

Karl H. E. Kroemer
Mechanical Engineer, Technical University Hannover, Germany:
Vor-Diplom (BS, 1957), Dipl.-Ing. (MS, 1960), Dr.-Ing. (PhD,
1965)

Dr. Kroemer was a research engineer at the Max Planck
Institute for Work Physiology, Germany, from 1960 to 1966.
Then he worked seven years as a research industrial engineer
in the Human Engineering Division of the Aerospace Medical
Research Laboratory, Wright-Patterson Air Force Base, Ohio.
In 1973, he was appointed director and professor, Ergonomics
and Occupational Medicine Divisions, in the Federal Institute
of Occupational Safety and Accident Research, Dortmund,
Germany. In 1976, he became professor of ergonomics and
industrial engineering at Wayne State University, Michigan.
In 1981, he began to serve as professor in Virginia Tech's
Department of Industrial and Systems Engineering and direc-
tor of the Industrial Ergonomics Laboratory in Blacksburg,
Virginia; and gained emeritus status in 1998.

He was a member of the Committee on Human Factors of the
National Research Council, National Academy of Sciences. The
Human Factors and Ergonomics Society and the Ergonomics
Society elected him Fellow. He worked as UN-ILO Expert on
Ergonomics in Bucharest, Romania, and in Mumbai, India.

Dr. Kroemer has authored and co-authored over 200 publications. His books include:

1975 *Engineering Anthropometry Methods*

1984 *Guide to the Ergonomics of Computer Workstations*

1997 *Engineering Physiology,* 3rd edition

1997 *Ergonomic Design of Material Handling Systems*

1997 *Fitting the Task to the Human*, translated into Serbo-Croatian and Portuguese

2001 *Office Ergonomics*, translated into Korean

2003 *Ergonomics*, amended 2nd edition

2006 *"Extra-Ordinary" Ergonomics*

2008 *Fitting the Human*, 6th edition

Contact Dr. Kroemer at kroemer@vt.edu.

The first page

Humans are similar—yet, they differ.

We are all alike. Our bodies follow the same design. We all have bones and muscles, the same circulatory and metabolic functions, and a similar brain layout.

Therefore, we can formulate principles and guidelines that apply to all people. Recognizing similarities in people is one trademark of ergonomics.

Each person is unique. Individuals are different; nobody is exactly like anybody else. We all differ from each other in appearances, in strengths, skills, interests, and expectations.

Therefore, work tasks and work systems must accommodate people of diverse sizes, powers, and wants. Recognizing differences among people is the other hallmark of ergonomics.

The ergonomic goal is to make work safe, efficient, satisfying, and even enjoyable.

Ergonomics is the application of scientific principles, methods, and data drawn from a variety of disciplines to the development of engineered systems in which people play significant roles. That design relies on the knowledge of human characteristics relevant to the system and on the understanding that the system exists to serve humans. User-oriented design acknowledges human variability. (See *http://iea.org* for more details.)

The human body

All humans have remarkably similar bodies: the head, two arms, and two legs attached to the trunk. Everybody uses the same set of bones and muscles that move the body. Respiration, circulation, nervous control, and other basic functions are uniform. Yet, the bodies of people who live in unique climates, consume special diets, or perform distinct activities over long periods of time evolve to adapt to the particular conditions. Thus, within the general similarity of humankind, there are differences in the bodies among groups of people. Furthermore, individuals differ from each other in size and capabilities.

We need to recognize and measure existing differences among people so that we can devise clothing and equipment, tasks, and procedures to fit their special needs. The following chapters provide related information.

- Chapter 1 Body sizes
- Chapter 2 Mobility
- Chapter 3 Muscular work
- Chapter 4 Body strength and load handling

CHAPTER ONE

Body sizes

Body sizes differ   We all experience changes in body size: quick growth during childhood, followed by a period of fairly constant dimensions during adulthood for about 20 to 40 or more years until final variations come with senescence. As a rule, men grow to be taller as adults than women do. During any of the age periods, some persons are smaller or bigger than their peers are and body proportions can differ widely among individuals.

1.1 Our Earth's populations

Height and weight   The most common way to describe the bodies of populations, and of individuals as well, is by stature (standing height) and body weight. Table 1.1 contains 1990 estimates of adult stature for 20 different regions of the earth. Such estimates give an overall impression about the differences in adults' body sizes, but designers need more precise anthropometric information. Variances, even though small compared to our similarities, must be put into exact numbers so that we can set up clothing tariffs, sizes of eyeglass frames, heights of workbenches, and cockpit dimensions to fit the various user groups.

1.2 Measurements

Anthropometrics around the globe   We can estimate body bulk, or we may ask persons about their body size, and if they know, they can tell us; however, with the current concerns about obesity in many countries, some respondents are likely to add a bit to their height and slightly reduce their weight. Therefore, actual measurements are in order.

Table 1.1   Estimates of average stature (in mm) in 20 regions of the Earth

	Females	Males
North America	1650	1790
Latin America		
Indian population	1480	1620
European and Negroid populations	1620	1750
Europe		
North	1690	1810
Central	1660	1770
East	1630	1750
Southeast	1620	1730
France	1630	1770
Iberia	1600	1710
Africa		
North	1610	1690
West	1530	1670
Southeast	1570	1680
Near East	1610	1710
India		
North	1540	1670
South	1500	1620
Asia		
North	1590	1690
Southeast	1530	1630
South China	1520	1660
Japan	1590	1720
Australia		
European extraction	1670	1770

Source: Adapted from Juergens, H.W., Aune, I.A., and Pieper, U. (1990). *International Data on Anthropometry, Occupational Safety and Health Series #65*, Geneva: International Labour Office.

Table 1.2 lists heights and weights of persons measured around the globe. Checking the entries in the table, we notice that often the measured samples are small, and that many regions and populations on earth are not represented. Yet, this table shows the real magnitudes of differences in average body sizes and weights among groups of people.

Table 1.2 Measured heights and weights: averages (and standard deviations) of adults in many populations

	Sample size	Stature (mm)	Weight (kg)
Algeria:			
Females (Mebarki and Davies, 1990)	666	1576 (56)	61 (13)
Brazil:			
Males (Ferreira, 1988; cited by Al-Haboubi, 1991)	3076	1699 (67)	nda[a]
Cameroon:			
Urban females	1156	1620	64
Urban males (35–44 years old) (Kamadjeu, Edwards, and Atanga, 2006)	558	1721	75
China:			
Females (Taiwan)	about 600	1572 (53)	52 (7)
Males (Taiwan) (Wang, Wang, and Liu, 2002)	about 600	1705 (59)	67 (9)
France:			
Females	5510	1625 (71)	62 (12)
Males (IFTH and Goncalves, personal communication, 2006)	3986	1756 (77)	77 (13)
Germany:			
Female army applicants	301	1674	64
Male army applicants (Leyk, Kuechmeister, and Juergens, 2006)	1036	1795	75

Continued

Table 1.2   Measured heights and weights: averages (and standard deviations) of adults in many populations (Continued)

	Sample size	Stature (mm)	Weight (kg)
India:			
Females	251	1523 (66)	50 (10)
Males	710	1650 (70)	57 (11)
(Chakarbarti, 1997)			
East-Ctr. India male farm workers (Victor, Nath, and Verma, 2002)	300	1638 (56)	57 (7)
South India male workers (Fernandez and Uppugonduri, 1992)	128	1607 (60)	57 (5)
East India male farm workers (Yadav, Tewari, and Prasad, 1997)	134	1621 (58)	54 (67)
Iran:			
Female students	74	1597 (58)	56 (10)
Male students (Mououdi, 1997)	105	1725 (58)	66 (10)
Ireland:			
Males (Gallwey and Fitzgibbon, 1991)	164	1731 (58)	74 (9)
Italy:			
Females	753*	1610 (64)	58 (8)
Females	386**	1611 (62)	58 (9)
Males	913*	1733 (71)	75 (10)
Males	410**	1736 (67)	73 (11)
*(Coniglio, Fubini, Masali et al., 1991)			
**(Robinette, Blackwell, Daanen et al., 2002)			
Japan:			
Females	240	1584 (50)	54 (6)
Males (Kagimoto, 1990)	248	1688 (55)	66 (8)
Netherlands:			
Females, 20–30 yrs old	68*	1686 (66)	67 (10)
Females, 18–65 yrs old	691**	1679 (75)	73 (16)
Males, 20–30 yrs old	55*	1848 (80)	81 (14)

Table 1.2 Measured heights and weights: averages (and standard deviations) of adults in many populations (Continued)

	Sample size	Stature (mm)	Weight (kg)
Males, 18–65 yrs old	564**	1813 (90)	84 (16)

* (Steenbekkers and Beijsterveldt, 1998)
** (Robinette, Blackwell, Daanen et al., 2002)

Russia:

	Sample size	Stature (mm)	Weight (kg)
Female herders (ethnic Asians)	246	1588 (55)	nda
Female students (ethn. Russians)	207	1637 (57)	61 (8)
Female students (ethn. Usbeks)	164	1578 (49)	56 (7)
Fem. factory workers (ethn. Russians)	205	1606 (53)	61 (8)
Fem. factory workers (ethn. Usbeks)	301	1580 (54)	58 (9)
Male students (ethn. Russians)	166	1757 (56)	71 (9)
Male students (ethn. Usbeks)	150	1700 (52)	65 (7)
Male factory workers (ethn. Russians)	192	1736 (61)	72 (10)
Male factory workers (ethn. mix)	150	1700 (59)	68 (8)
Male farm mechanics (ethnic Asians)	520	1704 (58)	64 (8)
Male coal miners (ethn. Russians)	150	1801 (61)	nda
Male construction workers (ethn. Russians) (Strokina and Pakhomova, 1999)	150	1707 (69)	nda

Saudi Arabia:

	Sample size	Stature (mm)	Weight (kg)
Males (Dairi, 1986; cited by Al-Haboubi, 1991)	1440	1675 (61)	nda

Singapore:

	Sample size	Stature (mm)	Weight (kg)
Males (pilot trainees) (Singh, Peng, Lim, and Ong, 1995)	832	1685 (53)	nda

Continued

Table 1.2   Measured heights and weights: averages (and standard deviations) of adults in many populations (Continued)

	Sample size	Stature (mm)	Weight (kg)
Sri Lanka:			
Females	287	1523 (59)	nda
Males	435	1639 (63)	nda
(Abeysecera, 1985; cited by Intaranont, 1991)			
Thailand:			
Females*	250	1512 (48)	nda
Females**	711	1540 (50)	nda
Males*	250	1607 (20)	nda
Males**	1478	1654 (59)	nda
* (Intaranont, 1991)			
** (NICE; cited by Intaranont, 1991)			
Turkey:			
Male soldiers	5108	1702 (60)	63 (7)
(Kayis and Oezok, 1991a,b)			
United Kingdom:			
Females	nda	1611	68
Males	nda	1746	81
(Erens, Primatesta, and Prior, 1999)			
United States:			
Females	about 3800	1625	75
Males	about 3800	1762	87
(Ogden, Fryar, Carroll et al., 2004)			
Midwest workers, with shoes and light clothes:			
Females	125	1637 (62)	65 (12)
Males	384	1778 (73)	84 (16)
(Marras and Kim, 1993)			
Male miners	105	1803 (65)	89 (15)
(Kuenzi and Kennedy, 1993)			
U.S. Army soldiers:			
Females	2208	1629 (64)	62 (8)

Table 1.2 Measured heights and weights: averages (and standard deviations) of adults in many populations (Continued)

	Sample size	Stature (mm)	Weight (kg)
Males (Gordon, Churchill, Clauser et al., 1989)	1774	1756 (67)	76 (11)
North American (Canada and United States) Females, 18–26 yrs. old	1255	1640 (73)	69 (18)
Males, 18–65 yrs. old (Robinette, Blackwell, Daanen et al., 2002)	1120	1778 (79)	86 (18)
Vietnamese, living in the United States:			
Females	30	1559 (61)	49
Males (Imrhan, Nguyen, and Nguyen, 1993)	41	1646 (60)	59

Source: Data adapted on 9 January 2008 from Kroemer (2006) who listed all sources except Erens, Primatesta and Prior, 1999; Kamadjeu, Edwards and Atanga, 2006; Leyk, Kuechmeister and Juergens, 2006.

Note: nda: no data available.

How to measure

The designer needs specific data, such as describing the head size for fitting helmets, or reach distance for devising a proper workspace. As a rule, such information cannot be deduced from stature or weight but must be specifically measured. Selecting a representative sample of the group of interest, and performing the measurements on the persons is usually a formidable task. The traditional technique is to take measurements with hand-held devices such as anthropometers, tapes, and gauges, and to record the data. An emerging technique is to record automatically the 3-D dimensions of the surface of the human body°; yet current information still relies almost exclusively on results of the hands-on approach. Figure 1.1 depicts 36 of the most commonly taken measurements and Table 1.3 describes these measurements and their use° in anthropometry and design.

Russian and Chinese adults

Table 1.4 is an example of the results of such detailed measures, taken between 1984 and 1986 on Russian students, 18 to

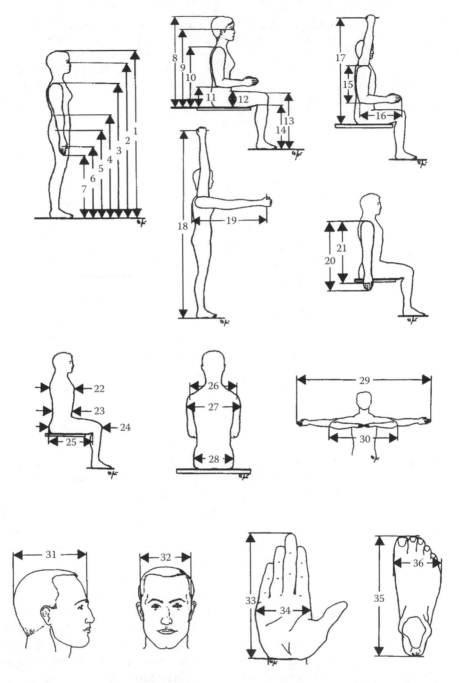

FIGURE 1.1   Illustration of measured body dimensions. (Adapted from Kroemer, K.H.E., Kroemer, H.J., and Kroemer-Elbert, K.E. (1997). *Engineering Physiology. Bases of Human Factors/Ergonomics*. New York: Wiley.)

Table 1.3 Descriptions of common body measures and their applications[a]

Dimensions	Applications
1. Stature	
The vertical distance from the floor to the top of the head, when standing.	A main measure for comparing population samples. Reference for the minimal height of overhead obstructions. Add height for more clearance, hat, shoes, stride.
2. Eye height, standing	
The vertical distance from the floor to the outer corner of the right eye, when standing.	Origin of the visual field of a standing person. Reference for the location of visual obstructions and of targets such as displays; consider slump and motion.
3. Shoulder height (acromion), standing	
The vertical distance from the floor to the tip (acromion) of the shoulder, when standing.	Starting point for arm length measurements; near the center of rotation of the upper arm. Reference point for hand reaches; consider slump and motion.
4. Elbow height, standing	
The vertical distance from the floor to the lowest point of the right elbow, when standing, with the elbow flexed at 90 degrees.	Reference for height and distance of the work area of the hand and the location of controls and fixtures; consider slump and motion.
5. Hip height (trochanter), standing	
The vertical distance from the floor to the trochanter landmark on the upper side of the right thigh, when standing.	Traditional anthropometric measure, indicator of leg length and the height of the hip joint. Used for comparing population samples.
6. Knuckle height, standing	
The vertical distance from the floor to the knuckle (metacarpal bone) of the middle finger of the right hand, when standing.	Reference for low locations of controls, handles, and handrails; consider slump and motion of the standing person.
7. Fingertip height, standing	
The vertical distance from the floor to the tip of the extended index finger of the right hand, when standing.	Reference for the lowest location of controls, handles, and handrails; consider slump and motion of the standing person.
8. Sitting height	
The vertical distance from the sitting surface to the top of the head, when sitting.	Reference for the minimal height of overhead obstructions. Add height for more clearance, hat, trunk motion of the seated person.

Continued

Table 1.3   Descriptions of common body measures and their applications[a]
(Continued)

Dimensions	Applications
9. Sitting eye height	
The vertical distance from the sitting surface to the outer corner of the right eye, when sitting.	Origin of the visual field of a seated person. Reference point for the location of visual obstructions and of targets such as displays; consider slump and motion.
10. Sitting shoulder height (acromion)	
The vertical distance from the sitting surface to the tip (acromion) of the shoulder, when sitting.	Starting point for arm length measurements; near the center of rotation of the upper arm. Reference for hand reaches; consider slump and motion.
11. Sitting elbow height	
The vertical distance from the sitting surface to the lowest point of the right elbow, when sitting, with the elbow flexed at 90 degrees.	Reference for the height of an armrest, of the work area of the hand and of keyboard and controls; consider slump and motion of the seated person.
12. Sitting thigh height (clearance)	
The vertical distance from the sitting surface to the highest point on the top of the horizontal right thigh, with the knee flexed at 90 degrees.	Reference for the minimal clearance needed between seat pan and the underside of a structure, such as a table or desk; add clearance for clothing and motions.
13. Sitting knee height	
The vertical distance from the floor to the top of the right kneecap, when sitting, with the knees flexed at 90 degrees.	Traditional anthropometric measure for lower leg length. Reference for the minimal clearance needed below the underside of a structure, such as a table or desk; add height for shoe.
14. Sitting popliteal height	
The vertical distance from the floor to the underside of the thigh directly behind the right knee; when sitting, with the knees flexed at 90 degrees.	Reference for the height of a seat; add height for shoe.
15. Shoulder–elbow length	
The vertical distance from the underside of the right elbow to the right acromion, with the elbow flexed at 90 degrees and the upper arm hanging vertically.	Traditional anthropometric measure for comparing population samples.

Table 1.3 Descriptions of common body measures and their applications[a] (Continued)

Dimensions	Applications
16. Elbow–fingertip length	
The distance from the back of the right elbow to the tip of the extended middle finger, with the elbow flexed at 90 degrees.	Traditional anthropometric measure. Reference for fingertip reach when moving the forearm in the elbow.
17. Overhead grip reach, sitting	
The vertical distance from the sitting surface to the center of a cylindrical rod firmly held in the palm of the right hand.	Reference for the height of overhead controls operated by a seated person. Consider ease of motion, reach, and finger/hand/arm strength.
18. Overhead grip reach, standing	
The vertical distance from the floor to the center of a cylindrical rod firmly held in the palm of the right hand.	Reference for the height of overhead controls operated by a standing person. Add shoe height. Consider ease of motion, reach, and finger/hand/arm strength.
19. Forward grip reach	
The horizontal distance from the back of the right shoulder blade to the center of a cylindrical rod firmly held in the palm of the right hand.	Reference for forward reach distance. Consider ease of motion, reach, and finger/hand/arm strength.
20. Arm length, vertical	
The vertical distance from the tip of the right middle finger to the right acromion, with the arm hanging vertically.	A traditional measure for comparing population samples. Reference for the location of controls very low on the side of the operator. Consider ease of motion, reach, and finger/hand/arm strength.
21. Downward grip reach	
The vertical distance from the right acromion to the center of a cylindrical rod firmly held in the palm of the right hand, with the arm hanging vertically.	Reference for the location of controls low on the side of the operator. Consider ease of motion, reach, and finger/hand/arm strength.
22. Chest depth	
The horizontal distance from the back to the right nipple.	A traditional measure for comparing population samples. Reference for the clearance between seat backrest and the location of obstructions in front of the trunk.

Continued

Table 1.3 Descriptions of common body measures and their applications[a] (Continued)

Dimensions	Applications
23. Abdominal depth, sitting	
The horizontal distance from the back to the most protruding point on the abdomen.	A traditional measure for comparing population samples. Reference for the clearance between seat backrest and the location of obstructions in front of the trunk.
24. Buttock–knee depth, sitting	
The horizontal distance from the back of the buttocks to the most protruding point on the right knee, when sitting with the knees flexed at 90 degrees.	Reference for the clearance between seat backrest and the location of obstructions in front of the knees.
25. Buttock–popliteal depth, sitting	
The horizontal distance from the back of the buttocks to back of the right knee just below the thigh, when sitting with the knees flexed at 90 degrees.	Reference for the depth of a seat.
26. Shoulder breadth (biacromial)	
The distance between the right and left acromions.	A traditional measure for comparing population samples. Indicator of the distance between the centers of rotation of the two upper arms.
27. Shoulder breadth (bideltoid)	
The maximal horizontal breadth across the shoulders between the lateral margins of the right and left deltoid muscles.	Reference for the lateral clearance required at shoulder level. Add space for ease of motion and tool use.
28. Hip breadth, sitting	
The maximal horizontal breadth across the hips or thighs, whatever is greater, when sitting.	Reference for seat width. Add space for clothing and ease of motion.
29. Span	
The distance between the tips of the middle fingers of the horizontally outstretched arms and hands.	A traditional measure for comparing population samples. Reference for sideway reach.
30. Elbow span	
The distance between the tips of the elbows of the horizontally outstretched upper arms when the elbows are flexed so that the fingertips of the hands meet in front of the trunk.	Reference for the lateral space needed at upper body level for ease of motion and tool use.

Table 1.3 Descriptions of common body measures and their applications[a] (Continued)

Dimensions	Applications
31. Head length	
The distance from the glabella (between the browridges) to the most rearward protrusion (the occiput) on the back, in the middle of the skull.	A traditional measure for comparing population samples. Reference for head gear size.
32. Head breadth	
The maximal horizontal breadth of the head above the attachment of the ears.	A traditional measure for comparing population samples. Reference for head gear size.
33. Hand length	
The length of the right hand between the crease of the wrist and the tip of the middle finger, with the hand flat.	A traditional measure for comparing population samples. Reference for hand tool and gear size. Consider manipulations, gloves, tool use.
34. Hand breadth	
The breadth of the right hand across the knuckles of the four fingers.	A traditional measure for comparing population samples. Reference for hand tool and gear size, and for the opening through which a hand may fit. Consider manipulations, gloves, tool use.
35. Foot length	
The maximal length of the right foot, when standing.	A traditional measure for comparing population samples. Reference for shoe and pedal size.
36. Foot breadth	
The maximal breadth of the right foot, at right angle to the long axis of the foot, when standing.	A traditional measure for comparing population samples. Reference for shoe size, spacing of pedals.
37. Weight (in kg)	
Nude body weight taken to the nearest tenth of a kilogram.	A traditional measure for comparing population samples. Reference for body size, clothing, strength, health, etc. Add weight for clothing and equipment worn on the body.

Source: Adapted from Kroemer, K.H.E. (2006). *"Extra-Ordinary" Ergonomics: How to Accommodate Small and Big Persons, the Disabled and Elderly, Expectant Mothers and Children.* Boca Raton, FL: CRC Press.

22 years of age. Table 1.5 gives anthropometric data on young adults in China (Taiwan). The literature contains compilations of data° describing people in Russia, France, Germany, Japan, the United Kingdom, and the United States yet, unfortunately, similarly detailed information remains missing on most populations, probably because it is so laborious and expensive to obtain. Thus, if we need information on other groups of people, we have to estimate the data or, better, do some measuring.

No "average person"

There is one serious problem with average data, such as shown in Tables 1.1, 1.2, and 1.4. We cannot design a glove, or workplace, for the "average person" because that is just a statistical phantom. If we did, the resulting glove or workplace would be too small for half the people, and too big for the others. Instead, it is necessary to consider specifically very big and tall people, so that they can fit into an airplane seat; or we must pay particular attention to small people, for example, to make sure that they can reach items stored on a high shelf. If doorways were fixed to suit people of average height, many persons passing through them would suffer bloodied foreheads, as they would strike the lintels.

The "normal" distribution

To avoid headaches, we need to get a bit of advice about the statistics° used to fit distribution curves to anthropometric measurements and, finally, compile them into tables. Figure 1.2 depicts a typical collection of anthropometric measurements, here the stature of a group of men. Statisticians call this symmetrical bell-shaped curve "normal" (or Gaussian): data cluster in the middle, and the farther away from the center, the fewer data points exist.

Mean and average

Such a normal distribution is easy to describe by two statistical values: one is the well-known *mean,* also often called *average.* The other descriptor is the *standard deviation*, a measure of the peakedness or flatness of the distribution. (Note that these descriptors do not apply to nonnormal distributions, which require more involved statistical treatments. There is a bit more about this in Chapter 4, which concerns muscle strength data.)

Percentiles

If we know the mean (m) and the standard deviation (SD), we can calculate the numerical value of any point, *percentiles* (p), in a normal distribution. The fifth percentile is often of design interest: 5% of all data are smaller; 95% are larger. Obviously, the mean (average) is the same as the fiftieth percentile: half the data lie below, and the other half above. The fifth percentile, $p5$ for short, is 1.65 standard deviations below the mean;

Table 1.4 Body size measures (in mm) taken between 1984 and 1986 on 681 Russian students, 18 to 22 years of age[a]

Dimension	Men				Women			
	5th %ile	Mean	95th %ile	SD	5th %ile	Mean	95th %ile	SD
1. Stature. Ethnic Russians (Moscow)	1664	1757	1849	56	1542	1637	1731	57
Usbeks (Tashkent)	1615	1700	1786	52	1498	1578	1658	49
2. Eye height, standing								
Ethnic Russians (Moscow)	1547	1637	1728	55	1433	1526	1618	57
Usbeks (Tashkent)	1496	1581	1665	51	1387	1463	1538	46
3. Shoulder height (acromion), standing								
Ethnic Russians (Moscow)	1351	1440	1529	54	1245	1334	1422	54
Usbeks (Tashkent)	1313	1391	1469	48	1217	1284	1371	47
4. Elbow height, standing								
Ethnic Russians (Moscow)	1004	1083	1162	48	941	1010	1080	42
Usbeks (Tashkent)	985	1042	1099	35	909	970	1031	37
5. Hip height (trochanter)	nda[b]	nda	nda	nda	nda	nda	nda	nda
6. Knuckle height, standing.								
Ethnic Russians (Moscow)	710	773	836	39	676	731	786	34
Usbeks (Tashkent)	676	734	792	35	632	687	742	33
7. Fingertip height, standing								
Ethnic Russians (Moscow)	nda	668	729	37	582	635	687	32
Usbeks (Tashkent)	579	635	691	34	546	599	652	32

Continued

Table 1.4 Body size measures (in mm) taken between 1984 and 1986 on 681 Russian students, 18 to 22 years of age[a] (Continued)

Dimension	Men				Women			
	5th %ile	Mean	95th %ile	SD	5th %ile	Mean	95th %ile	SD
8. Sitting height								
Ethnic Russians (Moscow)	860	912	964	32	806	859	911	32
Usbeks (Tashkent)	858	905	952	29	793	839	885	28
9. Sitting eye height								
Ethnic Russians (Moscow)	737	790	844	33	694	742	790	29
Usbeks (Tashkent)	737	784	830	28	676	723	771	29
10. Sitting shoulder height (acromion)	nda	nda	nda	nda	nda	nda	nda	nda
11. Sitting elbow height								
Ethnic Russians (Moscow)	202	243	284	25	196	236	275	24
Usbeks (Tashkent)	186	229	272	26	191	229	267	23
12. Sitting thigh height (clearance)								
Ethnic Russians (Moscow)	122	151	179	18	126	148	172	14
Usbeks (Tashkent)	120	143	165	14	114	142	170	17
13. Sitting knee height								
Ethnic Russians (Moscow)	520	562	603	25	487	527	567	24
Usbeks (Tashkent)	494	531	569	23	446	487	528	25
14. Sitting popliteal height								
Ethnic Russians (Moscow)	429	468	508	24	386	423	461	23
Usbeks (Tashkent)	400	430	460	18	366	398	430	20

15. Shoulder–elbow length	nda	nda	nda	nda	nda	nda	nda	nda
16. Elbow–fingertip length	nda	nda	nda	nda	nda	nda	nda	nda
17. Overhead grip reach, sitting								
Ethnic Russians (Moscow)	1199	1276	1354	47	1094	1169	1244	46
Usbeks (Tashkent)	1193	1256	1319	38	1085	1152	1219	41
18. Overhead grip reach, standing	nda	nda	nda	nda	nda	nda	nda	nda
19. Forward grip reach								
Ethnic Russians (Moscow)	697	759	821	38	641	702	763	37
Usbeks (Tashkent)	686	745	803	36	609	673	737	39
20. Arm length, vertical	nda	nda	nda	nda	nda	nda	nda	nda
21. Downward grip reach	nda	nda	nda	nda	nda	nda	nda	nda
22. Chest depth								
Ethnic Russians (Moscow)	207	245	nda	20	209	242	276	21
Usbeks (Tashkent)	211	244	276	20	200	233	265	20
23. Abdominal depth, sitting	nda	nda	nda	nda	nda	nda	nda	nda
24. Buttock–knee depth, sitting								
Ethnic Russians (Moscow)	561	610	660	30	536	584	631	29
Usbeks (Tashkent)	541	595	648	33	515	564	612	30
25. Buttock–popliteal depth, sitting								
Ethnic Russians (Moscow)	476	517	557	25	446	496	540	29
Usbeks (Tashkent)	459	504	550	28	423	472	520	29
26. Shoulder breadth (biacromial)								
Ethnic Russians (Moscow)	369	397	425	17	334	360	386	16
Usbeks (Tashkent)	349	377	404	17	320	347	373	16

Continued

Table 1.4 Body size measures (in mm) taken between 1984 and 1986 on 681 Russian students, 18 to 22 years of age[a] (Continued)

Dimension	Men				Women			
	5th %ile	Mean	95th %ile	SD	5th %ile	Mean	95th %ile	SD
27. Shoulder breadth (bideltoid)								
Ethnic Russians (Moscow)	416	458	492	23	377	412	446	21
Usbeks (Tashkent)	409	438	466	17	352	381	410	17
28. Hip breadth, sitting								
Ethnic Russians (Moscow)	323	362	410	23	334	372	411	23
Usbeks (Tashkent)	316	349	381	20	329	364	399	21
29. Span								
Ethnic Russians (Moscow)	1671	1782	1893	68	1516	1640	1763	75
Usbeks (Tashkent)	1640	1747	1855	66	1461	1579	1698	72
30. Elbow span								
Ethnic Russians (Moscow)	874	935	995	37	808	870	933	38
Usbeks (Tashkent)	842	909	976	41	781	837	894	34
31. Head length	nda	nda	nda	nda	nda	nda	nda	nda
32. Head breadth	nda	nda	nda	nda	nda	nda	nda	nda

33. Hand length								
Ethnic Russians (Moscow)	174	188	202	9	155	168	182	8
Usbeks (Tashkent)	175	188	201	8	nda	nda	nda	nda
34. Hand breadth								
Ethnic Russians (Moscow)	80	87	95	5	71	76	82	3
Usbeks (Tashkent)	82	89	96	4	73	79	87	4
35. Foot length								
Ethnic Russians (Moscow)	247	266	286	12	222	239	256	11
Usbeks (Tashkent)	242	260	279	11	220	237	254	10
36. Foot breadth								
Ethnic Russians (Moscow)	87	97	107	6	nda	88	95	4
Usbeks (Tashkent)	85	96	107	7	81	90	98	5
37. Weight (in kg)								
Ethnic Russians (Moscow)	57	71	85	9	49	60	73	7
Usbeks (Tashkent)	53	65	76	7	45	56	68	7

Source: Table adapted from Kroemer (2006). Data from Strokina, A.N. and Pakhomova, B.A. (1999). *Anthropo-Ergonomic Atlas.* Moscow: Moscow State University Publishing House.

[a] The measured sample consisted of 166 male and 207 female Russians from Moscow and 150 male and 164 female Usbeks from Tashkent.

[b] nda: No data available.

Table 1.5   Body size measures (in mm) taken between 1996 and 2000 on nearly 1200 persons in Taiwan, 25 to 34 years of age

Dimension	Men				Women			
	5th %ile	Mean	95th %tile	SD	5th %ile	Mean	95th %tile	SD
1. Stature	1608	1705	1801	59	1485	1572	1659	53
2. Eye height, standing	nda[a]	nda	nda	nda	nda	nda	nda	nda
3. Shoulder height (acromion), standing	1309	1396	1484	53	1204	1285	1367	50
4. Elbow height, standing	993	1059	1126	40	915	978	1040	38
5. Hip height (trochanter)	780	860	939	48	735	802	869	41
6. Knuckle height, standing	705	757	809	32	653	708	762	33
7. Fingertip height, standing	610	659	708	30	566	618	670	32
8. Sitting height	861	910	959	30	794	846	898	32
9. Sitting eye height	742	791	839	29	681	732	783	31
10. Sitting shoulder height (acromion)	560	602	645	26	516	561	605	27
11. Sitting elbow height	226	264	303	24	211	252	294	25
12. Sitting thigh height (clearance)	nda	nda	nda	nda	nda	nda	nda	nda
13. Sitting knee height	474	521	569	29	431	471	510	24
14. Sitting popliteal height	380	411	442	19	350	379	408	18
15. Shoulder–elbow length	308	338	369	19	280	309	339	18
16. Elbow–fingertip length	382	427	472	27	339	384	429	27
17. Overhead grip reach, sitting	1128	1208	1289	49	1033	1105	1177	44
18. Overhead grip reach, standing	1872	2002	2133	79	1721	1831	1942	67
19. Forward grip reach	650	710	770	36	597	651	705	33

20. Arm length, vertical	684	738	793	33	618	669	720	31
21. Downward grip reach	nda	nda	nda	nda	nda	nda	nda	nda
22. Chest depth	187	217	248	19	182	213	244	19
23. Abdominal depth, sitting	nda	nda	nda	nda	nda	nda	nda	nda
24. Buttock–knee depth, sitting	507	558	608	31	487	530	572	26
25. Buttock–popliteal depth, sitting	nda	nda	nda	nda	nda	nda	nda	nda
26. Shoulder breadth (biacromial)	323	369	415	28	282	324	366	25
27. Shoulder breadth (bideltoid)	422	460	499	23	367	406	445	24
28. Hip breadth, sitting	315	360	404	27	316	353	390	23
29. Span	1625	1738	1852	69	1469	1571	1672	62
30. Elbow span	820	894	968	45	737	801	866	39
31. Head length	185	197	209	7	176	187	198	6
32. Head breadth	154	167	181	8	146	161	175	9
33. Hand length	168	183	199	10	154	167	181	8
34. Hand breadth	77	86	94	5	68	75	82	4
35. Foot length	nda	nda	nda	nda	nda	nda	nda	nda
36. Foot breadth	nda	nda	nda	nda	nda	nda	nda	nda
37. Weight (in kg)	53	67	81	9	40	52	64	7

Source: Table adapted from Kroemer (2006). Data from Wang, M.J.J., Wang, E.M.Y., and Lin, Y.C. (2002). *Anthropometric Data Book of the Chinese People in Taiwan.* Hsinchu, ROC: The Ergonomics Society of Taiwan.

[a] nda: No data available.

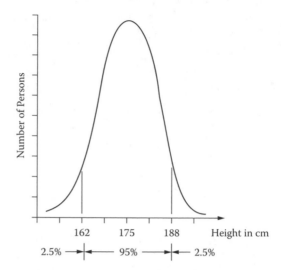

FIGURE 1.2　Typical distribution of anthropometric data

conversely, $p95$ is 1.65 SD (standard deviations) above m. Table 1.6 contains multiplication factors needed to calculate various percentage points in a normal distribution.

Hand sizes　　For devising gloves, hand tools, or hand-operated controls we need information about hand sizes. Figure 1.3 shows relevant measurements; Table 1.7 contains available data. The table shows that for many groups of people, hand data are missing, so we must either measure or estimate the missing numbers.

Further advice　　Collections of anthropometric data provide a wealth of information to the skilled statistician, but any laypersons including engineers or designers can use the data as well, for example, to calculate percentiles, cut-off values, adjustment ranges, and so on of homogeneous or composite population groups. Another task is to estimate data on groups of people, such as to fill in the vacant cells in Table 1.7. In doing so, it is advantageous to employ the advice given in books written for ergonomists and human factors engineers, listed under Further Reading.

1.3　Designing to fit the body

People differ　　Nearly any sample of all the adults on Earth can fit into an airplane, or use the same hand tools, if these products are sized well. Yet, individuals within the species vary from each other;

Table 1.6 Values of factor k to calculate percentage points in a normal distribution

k	Percentile p located below the mean m $p = m - k \times SD$	Percentile p located above the mean m $p = m + k \times SD$
4.25	0.001	99.999
2.33	1	99
2.06	2	98
1.96	2.5	97.5
1.88	3	97
1.65	5	95
1.28	10	90
1.04	15	85
1.00	16.5	83.5
0.84	20	80
0.67	25	75
0	50	50

we need differently sized shoes to fit our individual feet. Even among seemingly similar groups, body sizes or body segment measurements can differ significantly. For example, on average, in the United States agricultural workers are shorter by an average of 2.5 cm in height than other workers. Female American agricultural and manufacturing workers have larger waist circumferences than those in other occupations; firefighters, police,

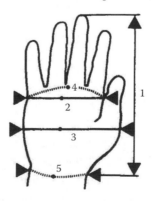

FIGURE 1.3 Common measurements of hand and wrist

Table 1.7   Hand and wrist sizes (in mm)

Hand measures	Population	Men Mean	Men SD	Women Mean	Women SD
1. Length	British	180	10	175	9
	British, estimated 1986	190	10	175	9
	Chinese, Taiwan	183	10	167	8
	French	190	nda[a]	173	nda
	Germans	189	9	174	9
	German soldiers 2006	191	nda	176	nda
	Japanese	nda	nda	nda	nda
	Russians, Moscow	188	9	168	8
	U.S. soldiers	194	10	181	10
	U.S. Vietnamese	177	12	165	9
2. Breadth at knuckles	British	85	5	75	4
	Chinese, Taiwan	86	5	75	4
	French	86	nda	76	nda
	Germans	88	5	78	4
	Japanese	nda	nda	90	5
	Russians, Moscow	87	5	76	3
	U.S. soldiers	90	4	79	4
	U.S. Vietnamese	79	7	71	4
3. Maximal breadth	British	105	5	92	5
	Chinese	nda	nda	nda	nda
	French	nda	nda	nda	nda
	Germans	107	6	94	6
	Japanese	nda	nda	nda	nda
	Russians	nda	nda	nda	nda
	U.S. soldiers	nda	nda	nda	nda
	U.S. Vietnamese	100	6	87	6
4. Circumference at knuckles	British	nda	nda	nda	nda
	Chinese	nda	nda	nda	nda
	French	nda	nda	nda	nda
	Germans	nda	nda	nda	nda
	Japanese	nda	nda	nda	nda
	Russians	nda	nda	nda	nda
	U.S. soldiers	214	10	186	9
	U.S. Vietnamese	nda	nda	nda	nda

Table 1.7 Hand and wrist sizes (in mm) (Continued)

Hand measures	Population	Men		Women	
		Mean	SD	Mean	SD
5. Wrist circumference	British	nda	nda	nda	nda
	Chinese	nda	nda	nda	nda
	French	nda	nda	nda	nda
	Germans	nda	nda	nda	nda
	Japanese	nda	nda	nda	nda
	Russians	nda	nda	nda	nda
	U.S. soldiers	174	8	151	7
	U.S. Vietnamese	163	15	137	18

Source: Data taken from Imrhan et al. (1993), Kroemer (2006), and Leyk et al. (2006). Adapted from Kroemer, K.H.E. (2006). *"Extra-Ordinary" Ergonomics: How to Accommodate Small and Big Persons, the Disabled and Elderly, Expectant Mothers and Children*. Boca Raton, FL: CRC Press.

[a] nda: No data available.

and guards are taller and heavier (males by 7 kg, females over 10 kg) than persons in all other occupations[o].

Design principles

Before we begin to design a glove, helmet, or other object that must fit its user exactly, we must decide which range of relevant body sizes we want to accommodate. For this aim, we have a choice among five approaches.

1. *Custom-fit each individual.* This is a laborious and expensive solution, justifiable in a few exceptional cases.

2. *Have several fixed sizes.* This can be a reasonable solution, but all sizes must be available and "between-sized" users may not be accommodated well.

3. *Make it adjustable.* This is usually the approach that provides the best fit to all people, but the adjustment features must be easy to use.

4. *Design for the extreme bodies.* This is the appropriate solution when you want to assure that any individual:

   • Can operate a gadget, so locate an emergency stop button within the shortest reach; or

   • Can fit through an opening, so make a door or escape hatch wide enough for even the largest person; or

- Cannot pass through an opening behind which danger lurks, so make railings or safety guards at machinery tight.

5. *Select those persons whose bodies fit the existing design.* This is a "last and bad resort" if we failed to achieve the fundamental principle of good design; all intended users should be able to employ our design effectively and efficiently.

Fit a range

Solutions 2 and 3 are the most common. Here, we need to select a range of body sizes that we intend to accommodate with our design. For that goal, we must select the extremes of those body sizes that establish the endpoints of the fit range. Often, we will aim to fit the persons bigger than the smallest 5% and smaller than the biggest 5%; in other words, we are accommodating the central 90% of a group. In doing so, we exclude 10%, half of them very small and the others very big.

Select design limits

The design endpoints, the minimum and the maximum of a range, depend on the design purpose and therefore must be selected carefully. We may decide to divide our designs into several sizes (Solution 2), each appropriate for a subgroup of all users. This is a routine approach for ready-made clothing: clothes come assembled in size clusters. Within each range, adjustment features (Solution 3) can provide further fitting: good examples are shoes with laces and the office chair seats that can be raised and lowered.

Statics and dynamics

The measures shown in Figures 1.1 and 1.2 and in Tables 1.1 through 1.4 are dimensions taken on the body while standing or sitting still. In reality, hardly anybody is stiffly static at work; we usually move about. To design for movement, the engineer needs to modify the static data. The recommendations in Table 1.8 can help accomplish this task.

Table 1.8 Guidelines for the conversion of standard measuring postures to functional stances and motions

To consider	Do this
Slumped standing or sitting	Deduct 5–10% from appropriate height measurements.
Relaxed trunk	Add 5–10% to trunk circumferences and depths.
Wearing shoes	Add approximately 25 mm to standing and sitting heights; more for high heels.
Wearing light clothing	Add about 5% to appropriate dimensions.
Wearing heavy clothing	Add 15% or more to appropriate dimensions. (Note that mobility may be strongly reduced by heavy clothing.)
Extended reaches	Add 10% or more for strong motions of the trunk.
Use of hand tools	Center of handle is at about 40% of hand length, measured from the wrist.
Comfortable seat height	Add or subtract up to 10% to or from standard seat height.

Source: Adapted from Kroemer, K.H.E. (2006). *"Extra-Ordinary" Ergonomics: How to Accommodate Small and Big Persons, the Disabled and Elderly, Expectant Mothers and Children.* Boca Raton, FL: CRC Press.

Summary

To fit equipment and tasks to persons of various body sizes requires (a) anthropometric data and (b) proper procedures. Data on many populations are available; missing information may be estimated or, better, measured following standard procedures. Design procedures often involve the selection of percentile values that serve as lower and upper limits of accommodation ranges.

Fitting steps

Step 1: Determine which body dimensions are important for your design.

Step 2: Decide on the range(s) to be fitted, on cut-off point(s).

Step 3: Design, then test. Modify as necessary.

Further reading

KROEMER, K.H.E. (2008). Anthropometry and biomechanics: Anthromechanics. In Kumar, S. (Ed.) *Biomechanics in Ergonomics* (2nd Edn) Chapter 2. Boca Raton, FL: CRC Press.

MARRAS, W.S. and KARWOWSKI, K. (EDS.) (2006). *The Occupational Ergonomics Handbook* (2nd Edn). Boca Raton, FL: CRC Press.

ROBINETTE, K.M. (ED.) (2009). *Computer Aided Anthropometry for Research and Design.* Boca Raton, FL: CRC Press (in press).

Notes

The text contains markers, °, to indicate specific references and comments, which follow.

1.2 Measurements: Techniques: Gordon et al. 1989, Landau et al. 2000, Lohman et al. 1988, Robinette 2009, Robinette et al. 2003, Roebuck 1995.

Compilations of data: see, for example, Kroemer 2006.

Measurements and their use: ISO Standard 7250, Kroemer 2006.

1.1.2 Statistics: Marras et al. 2006, Appendix in Pheasant et al. 2006, Roebuck 1995.

1.1.3 Designing to fit the body: Differences among occupational groups: Hsiao et al. 2002.

CHAPTER TWO

Mobility

Ancient experiences in agriculture, forestry, fishing, and other traditional work have taught people how to do the embedded tasks best. However, new jobs, tools, and workplaces in modern industry and commerce, in transportation on land, in water, air, and space need to be laid out purposefully to suit the human body and mind. The foundation for human-factored designs is the insight that our body functions best in motion, not in a maintained static stance.

2.1 Work in Motion

Made for motion

We are continually changing our body's configuration while we walk or sit, even when we sleep. If injury or disease imposes a fixed body position, circulatory and metabolic functions become impaired; people who must lie in place develop bedsores. Holding still is tiring, almost impossible for just an hour. Apparently, the human body functions best in motion. Therefore, we should design our equipment and tasks for movement.

No static templates

However, it is convenient to measure the human body while the person holds still in a defined upright static posture, standing or sitting as Figure 1.1 in the first chapter of this book shows. Unfortunately, such artificial measuring stances such as shown in Figure 2.1 became misleading templates for workstation layout; these not only created the false image of a static operator, but they also gave the wrong impression that being stiff upright is desirable or healthy. People like to move, not to stay still; if left alone, they will sit any way they want, as sketched in Figure 2.2. To design for movement° is not difficult, as we will see later in this chapter.

31

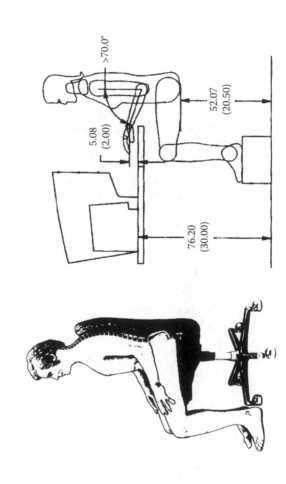

FIGURE 2.1   Unrealistic circa 1988 depictions of "orthopedically good" sitting and of a computer operator at work

FIGURE 2.2 People move and sit as they want. (Adapted from Kroemer, K.H.E. and Kroemer, A.D. (2005). Office Ergonomics, authorized translation into Korean. Seoul: Kukje.)

Excessive motions

Of course, overdoing motions can lead to trouble. Highly repetitive movement requirements° have been associated with hand/wrist/arm problems since the early 1700s. Excessive demands on the limited motion capabilities of the spinal column can easily result in overloading, especially if twist rotation of the torso combines with bending; low back pain° seems to have been with humankind since the earliest times.

2.2 Body joints

Extensive leg and arm mobility

Legs provide us with powerful mobility: these long body members rotate about their articulations with the trunk at the hip joints, which provide wide-ranging angular freedom°. Simpler angular motions occur in the knee joints. Foot angle changes in the ankles are small but important for balance and subtle actions. Our arms provide us with long reaches and extensive mobility in the shoulder and elbow joints. Thumbs and fingers are able to perform complex finely controlled motions via wrist and digit articulations.

Rotations in body joints

Hip and shoulder joints consist of a fixed panlike bone structure within which the mobile globe-shaped part (of thigh and upper arm) can rotate. The technical analogy is a mechanical ball joint that can move about what the engineer calls three "axes of rotation" (has three "degrees of freedom" mobility): the upper leg and arm can rotate (1) fore–aft as well as (2) left–right and they can (3) twist. The knee and elbow are simpler joints, having only one axis of rotation, and so the lower leg and forearm can only swing forth and back in their hinge-type joints. The ankle articulation is another ball-joint type but with very narrow excursions in the three axes.

Hand mobility

2.2.1 The hand   The wrist gives the hand wide-ranging mobility, in three axes. (To be exact, the ability to twist is actually within the forearm.) The 27 bones of the hand provide the solid structure; Figure 2.3 shows them and names their articulations.

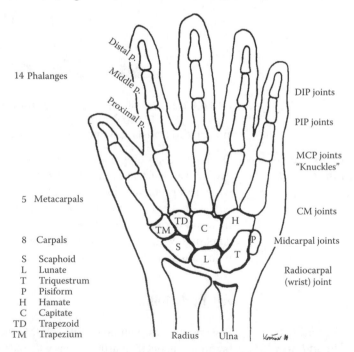

FIGURE 2.3   Bones and joints of the hand. (Adapted from Kroemer, K.H.E. and Kroemer, A.D. (2001). *Office Ergonomics.* London: Taylor & Francis.)

Whereas the main body of the hand is only slightly deform-able among its eight carpal bones, the attached five digits (one thumb and four fingers) provide great dexterity.

Hand digits   As Figure 2.3 shows, the five digits attach to the main body of the hand by their metacarpal bones in the CM articulations (carpal–metacarpal joints), which have 2-axes mobility. The thumb is particularly mobile in its CM joint, but it has only two final segments (phalanges) whereas the four fingers each have three phalanges. The knuckles (metacarpal–phalangeal MCP joints) provide two axes of mobility to the fingers but their interphalangeal (IP) joints, proximal (PIP), and distal (DIP), are simple 1-axis hinges.

Movers of the hand   The importance of the hand for so many of our tasks warrants a more detailed discussion. Some of the muscles that bend (flex) and straighten (extend) the hand are located within the hand. These are the intrinsic muscles, which control most of the finer aspects of manipulations. Extrinsic muscles are in the forearm, from where they control most of the powerful hand flexion and extension activities by pulling on tendons that cross the wrist. Figures 2.4 and 2.5 show the extensor and flexor tendons. Sheaths encapsulate the tendons, keeping them in place while providing lubrication for their gliding.

Carpal tunnel   At the wrist, where only limited space is available for passage of the tendons, bending can create increased pressure between a tendon and its sheath. This usually does not affect the exten-sor tendons at the back of the hand, but can be a problem for the flexors on the other side; see Figure 2.6. A particularly critical region is the carpal tunnel, a narrow opening at the base of the hand formed of the carpal bones and a strong transverse ligament at the palm side of the hand. The nine tendons of the extrinsic muscles that flex the hand's digits must pass through this tight tunnel together with the median nerve and blood vessels. The median nerve is a cord about the size of a pencil containing thousands of nerve fibers supplying "feel" to the thumb, index, and middle fingers, and to part of the ring finger. If swelling occurs in this tightly packed carpal tunnel (such as by inflammation that often results from highly repetitive digit actions, for example, on keyboards), the pressure increases. Resulting compression can make tendon movements difficult and painful and affect nervous feedback and control through

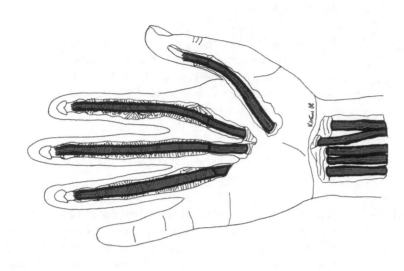

FIGURE 2.5 Digit flexor tendons that bend the digits of the hand. (Adapted from Kroemer, K.H.E. and Kroemer, A.D. (2001). *Office Ergonomics*. London: Taylor & Francis.)

FIGURE 2.4 Digit extensor tendons that straighten the digits of the hand. (Modified from Kroemer, K.H.E. and Kroemer, A.D. (2001). *Office Ergonomics*. London: Taylor & Francis and from Putz-Anderson, V. (1988). *Cumulative Trauma Disorders*. London: Taylor & Francis.)

Sheaths

Tendon

Carpal ligament

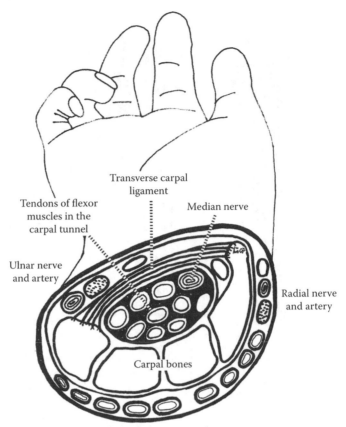

FIGURE 2.6 Cross-section of the right hand, distal from the wrist. Carpal bones and the transverse carpal ligament form the carpal tunnel through which pass digit flexor tendons and the median nerve (Modified from Kroemer, K.H.E. and Kroemer, A.D. (2001). *Office Ergonomics*. London: Taylor & Francis.)

the median nerve, causing the carpal tunnel syndrome with temporary disability, possibly even permanent injury°.

Tendon sheaths   The sheaths of the tendons have complex designs, depending on their locations and purposes. Besides providing lubrication to facilitate gliding of the tendons, sheaths also constitute attachments to the bones and provide "pulleys" at which the tendons pull to articulate the sections of the digits against each other to flex or extend them, as sketched, much simplified, in Figures 2.7 and 2.8. The complex design of the hand with its

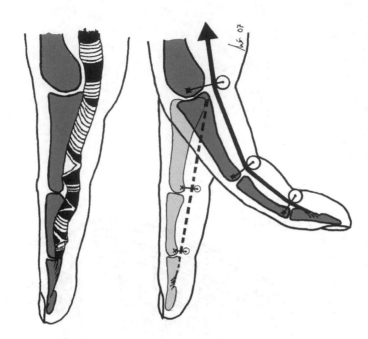

FIGURE 2.8 Ringlike and cruciform ligament attachments to the bones keep the flexor tendon in place and allow it to bend the finger by pulling. (Adapted from Kroemer, K.H.E. and Kroemer, A.D. (2001). *Office Ergonomics*. London: Taylor & Francis.)

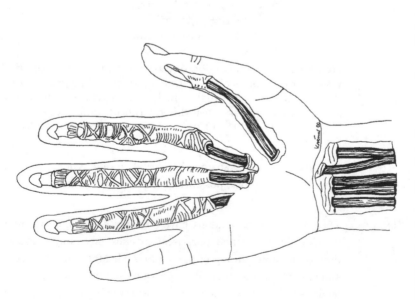

FIGURE 2.7 Sheaths guide digit flexor tendons. (Adapted from Kroemer, K.H.E. and Kroemer, A.D. (2001). *Office Ergonomics*. London: Taylor & Francis.)

multiple functions provides wide-ranging, forceful mobility with fine manipulative control.

2.2.2 The spine The spinal column supports the trunk and head. It is essentially a stack of 24 bones, vertebrae, on top of the fused tailbone, the sacrum. Seen from the front or back, the healthy spine is essentially straight, but viewed from the side, appears bent in a series of flat curves, as depicted in Figure 2.9. At the neck is a forward bend, called a lordosis; at chest height, the curvature points backward, called a kyphosis; below, in the lumbar region, is another lordosis.

Spine mobility   Compared to the hand, the joints and motion capabilities of the spinal column are quite different. Each spinal bone, or vertebra, sits on top of an intervertebral disc, which consists of a viscous fluid enclosed in tough fibrous ring. The disc acts as a cushion that absorbs shocks; the elastic pad allows the vertebrae above and below small changes in their angles of tilt against each

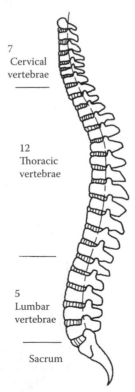

7
Cervical
vertebrae
———

12
Thoracic
vertebrae

———

5
Lumbar
vertebrae

———

Sacrum

FIGURE 2.9   Curvature of the spinal column, seen from the left side

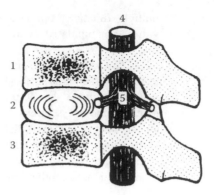

FIGURE 2.10 Side view, from the left, of a disc (2) between the main bodies of two vertebrae (1, 3); behind them is the spinal cord (4) with its nerve extensions (5)

other. However, the small angular displacements add up over the stack of discs, providing the whole spinal column considerable twist, fore–aft, and left–right bending capabilities. Figure 2.10 shows, schematically, a disc between the main bodies of two adjacent vertebrae. It also shows the spinous processes of the vertebrae, and the spinal cord with its emanating extensions ("nerve roots").

Figure 2.11 is a 3-D sketch of a vertebra: behind the main body is an opening, the vertebral foramen, through which the spinal cord runs up and down. That protective structure has three protrusions, the spinal and the two transverse processes (SP, TP), pointing backward, left, and right. These act as lever

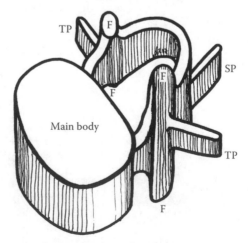

FIGURE 2.11 3-D sketch of a vertebra

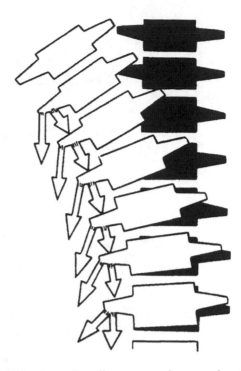

FIGURE 2.12 Lateral pulls on vertebrae make them tilt. (Adapted from Kapandji, I.A. (1982). *The Physiology of the Joints*. Amsterdam: Elsevier.)

arms for ligaments and muscles, which attach to them and, by their combined actions, stabilize or bend the spinal column; see Figure 2.12.

Figure 2.13 depicts the five lumbar vertebrae atop the sacrum, seen from the left. Together with Figure 2.11, this schematic illustrates that each vertebra has six bearing surfaces with the vertebrae above and below: two interfaces at the bottom and top of the main body with cushioning discs; but there are no cushions at the facets (F in Figure 2.11) which provide four bony articulations, two superior and two inferior surfaces. Thus, the design of the spinal joints provides complex yet limited mobility.

Limited trunk flexibility

With only small displacement between the bones of the spine provided by deformation of the discs, the head and neck can twist and bend, fore–aft and sideways, in the upper section of the spinal column, within fairly narrow limits. The larger movements of the trunk, in bending as well as twisting, occur mostly in the lower parts of the spine, especially in the lumbar

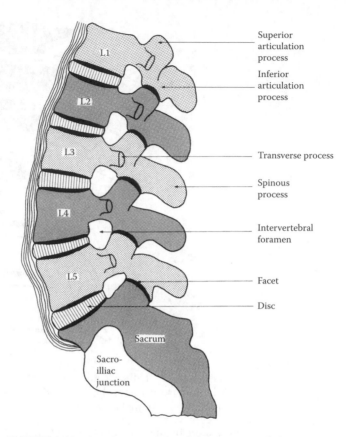

FIGURE 2.13 Lumbar section of the spine, with bearing surfaces drawn in heavy lines. (Adapted from Kroemer, K.H.E., Kroemer, H.J., and Kroemer-Elbert, K.E. (1997). *Engineering Physiology. Bases of Human Factors/Ergonomics* (3rd Edn). New York: Wiley.)

section. The ability of the spine to transmit large forces, mostly in compression, is remarkable. Yet, overloading can cause damage, most often to the spinal discs in the lumbar region°, which is frequently the location of discomfort, pain, and injury because it must transmit substantial forces and torques to and from the upper body.

Skeletal adjustments for pregnancy

When our prehuman ancestors started to walk on two legs instead of on four, that new ambulation caused several skeletal adjustments: vertebrae increased in number and thickness to support the upper body, and the spine changed its curvature. In women, pregnancy adds bulk to the abdomen which shifts the center of mass forward; to counteract this, the mother-to-be leans back, which increases the curve of the lumbar spine. Exaggerating the lumbar curve creates problems: vertebrae are more likely to slip against each other, which can cause back pain or fracture. The female spine° developed several features that help to prevent that damage. In men, the curve in the lower back spans two vertebrae, but three in women, which distributes strain over a wider area. Furthermore, the facet joints are larger in women than in men and the joints are oriented at a slightly different angle, so that they can resist the higher force and better brace the vertebrae against slipping.

2.3 Designing for mobility

Given that we have so many ways to move, it is convenient to describe the rotations in each involved body articulation separately. Figure 2.14 illustrates body joint displacements and their traditional descriptions.

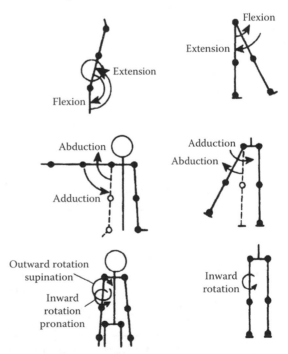

FIGURE 2.14 Displacements in body joints. (Adapted from Kroemer, K.H.E., Kroemer, H.J., and Kroemer-Elbert, K.E. (1997). *Engineering Physiology. Bases of Human Factors/ Ergonomics* (3rd Edn). New York: Wiley.)

Table 2.1 Mobility (in degrees) of 100 female and 100 male students of physical education

		5th Percentile		95th Percentile	
Joint	Movement	Females	Males	Females	Males
Neck	Ventral flexion	34.0	25.0	69.0	60.0
	Dorsal flexion	47.5	38.0	93.5	74.0
	Right rotation	67.0	56.0	95.0	85.0
	Left rotation	64.0	67.5	90.0	85.0
Shoulder	Flexion	169.5	161.0	199.5	193.5
	Extension	47.0	41.5	85.0	76.0
	Adduction	37.5	36.0	67.5	63.0
	Abduction	106.0	106.0	139.0	140.0
	Medial rotation	94.0	68.5	127.0	114.0
	Lateral rotation	19.5	16.0	54.5	46.0
Elbow	Flexion	135.5	122.5	160.5	150.0
	Supination	87.0	86.0	130.0	135.0
	Pronation	63.0	42.5	99.0	86.5
Wrist	Extension	56.5	47.0	87.5	76.0
	Flexion	53.5	50.5	89.5	85.0
	Adduction	16.5	14.0	36.5	30.0
	Abduction	19.0	22.0	37.0	40.0
Hip	Flexion	103.0	95.0	147.0	130.0
	Adduction	27.0	15.5	50.0	39.0
	Abduction	47.0	38.0	85.0	81.0
	Medial rotation (prone)	30.5	30.0	58.5	62.5
	Lateral rotation (prone)	29.0	21.5	62.0	46.0
	Medial rotation (sitting)	20.5	18.0	43.5	43.0
	Lateral rotation (sitting)	20.5	18.0	45.5	37.0
Knee	Flexion (standing)	99.5	87.0	127.5	122.0
	Flexion (prone)	116.0	99.5	144.0	130.0
	Medial rotation	18.5	14.5	44.5	35.0
	Lateral rotation	28.5	21.0	58.5	48.0
Ankle	Flexion	13.0	18.0	33.0	34.0
	Extension	30.5	21.0	51.5	51.5
	Adduction	13.0	15.0	34.0	38.0
	Abduction	11.5	11.0	36.5	30.0

Source: Adapted from Staff, K.R. (1983) *A comparison of range of joint mobility in college females and males*, unpublished master's thesis, Texas A&M University, College Station.

Actual mobility

As we move about, we usually combine movements in several body joints to generate the flexibility needed. Table 2.1 lists mobility measurements taken on physical education students, 100 females and 100 males. Most ordinary people are likely to be less flexible but, of course, actual ranges of motion depend very much on health, fitness, training, skill, age, and any disability. Furthermore, dissimilar measuring techniques and instructions to the clients can cause great diversity in reported results of mobility ranges of groups of people°.

2.4 Workspaces

In our everyday activities, at work or leisure, we like to walk about; whether we stand or sit, changes in posture are essential. Being forced to stand or sit still for long periods is difficult to tolerate.

Preferred motions

Preferred motion ranges of the feet or hands depend on habits and skills, on workplace layout, and on the dominant task requirements° such as strength, speed, accuracy, or vision. Clearly, there is not just one work envelope; rather, different people prefer to do different tasks in different workspaces.

Everyday motion ranges

"Convenient" mobility is somewhere within the overall array, such as listed in Table 2.2, but is not always in the middle of the range. Some preferred motions occur near the mean mobility of a body joint, but often motions take place close to a joint's flexibility limits. For example, a person walking around on a job site or standing often has the hips and knees almost fully extended, near 180 degrees. However, when the person sits on a chair, both angles cluster around 90 degrees. Table 2.2 lists motion ranges associated with everyday activities.

Who uses chairs?

However, in many parts of the earth, working habits differ: for example, chairs are in wide use in the so-called western countries, such as in Europe, the Americas, and Australia/New Zealand. In other regions, persons often hover close to the ground, such as sketched in Figure 2.15. Furthermore, work tasks may be quite different and require special skills and motions. Unfortunately, few ergonomic recommendations for non-western conditions are at hand.

Reach envelopes

In general, preferred work areas of the hands and feet are before the trunk of the body, within curved envelopes that reflect the mobility of the arms and legs. Thus, it appears convenient to describe

Table 2.2 Mobility ranges at work

Body joints/parts	Angles/positions	
Shoulder	Mostly mid-range, upper arm often hanging down	
Elbow	Mostly mid-range, at about 90 degrees	
Wrist	Mostly mid-range, about straight	
Neck/Head	Mostly mid-range, about straight	
Back	Near complete stretch, about erect	
	When walking or standing	**When sitting**
Hip (side view)	Near complete stretch, at about 180 degrees	Mostly mid-range, at about 90 degrees
Knee	Near extreme stretch, slightly less than or near 180 degrees	Mostly mid-range, at about 90 degrees

Source: Adapted from Kroemer, Kroemer, and Kroemer-Elbert (1997). *Engineering Physiology: Bases of Human Factors/ Ergonomics* (3rd Edn). New York: VNR–Wiley.

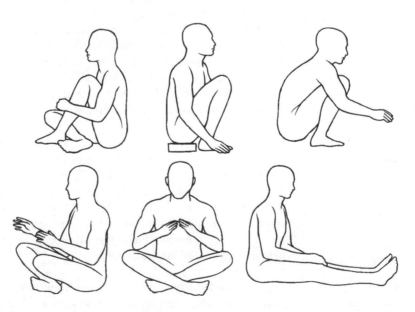

FIGURE 2.15 Non-western work postures

these reach envelopes for hands and feet as partial spheres around the temporary locations of body joints. In reality, however, several parts of the body may move at the same time, and force or vision requirements may establish specific constraints.

Hand workspace

Movement of the forearm in the elbow joint, close to an approximately 90-degree bend, combined with some rotation in the shoulder, determines the envelope for fine manipulation and other light work; this is the shaded space in Figure 2.16. Supporting the elbows, as shown in Figure 2.17, may facilitate precision work. Full movements of the upper and forearms

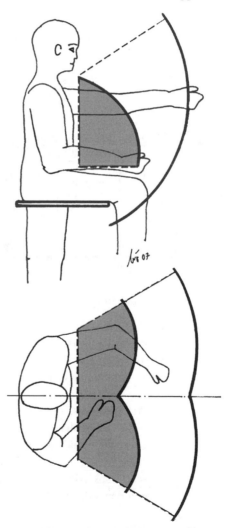

FIGURE 2.16 Workspace of the hands

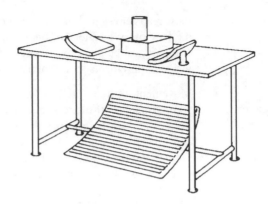

FIGURE 2.17 Workstation with support for the elbows and feet

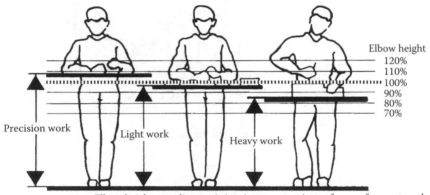

Elbow height, standing or sitting, is an appropriate reference for most work

FIGURE 2.18 Work surface height of the hands

allow handwork from about hip to shoulder height in a wide space, mostly in front of the body. The main determinant of the proper height of the workspace is the height of the person's elbow, standing or seated; see Figure 2.18.

Easy foot actions

When a person sits, foot activities usually occur in a shallow area below the knees, shaded in Figure 2.19. The operator can cover this area with small displacements in knee and ankle joints, both near 90 degrees. This is the best space for foot activities that require exertion of only small force and energy, usually by pushing down the whole foot, or just the heel or toes.

Strong foot push

In contrast, foot controls that demand substantial forward-push force from the seated operator (such as the brake pedal in a

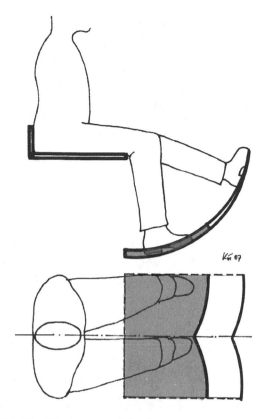

FIGURE 2.19 Workspace of the feet

vehicle) must be located higher and farther away so that the knee angle is around 150 to 160 degrees. However, the designer needs to pay attention to two facts: the almost fully extended leg does not allow for much displacement of the foot; furthermore, a strong backrest is necessary to counteract the pedal push.

No standing on one foot

Frequent operation of pedals is only suitable for a seated operator: a standing person obviously cannot use both feet simultaneously, and even use of only one foot would require balancing the whole body on the other foot; although birds can stand on one foot for a long time, humans cannot do so easily.

Summary

Humans are unable to maintain any given body position over long periods; we need to move our bodies. Designing to fit suitable motion ranges instead of fixed postures is not difficult. Human articulations have known movement limits within which we prefer certain regions for tasks that require strength exertion, fast motions, or exact and enduring work.

Fitting steps

Step 1: Determine which specific task must be done.

Step 2: Decide on how to do the work, with hand or foot, and where in the movement space.

Step 3: Design, then test. Modify as necessary.

Further reading

DELLEMAN, N.J., HASLEGRAVE, C.M., and CHAFFIN, D.B. (EDS). (2004). *Working Postures and Movements*. Boca Raton, FL: CRC Press.

MARRAS, W.S. and KARWOWSKI, K. (EDS.) (2006a). *The Occupational Ergonomics Handbook* (2nd Edn), *Fundamentals and Assessment Tools for Occupational Ergonomics*.

MARRAS, W.S. and KARWOWSKI, K. (EDS.) (2006b) *The Occupational Ergonomics Handbook* (2nd Edn), *Interventions, Controls, and Applications in Occupational Ergonomics*, Boca Raton, FL: CRC Press.

Notes

The text contains markers, ° , to indicate specific references and comments, which follow.

2.1 Work in motion:

Design for movement: See Kroemer et al. 2001, 2003; Kroemer 2006.

Repetitive movement requirements: See Armstrong 2006, Arndt et al. 2001, Ramazzini 1713, Kroemer 2001, Violante et al. 2000, Wright 1993.

Low back problems: See Snook 2001.

2.2 Body joints:

Articulations of the body: See Chaffin et al. 2006, Kapandji 1988, Oezkaya et al. 1991, Walji 2008.

2.2.1 The hand, carpal tunnel: See Freivalds 2006, Hughes et al. 2008, Kroemer 2008, Walji 2008.

2.2.2 The spine, low back problems: See Marras 2008, Violante 2000.

Female and male spine features: Already present in *Australopithecus* fossils; see Whitcome et al. 2007.

2.3 Designing for mobility:

Diversity in mobility measurements: See, for example, data on NASA astronauts at http://msis.jsc.nasa.gov/ and Wu et al. 2002.

2.4 Workspaces: Dominant strength, speed, accuracy or vision requirements: see Kroemer et al. 1997, 2003.

CHAPTER THREE

Muscular work

Muscles are the human's "natural engines". Muscle attaches to a bone and then extends across one or two body joints to another bone. By contracting, the muscle pulls on the body's internal bone framework. That pull can change the angle between bones, setting body segments into motion or stabilizing their position.

Basic physics explains the relations between common strength measurements:

> Force (in Newton, *N*) divided by transmitting cross-section area is *tension* (usually in *N/mm²*) or, in the opposite direction, *compression* or *pressure*.
>
> Force multiplied by lever arm is *moment* or *torque* (usually in *Nm*).
>
> The integral of force and distance traveled is *work*.
>
> Work divided by time is *power*.

3.1 Physiological principles

Muscle components

The human body has several hundred skeletal muscles, known by their Latin names; for example, *m. biceps brachii* can flex the elbow. Every muscle consists of bundles of fibers, sketched in Figure 3.1. Tissue that embeds blood vessels and nerves wraps and permeates the muscle. At the muscle ends, tissues combine to form tendons, which are like cables that extend to and attach to bones°.

Power to the muscle

Thousands of individual muscle fibers run in bundles, essentially parallel, along the length of the muscle. Inside the bundles

53

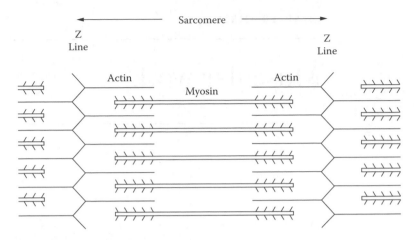

FIGURE 3.1 Basic structures of skeletal muscle. (Adapted from Kroemer, K.H.E., Kroemer, H.J. and Kroemer-Elbert, K.E. (1997). *Engineering Physiology. Bases of Human Factors/Ergonomics*. New York: Wiley.)

are hundreds of mitochondria, the muscle's "power factories". These are cells specialized to liberate chemically stored energy (in ATP, adenosine triphosphate, and CP, creatine phosphate°) that enables the muscle to contract. Blood supply is essential for muscle function because it supplies energy and oxygen and removes the byproducts of metabolic processes (see Chapter 10) such as heat, water, and carbon dioxide. Nerve signals trigger the actions of the muscle and control their intensity.

Contractile microstructure

Contraction is the only *active* action that a muscle can take. Elongated protein molecules, side-by-side threadlike filaments, temporarily adhere to each other when triggered by nervous system signals. One kind of filament, called myosin, has projections that protrude toward the surrounding type of filament, called actin; see Figures 3.1, 3.2 and 3.3. When the muscle is at rest, the projections remain relaxed and do not transmit forces

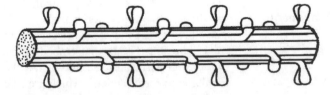

FIGURE 3.2 Scheme of the clublike "cross-bridges" of myosin. (Adapted from Herzog, W. (2008). Determinants of muscle strength. In Kumar, S. (Ed.) *Biomechanics in Ergonomics* (2nd Edn) Chapter 7. Boca Raton, FL: CRC Press.)

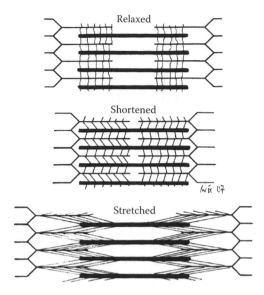

FIGURE 3.3 Scheme of a sarcomere relaxed, shortened and stretched. (Adapted from Kroemer, K.H.E., Kroemer, H.J., and Kroemer-Elbert, K.E. (1997). *Engineering Physiology. Bases of Human Factors/Ergonomics*. New York: Wiley.)

between the actin and myosin filaments. However, when the muscle receives a signal to contract, the myosin's projections activate and attach to the actins (like cross-bridges), and, in ratcheting motions, try to pull the actin rods along the myosin. If that pull is stronger than any opposing force, the ends of actin filaments slide, which shortens the sarcomere, the distance between adjacent z-lines, and hence the muscle. If the pull is just as strong as an opposing force, the lengths of sarcomere and muscle remain unchanged: this is called an isometric contraction. If the contraction pull is weak, then the sarcomere can become lengthened by the external force.

Striated skeletal muscle

This shortening or stretching is visible under the microscope when we look at the z-lines. (The letter z stands for the German *zwischen*, between.) These lines appear as dark stripes that run across the muscle fibers where the fixed ends of actin filaments connect. When a contraction shortens the muscle, the z-lines are pulled closer to each other. The striping, caused by the z-lines and other bands crossing skeletal muscle, provides its second name, striated (striped) muscle. The larger the number of actin–myosin bundles that exist side by side in a muscle, the more strength it can exert. Hence, the strength of a muscle is proportional to its cross-sectional area.

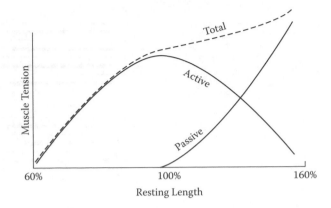

FIGURE 3.4   Active, passive and combined tensions at different muscle lengths. (Adapted from Kroemer, K.H.E., Kroemer, H.J., and Kroemer-Elbert, K.E. (1997). *Engineering Physiology. Bases of Human Factors/Ergonomics.* New York: Wiley.)

Smooth and cardiac muscles

The human body also has two other kinds of muscle: smooth muscle surrounding vessels, mostly as sphincters that control the opening of blood vessels; the other kind, cardiac muscle, operates the heart. Smooth and cardiac muscles have characteristics that differ from those of skeletal muscle°.

Sarcomere

The distance between adjacent z-lines is the sarcomere. At rest, the length of a sarcomere is approximately 250 Ångstroms. (Because 1 Å = 10^{-10} *m*, about 40,000 sarcomeres can lie in series within 1 *mm* of muscle fiber length.) In full contraction, the length of a sarcomere, and hence the length of a muscle, can shorten to about 60% of its resting length. The muscle's contractile force is largest at about resting length and decreases as the muscle shortens, as sketched in Figure 3.4.

Muscle tension

However, a strong force external to the muscle (such as due to gravity or stemming from the action of another muscle) can lengthen it beyond its relaxed length. That elongation can reach about 160% of resting length, from which point on structural damage, eventually even breakage, occurs. The farther the stretch, the more the muscle resists. Thus, the overall tension inside the muscle is the result of adding the active and passive tensions, illustrated in Figure 3.4. By experience, we use the prestretch "trick" by combining active and passive tension to increase our muscular power, such as when moving hand and arm back behind the shoulder in order to throw an object with force.

Co-contraction   Muscles usually appear in pairs: one muscle turns a bone around an articulation in one direction while another muscle turns the opposite way. The elbow provides a good example: pull of the *biceps* muscle reduces the elbow angle and *triceps* pull increases it, as Figure 3.5 illustrates. The drawing also shows that one "head" of the triceps muscle reaches across the shoulder joint and attaches to the shoulder blade and the other two attachments are to the femur bone. The biceps has two heads and attachments, as its name indicates. The seemingly simple muscle system around the elbow is actually quite complex, because two more muscles assist the biceps in reducing the elbow angle: the *m. brachialis*, connecting the humerus and ulna bones, and the *m. brachioradialis*, linking the humerus and radius bones. When muscles contract simultaneously to pull

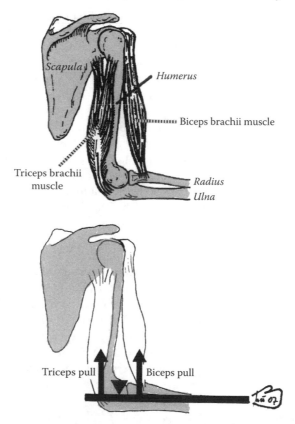

FIGURE 3.5   Triceps and biceps muscles control elbow extension and flexion. (Adapted from Kroemer, K.H.E., Kroemer, H.J., and Kroemer-Elbert, K.E. (1997). *Engineering Physiology. Bases of Human Factors/Ergonomics*. New York: Wiley.)

in the same direction, their combined effort obviously increases the force on the attached bone(s); however, co-contraction of two opposing muscles (such as biceps and triceps) generates a counterplay of forces that helps to control segment movement.

3.2 Dynamic and static efforts, strength tests

As already discussed in Chapter 2, the human body functions best in motion rather than when immobile. However, it is easier to assess certain traits of the body when it remains motionless in a defined position. Thus, on persons holding still, body sizes are simple to measure (see Chapter 1); likewise, it is relatively simple to assess muscular strength when muscle lengths remain constant.

Contraction and motion

There is much interplay between the tension (force, energy, power) a muscle can develop and how it moves while contracting. The simplest case is no motion: although its myofilaments attempt to contract, the overall length of the muscle does not change (and the involved body segments stay in place). Physiologists call this condition *isometric*; obviously, physicists would call it *static*. If the muscle succeeds in shortening its length, this case of contraction is called *concentric*. Such shortening can occur only if the muscle is stronger than the resisting force. However, if the opposing force overcomes the muscle while it attempts to contract, the muscle becomes lengthened. This condition is called *eccentric*. (This shows that the term *contraction* does not necessarily mean that the muscle shortens.)

Dynamic muscular efforts are more complicated to describe than static contractions and they are more difficult to control in experiments. In dynamic activities, muscle length changes, and therefore involved body segments move. This results in displacement. The time derivatives (velocity, acceleration, and jerk) of displacement are of importance° both for the internal muscular effort and for its external effect: for example, change in velocity determines impact and force, as per Newton's second law.

Anthro-mechanics

The application of physics principles to biological events is called *biomechanics*, or, when applied to the human body, *anthromechanics*. Overlapping separate traditional disciplines of knowledge is typical for ergonomics/human factors engineering, as exemplified in this book.

Athletics and sports

We can measure muscle strength in different ways. One kind of assessment is common in athletic and sports events with their immense, even explosive efforts, usually performed during in competition with other persons selected for their abilities and competitive spirits. Such conditions are impossible to control and reproduce in the laboratory, and therefore the feats are scored in terms of records achieved, weights lifted, distances covered, and matches won. However, these muscular performances have little bearing on everyday conditions at work.

Work requirements

At work, extreme efforts are rarely required. Instead, moderate muscular labor is usually in demand throughout the working time, such as eight hours a day. These efforts are often intermittent, yet possibly repeated often during the work shift. Also, the persons doing such work commonly include individuals of different sizes and physical capabilities. For all these reasons, the required muscular exertions are usually on a fairly low level so that all persons can do them over the duration of the work shift.

Controlled strength tests

To determine work-related muscular capabilities°, strength tests are done routinely in a controlled laboratory environment, where subjects are asked to perform specific exertions; this may be an elbow flexion where the biceps muscle is dominant, or a whole-body lift that involves many muscles. These tests follow carefully planned test protocols°. The peak force (or torque, moment) observed during the test performance, which usually lasts three to five seconds, is commonly taken as the strength measurement. Although this laboratory approach does not reflect true work conditions, as in an industrial or agricultural environment, the controlled procedure allows standardization and provides some test–retest reliability.

Testing muscle strength

The strength of a muscle can be measured objectively and reliably if removed from the body, as has been done in the past with animal models. A less radical approach is to instrument a muscle, for example with electrodes, which pick up nervous excitation signals. Such tests have shown that the strength of exertion depends on the muscle length at which the exertion takes place. Figure 3.3, above, illustrates how active and passive tension within the muscle change with its length. The actual length at which a muscle performs depends on the body posture at that instant; for example, the strength of elbow extension depends on the angle between upper arm and forearm, which determines the length of the triceps muscle. Furthermore, the actual strength capabilities of muscles depend on several factors, such as size (thickness), shape (pennation), fiber composition (fast or slow), and state of training or fatigue.

MVC

Clearly, it is not permissible to measure human muscle strength by forcing muscle fibers to perform a truly maximal effort, at a tension where damage might occur. Therefore, all strength measures rely on subjects giving their best effort; the literature calls this MVC, maximum voluntary contraction. The subjects regularly receive the cautionary admonition not to hurt themselves in that test. This means that all maximal strength data reported in the literature depend on the subjects' motivation; their will to perform, in turn, depends on various physiological, psychological, and environmental conditions that exist at the moment of measurement.

Influencing test scores

The actual results of strength tests depend on a number of variables. These include age and gender of the subject; health, nutritional and hydration status; fitness, training, skill, and experience; rewards and possible drawbacks such as pain or injury; and, of course, the subjects' willingness to participate and perform truly maximal voluntary contractions. Actions of the experimenters, such as instructions, encouragement, even feedback of results, can easily influence the subjects' ego involvement and hence the test scores°. Table 3.1 lists conditions that may increase or decrease muscular performance.

Situational conditions

As just mentioned, muscle length, and hence its strength, depend on so-called situational variables°: one is posture; another is external body support (more on this topic in the next chapter). For example, when standing upright to exert a horizontal pull with one hand, that pull is weakest when you keep your feet side by side; the pull becomes stronger with one foot placed forward, and strongest if you can grasp with the free hand an external structure that provides support. Another example is to press with a foot on a pedal when we sit, such as when operating the brake in a vehicle: that push becomes strongest if we can extend the leg almost fully while we brace ourselves against the backrest of the seat.

Test instruments

Another situational condition that determines the strength test score is the instrumentation used to record the effort. By far most test instruments are of the static type, not of the dynamic sort: the handles of a grip strength tester, a force platform, and other so-called dynamometers do not give, at least not appreciably, while the test subject squeezes, pushes, or pulls. With no motion, muscular exertion is isometric; as stated above, it is the

Table 3.1 Conditions likely to increase or decrease maximal muscular performance

	Likely effect on strength exertion
Feedback of results to subject	Increase
Instructions on how to exert	Increase
Arousal of ego involvement, aspiration	Increase
Pharmaceutical agents, drugs	Increase
Startling noise, subject's shout or grunt	Increase
Hypnosis	Increase
Setting of goals, incentives	Increase or decrease
Competition, contest	Increase or decrease
Verbal encouragement	Increase or decrease
Spectators	Increase or decrease
Deception	Decrease
Fear of injury	Decrease

Source: Adapted from Kroemer, K.H.E., Kroemer, H.J., and Kroemer-Elbert, K.E. (1997). *Engineering Physiology. Bases of Human Factors/Ergonomics.* New York: Wiley.

physiological term indicating that muscle length remains constant, which is equivalent to the physics descriptor, static. True dynamometers, which measure while muscle length changes, are much less common; most are expensive, complex, and difficult to use. However, objects of constant mass provide one simple yet effective means for dynamic measurements: moving a load, called *isoinertial* testing, supplies information on the ability to push and pull, lift and lower loads, discussed in Chapter 4. However, there remains a general need for research on complex, fast dynamic strength exertions°.

3.3 Fatigue and recovery

Muscle fatigue° is a subjective experience signaling that one is becoming unable to continue or repeat a muscular effort. The onset of fatigue depends on the muscle fiber types involved in different kinds of muscular exertions, such as dynamic or static, sustained or brief. The underlying physiological reasons for fatigue are manifold: they include restriction of blood flow through muscle tissue due to intramuscular pressure, depletion of energy sources available (especially ATP and CP) in the muscle, formation of lactic acid (a byproduct of the energy conversion process), and other events caused by a continuous muscular effort. These events especially impede the processes associated with formation and detachment of cross-bridges. Furthermore, so-called central fatigue may occur in the nervous control system, associated with one's sense of effort and motivation; however, in competition, a new emotional drive to perform can temporarily overcome muscular fatigue.

Fatigue ensues when an individual tries to exert a larger effort than she or he is capable of producing. The required exertion can be too large because of extreme magnitude, unwarranted duration, or an undue combination of amount and time. The two variables interact with each other: the longer the effort, the smaller the possible amount of exertion, and vice versa. The limiting organ may not be one that does the actual task, but may be in an auxiliary subdivision of the body; a typical situation is working overhead, requiring the arms to extend upward to reach the work site, and the neck bent backward to see the work, as sketched in Figure 3.6. In this case, the shoulder

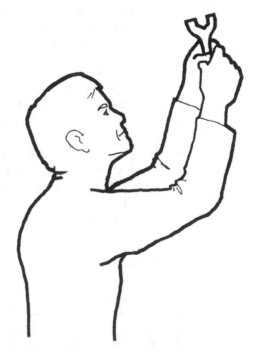

FIGURE 3.6 Fatiguing overhead work. (Adapted from Nordin, M., Andersson, G.B.J., and Pope, M.H. (1997). *Musculosketal Disorders in the Workplace: Principles and Practice*. St. Louis: Mosby.)

muscles keeping the arms up, and the muscles bending the neck, are likely to fatigue earlier than the muscles doing the actual work.

Avoid fatigue

Fortunately, simply discontinuing what caused muscular fatigue by taking a break and resting regularly lead to complete recovery. The benefit of the feeling of fatigue is prevention of serious damage to muscles. However, to the ergonomist/human factors designer, the incidence of fatigue indicates the need to improve the task conditions. The occurrence of fatigue depends on the magnitude and duration of effort compared to the capability of the afflicted muscle; hence, muscle training and skill development are subjective counteractions, effective to some extent. However, the proper ergonomic approach is to design out any work requirements that generate fatigue. The human engineering solution is fitting the task to the human.

3.4 Use of muscle strength data in design

Factors affecting strength

As mentioned, the actual strength (usually defined in terms of amount, direction, and duration) that people will generate depends decidedly on individual factors including gender, age, training, fitness, skill, experience, and motivation. Employing muscles in suitable body motion or position while bracing the body against a support structure are examples of situational factors, which also definitely affect the strength that persons exert. Figures 3.7 and 3.8 illustrate such effects. The next chapter contains more data and discusses their use in design in considerable detail.

Sources of data

The literature provides a wealth of static muscle strength data°. Sources of such information are especially journals° because they contain the newest information on human factors engineering/ergonomics issues. International and national agencies, insurance companies, and specialized research groups° can provide answers to specific inquiries. Textbooks and handbooks, encyclopedias°, and military and industry standards° can provide comprehensive overviews of accumulated data. (Push/pull, lift/lower, and carry activities are the specific topics of the next chapter.)

Naturally, all data published in the literature describe the strength capabilities of that specifically measured population. The users of the tools and equipment whom the designer has in mind may differ from a previously measured group. If

Force-plate[1] height	Distance[2]	Force, N Mean	Force, N SD
Percent of shoulder height*		With both hands	
50	80	664	177
50	100	772	216
50	120	780	165
70	80	716	162
70	100	731	233
70	120	820	138
90	80	625	147
90	100	678	195
90	120	863	141
Percent of shoulder height*		With one shoulder	
60	70	664	177
60	80	772	216
60	90	780	165
70	60	716	162
70	70	731	233
70	80	820	138
80	60	625	147
80	70	678	195
80	80	863	141
Percent of shoulder height*		With both hands	
70	70	623	147
70	80	688	164
70	90	586	132
80	70	545	127
80	80	543	123
80	90	533	81
90	70	433	95
90	80	448	93
90	90	485	80
100 Percent of shoulder height	Percent of thumb-tip reach*	With both hands	
	50	581	143
	60	667	160
	70	981	271
	80	1285	398
	90	980	302
	100	646	254
		With the preferred hand	
	50	262	67
	60	298	71
	70	360	98
	80	520	142
	90	494	169
	100	427	173
100 Percent of shoulder height	Percent of span	With either hand	
	50	367	136
	60	346	125
	70	519	164
	80	707	190
	90	325	132

FIGURE 3.7 Average horizontal push forces, in N, exerted by standing military men with hand, shoulder, or back. Legend: (1) height of the center of the 20 cm high, 25 cm wide, force plate; (2) horizontal distance between the surfaces of the force plate and the opposing bracing structure; (*) and see the authropometric definitions in Chapter 1. (Adapted from AMRL-TR-70-114 (1971). W-P AFB, OH: Aerospace Medical Research Laboratory.)

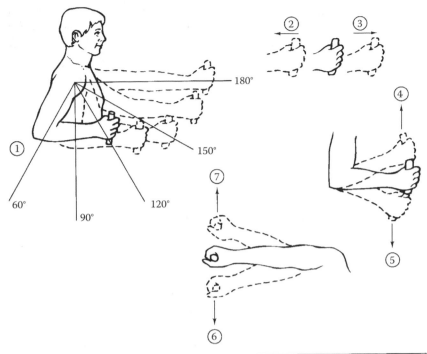

FIGURE 3.8 Fifth-percentile arm forces, in N, exerted by sitting military men with their arms in various positions. (Adapted from MIL HDBK 759 (1981), Philadelphia: Naval Publications and Forms Center.)

Fifth-percentile arm strength (N) exerted by sitting men													
(1)	(2)		(3)		(4)		(5)		(6)		(7)		
Degree of elbow flexion (deg)	Pull		Push		Up		Down		In		Out		
	Left	Right	L	R	L	R	L	R	L	R	L	R	
180	222	231	187	222	40	62	58	76	58	89	36	62	
150	187	249	133	187	67	80	80	89	67	89	36	67	
120	151	187	116	160	76	107	93	116	89	98	45	67	
90	142	165	98	160	76	89	93	116	71	80	45	71	
60	116	107	96	151	67	89	80	89	76	89	53	71	

differences are expected, it may be necessary to run a set of tests, because, in order to fit design to the human, it is necessary to know the relevant human capabilities.

Summary

The literature contains many sources of information on maximal human muscle strength. Most of it concerns static exertions. The ergonomist must be cautious when applying published data because they may not apply to the prospective users and rely instead on circumstances different from those during actual use.

Fitting steps

Step 1: Assess which kind and magnitude of body strength is necessary to do a specific task.

Step 2: Determine whether people can do the required exertion. If not, reset the task.

Step 3: Design, then test. Modify as necessary.

Further reading

KUMAR, S. (ED.) (2004). *Muscle Strength*. Boca Raton, FL: CRC Press.
ASTRAND, P.O., RODAHL, K., DAHL, H.A., and STROMME, S.B. (2004). *Textbook of Work Physiology. Physiological Bases of Exercise* (4th Edn). Champaign, IL: Human Kinetics.
WINTER, D.A. (2004). *Biomechanics and Motor Control of Human Movement* (3rd Edn). New York: Wiley.

Notes

The text contains markers, °, to indicate specific references and comments, which follow.

3.1 Physiological principles:

Tendons attach like cables to bones: Chaffin et al. 2006; Kroemer 1998; Kumar 2004, 2008; Nordin et al. 1997.

Chemically stored energy: Astrand et al. 2004, Winter 2004.

Smooth, cardiac and skeletal muscle: Astrand et al. 2004.

3.2 Dynamic and static efforts, strength tests:

Time derivatives of displacement are important: Marras et al. 1993.

Work-related strength tests: Kroemer 1999.

Strength test protocols: Chaffin et al. 2006, Kroemer et al. 1997, 2003; Marras et al. 1993.

Experimenters can influence test scores: Kroemer et al. 1997.

Situational variables: Daams 1993, 2001; Kroemer et al. 1971.

Dynamic strength tests: Dempsey 2006, Kumar 2004, Marras et al. 1993, Winter 2004.

3.3 Fatigue and recovery: Astrand et al. 2004; Chaffin et al. 2006; Kumar 2004, 2008; Winter 2004.

3.4 Use of muscle strength data in design:

Literature: Journals: for example, Applied Ergonomics, Ergonomics, Ergonomics in Design, Human Factors.

Textbooks and handbooks: for example, Chengular et al. 2003, Konz et al. 2000, Kroemer et al. 2003; Marras et al. 2006; Peebles et al. 1998, 2000; Phillips 2000; Univ. Nottingham 2002.

International and national agencies and research groups: for example, Dept. of Trade and Industry, London; Fraunhofer Gesellschaft in Germany; International Labour Office, Geneva, Switzerland; Liberty Mutual Insurance Company, Framingham, MA, USA.

Encyclopedias: for example, Karwowski 2001; Marras and Karwowski 2006, Salvendy 1997.

Standards: U.S. military standards such as 759, 1472; NASA 3000; Australian and ISO standards.

CHAPTER FOUR

Body strength and load handling

Individual strength factors

The actual strength that people generate depends primarily on two clusters of factors, already mentioned in the foregoing chapter, whose effects overlap and influence each other°. The first group consists of *individual factors*, generally related to body build, gender, age, training, and fitness; specific muscle properties are

- Muscle cross-section area
- The length of the muscle
- Speed of motion (dynamic or static exertion)
- Fatigue
- Neurophysiological excitation

Situational strength factors

The second group of factors is *situational:* these are conditions that affect—often strongly so—the amount of body strength that a person will or can apply. Situational factors include

- Motivation, ego involvement, will to succeed or fear of injury
- Skill and experience
- Body motion, or lack thereof
- Ability to brace the body against a support structure
- Body posture, for example, allowing the use of strong muscles at advantageous leverage
- The body site used for exertion, such as one foot or both feet, shoulder, hand or hands
- Coupling between body and object

Obviously, both groups of factors are of great importance for the human factors engineer who wants to make muscle use easy; the last five situational factors are of particular concern to the designer of work equipment and tools.

4.1　Static and dynamic strength exertions

Body motion, or lack thereof, is a major factor that determines the force, torque, work, power, or impulse transmitted from the body to an external object°. When the body does not move, the muscles involved in this "static" strength exertion do not change in length. In physiological terms, this called an *isometric* muscle contraction. In physics terms, all forces acting within the system are in equilibrium, as Newton's first law requires.

Static exertion

The static condition is theoretically simple and experimentally easily controllable. It allows rather simple measurements of muscular effort. For a long time, assessments of human muscle strength routinely used isometric muscle contractions. Therefore, most of the information available on body strength describes outcomes of static (isometric) testing. Accordingly, nearly all tables on body segment strength in the human engineering and physiological literature contain static data. One exception involves *isoinertial* tests, where a person moves a constant external mass. This procedure often serves to determine lifting and lowering abilities, discussed in more detail below.

Static or dynamic?

Is the real-world strength exertion static or dynamic? This question is of great concern to the human factors engineer who intends to design for easy use of muscles:

- If it is *static*, information about isometric strength capabilities is obviously appropriate. As a rule, strength exerted in movement is of lower magnitude than the static strength values measured in positions located on the path of motion.

- If the motion is *very slow*, especially when eccentric, data on isometric strength measured at points on the path usually yield a reasonable estimate (although usually a bit high) of the maximally possible exertion. ("Eccentric" means that the muscle actually lengthens during the exertion.)

- Isometric data do not estimate *fast exertions* well, especially if they are concentric and of the ballistic/impulse type, such as throwing or hammering. ("Concentric" means that the muscle shortens during the exertion.)

- If the *dynamic action is fast or variable*, special experiments° may be necessary to measure the strength capability that is specific to the demanded kind of exertion.

4.2 Maximal or minimal strength exertion

Strong and weak operators

A designer who wants to consider operator strength has to make a series of decisions when developing a new process or device. The first decision follows from the answer to the question, "Is a strong or a weak strength exertion the critical design factor?" The structural integrity of the object must be above the maximal strength that any user might apply, because even the strongest operator should not break a handle or a pedal. Therefore, the design value should be, with a safety margin, above the highest perceivable strength application. The other extreme is the minimal strength expected from the weakest operator, which still yields the desired result; an example is successfully operating a brake pedal in a vehicle under the worst foreseeable circumstances, such as failure of the power booster.

No average user

Consequently, the designer must carefully determine both the minimum and the maximum of expected operator strength exertions. Both values usually bracket the design range whereas *average user* strength usually has no value in design (as already discussed in Chapter 1 of this book). Determining the upper and lower design limits requires that we know the actual distribution of strength data. With this knowledge, we determine the lower cut-off so that nearly everyone can operate the equipment; a common cut-off is the 5th percentile strength, meaning that only 5% of the users are weaker than expected. Another common procedure is to select the 95th percentile as the upper value; however, this may not be high (safe) enough, because the strongest 5% of users can still break the equipment.

MAX and MIN values

It is easy to calculate percentile points in a normal (Gaussian) data collection, which appears in the well-known symmetrical bell-shaped distribution, often encountered with anthropometric data; see, for example, Figure 1.2 in Chapter 1. Many strength data, however, are not so evenly distributed but show irregular (nonparametric) accumulations. In this case, regular statistical procedures do not apply; yet, we can still determine, simply by "eye-balling" or other estimates, or by formal non-parametric procedures°, where we want to set our minimal and maximal design limits.

4.3 Hand strength

Most tasks, at work and in our private lives, involve grasping, holding, turning, pressing, pushing, pulling, and many other manipulations done with our digits, on objects held between the thumb and four fingers and the palm of the hand. Given that importance of our ability to handle objects, a large number of publications exist discussing hand capabilities°, especially forces.

Intrinsic and extrinsic muscles

Hand force depends on the combined effect of intrinsic and extrinsic muscles. Intrinsic muscles and their tendons are completely contained in the hand whereas extrinsic muscles are located in the forearm; they extend their tendons across the wrist joint to the hand with its digits. Most intrinsic muscles contribute to finely controlled actions of the digits, whereas the extrinsic muscles primarily generate large forces for moving the whole hand about the wrist joint and for exercising the individual digits. As with all other skeletal muscles of the human body, they can be strengthened by use, weakened by disuse, and they can suffer fatigue and recover from it.

Types of hand tasks

There are three major types of requirements in hand tasks: for accuracy, for strength exertion, and for displacement. One may divide hand tasks further in this manner:

- *Fine manipulation of objects*, with little displacement and force. Examples are writing by hand, assembly of small parts, adjustment of controls.
- *Fast movements to an object*, requiring moderate accuracy to reach the target but fairly small force exertion there. An example is the movement to a switch and its operation.
- *Frequent movements between targets*, usually with some accuracy but little force, such as in an assembly task, where small parts must be taken from bins and assembled; another example is typing on a keyboard.

1. Finger Touch: One finger touches an object without holding it.

2. Palm Touch: Some part of the inner surface of the hand touches the object without holding it.

3. Finger Palmar Grip (Hook Grip): One finger or several fingers hook(s) onto a ridge, or handle. This type of finger action is used where thumb counterforce is not needed.

4. Thumb-Fingertip Grip (Tip Grip): The thumb tip opposes one fingertip.

5. Thumb-Finger Palmar Grip (Pinch or Plier Grip): Thumb pad opposes the palmar pad of one finger (or the pads of several fingers) near the tips. This grip evolves easily from coupling #4.

6. Thumb-Forefinger Side Grip (Lateral Grip or Side Pinch):Thumb Opposes the (radial) side of the forefinger.

7. Thumb-Two-Finger Grip (Writing Grip): Thumb and two fingers (often forefinger and index finger) oppose each other at or near the tips.

8. Thumb-Fingertips Enclosure (Disk Grip): Thumb pad and the pads of three or four fingers oppose each other near the tips (object grasped does not touch the palm). This grip evolves easily from coupling #7.

9. Finger-Palm Enclosure (Collet Enclosure): Most or all of the inner surface of the hand is in contact with the object while enclosing it.

10. Power Grasp: The total inner hand surface is grasping the (often cylindrical) handle which runs parallel to the knuckles and generally protrudes on one or both sides from the hand.

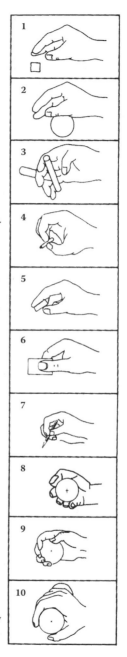

FIGURE 4.1 Couplings between hand and handle. (Adapted from Kroemer, K.H.E. (1986). Coupling the hand with the handle. *Human Factors, 28*(3), 337–339.)

- *Forceful activities with moderate displacement* such as with many assembly or repair activities, for example, when turning a hand tool against resistance.

- *Forceful activities with large displacements* such as hammering.

Grips and grasps

Tools serve as extensions of the hands, fortifying and protecting them, in many tasks. Depending on the nature of the job, the objects are held in different ways between the digits and the palm of the hand. Some hand tools require a fairly small force but precise handling, such as surgical instruments, screwdrivers used by optometrists, or writing utensils. Commonly, the manner of holding these tools is called the "precision grip". Other instruments must be held strongly between large surfaces of the fingers, thumb, and palm in what is often called the "power grasp." Yet, there are many transitions from merely touching an object with a finger (such as pushing a button) to pulling on a hooklike handle, from holding small objects between the fingertips to transmitting large energy from the hand to the handle. One attempt to classify the couplings between hand and object systematically is shown in Figure 4.1.

Hand tool design

Shaping the tool handle° to provide high friction (often helped by wearing a glove), even mechanical interlocking between hand and handle (by grooves, bulges, and serrations), facilitates secure holding and transfer of energy. Proper tool design and use should keep the wrist straight, not bent, to avoid overexertion of connective tissues (muscles, tendons, tendon sheaths) and especially to prevent compression of the median nerve in the carpal tunnel at the base of the hand. The carpal tunnel syndrome (see Chapter 2) is a painful and disabling affliction that often results from repetitive motions of the hand, such as in assembly work or keyboard use.

The natural grasp centerline of a straight handle is not perpendicular to the forearm axis, but at an angle of about 60 to 70 degrees; see Figure 4.2. For example, use of common straightnose pliers often requires a strong bend in the wrist, and neither the direction of thrust nor the axis of rotation correspond with those of the hand and arm. This often results in wrist bending, which reduces the force that can be applied. Bending the tool, not the wrist improves that situation, as shown in Figure 4.3.

"Lefties"

Most women and men are right-handed. Some tools fit only the right hand but many tools can be used with either the left or right hand. About one of ten persons prefers to use the left hand

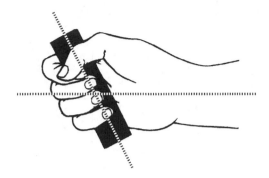

FIGURE 4.2 Angle between forearm and the hand grasp center line

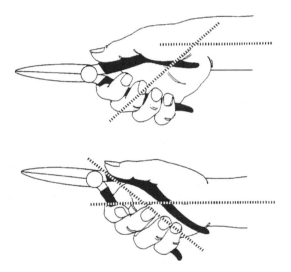

FIGURE 4.3 "Bend the tool, not the wrist." (Adapted from Kroemer, K.H.E., Kroemer, H.B., and Kroemer-Elbert, K.E. (2003, amended reprint of the 2001 edition). *Ergonomics: How to Design for Ease and Efficiency* (2nd Edn), Upper Saddle River, NJ: Prentice-Hall/Pearson Education.)

and has better skills and more strength available there. Thus, it is advisable to provide them, if needed, tools specifically designed for use with the left hand.

4.4 Whole body strength

Very few jobs require use of only one finger; most hand tasks involve hand, arm, and shoulder as well. Observing the static strength exerted at various arm positions, as shown in Figure

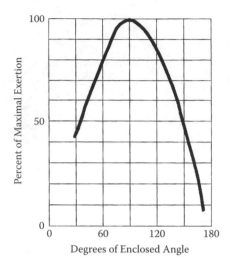

FIGURE 4.4 Effects of elbow angle on static torque exertion. (Based on diagrams in Kroemer, K.H.E., Kroemer, H.B., and Kroemer-Elbert, K.E. (2003, amended reprint of the 2001 edition). *Ergonomics: How to Design for Ease and Efficiency* (2nd Edn), Upper Saddle River, NJ: Prentice-Hall/Pearson Education.)

4.4, reveals that the measured torque varies systematically with the enclosed elbow angle.

Pushing and pulling, lifting, and lowering loads are examples of strength exertions that involve much of the whole body. Thus, in most cases, a series of body segments engages to perform a job. The chain of strength-transmitting segments starts at the hand on the steering wheel, at the foot on the brake pedal, at the shoulder pushing against a stuck car, and then runs through all body parts involved to the surface that provides reaction force, usually the seat when sitting, or the ground when standing. Figure 4.5 depicts the chain of strength-transmitting body segments. Every involved body segment must be able to perform its duty; if one fails, the task cannot be done. Furthermore, (according to Newton's third law) the reaction force must be sufficient to counteract the force exerted by hand, foot, or shoulder.

Foot strength If a person must stand at work, fairly little force and only infrequent operation of foot controls should be required because, during these exertions, the operator must support the body solely on the other leg. For a seated operator, however, operation of foot controls is much easier because the seat carries the

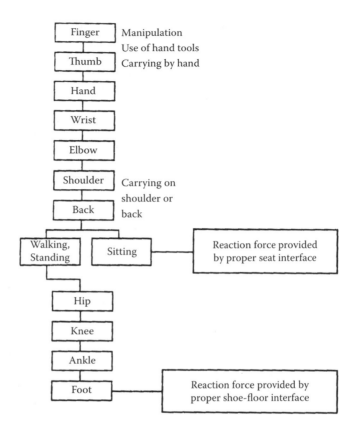

FIGURE 4.5 Chain of strength-transmitting body segments. (Adapted from Kroemer, K.H.E., Kroemer, H.J., and Kroemer-Elbert, K.E. (1997). *Engineering Physiology: Bases of Human Factors/Ergonomics* (3rd Edn), New York, Van Nostrand Reinhold–Wiley.)

body. Thus, the feet can move more freely and, given suitable conditions, can exert large forces and energies, such as when pedaling a bicycle. Bicycle pedals are best set underneath the body, so that the body weight above them provides the reactive force to the foot force transmitted to the pedal. Placing the pedals higher and forward makes body weight less effective for generation of reaction force, but if a suitable backrest is present against which buttocks and lower back can press, the feet can push forward on the pedals. That is the design principle in so-called recumbent bicycles and, similarly, in automobiles.

Foot thrust   Foot movements are relatively slow and clumsy compared to hand motions, with rather large leg masses involved. Yet, with proper seat support, our feet can exert large forces in the down

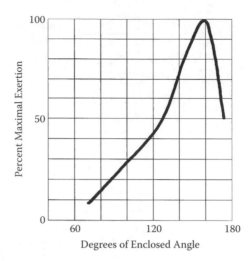

FIGURE 4.6 Effects of knee angle on static pedal push force. (Based on diagrams in Kroemer, K.H.E., Kroemer, H.B., and Kroemer-Elbert, K.E. (2003, amended reprint of the 2001 edition). *Ergonomics: How to Design for Ease and Efficiency* (2nd Edn), Upper Saddle River, NJ: Prentice-Hall/Pearson Education.)

and fore directions, a fact utilized in the early years of automotive design (and, surprisingly, still used today) when the driver had to stomp hard on pedals to operate brake and clutch, actions that are now, in modern vehicles, much less demanding. A nearly fully extended leg can generate a very large foot thrust force, but only along the direction of the lower leg. If there is no seat providing "butt and back" support, or the angle enclosed at the knee becomes smaller or larger than about 160 degrees, the pedal force diminishes dramatically, as Figure 4.6 demonstrates. Thus, there is very small space available within which the designer can place the pedal. This case is a striking example of how the success of a body exertion depends not only on the existent muscle strength, but also on how the seat provides the needed reactive support and how muscle strength gets transmitted along the hip–knee–ankle chain. An improper seat, a frail hip, knee, or ankle, or even a bad coupling at the shoe may all make for a weak kick.

Pulling and pushing

In the previous chapter, Figures 3.7 and 3.8 showed examples of one-time static push and pull forces measured under experimental conditions set up so that the subjects kept their bodies still while they exerted one-time maximal efforts. In reality, however, usually people actually move objects by pushing and pulling on them; these activities include body motions and are often repetitive.

Psychophysical tests

To design and control experiments that represent such realistic dynamic strength exertions are more complex than those needed for static tests°. In psychophysical assessments°, subjects make judgments based on their personal perception of the strenuousness of the task, thus integrating internal feedback about various physiological (muscular, circulatory, metabolic, etc.) body functions° involved in the effort. In the tests, subjects determine the amount of energy or strength that they are willing to exercise for a given time, say 5 minutes or 8 hours, without endangering or exhausting themselves. The experimenters control essential test conditions but do allow the subjects enough freedom to change other variables to make the testing conditions similar to actual work.

Pushes and pulls on the job

Push and pull strength exerted on the job depend, of course, both on the worker's body condition and on the work situation, such as one- or two-handed application, on whether one has to set the object into motion or keep it moving, and over how far a distance one moves the object. Table 4.1 provides information about push forces°, which male and female U.S. workers were willing to apply while working as hard as they could without straining themselves and without becoming unusually tired, weakened, overheated, or out of breath.

Carrying

Carrying loads, often in the hands, is an everyday task. However, it can easily overload a person if the carrying is done in awkward postures and with heavy loads. Snook and Ciriello used their psychophysical method, mentioned above, to determine the loads that American workers would voluntarily carry. Tables 4.2 and 4.3 show excerpts from their results.

Load carrying can be done in many ways, not only in the hands, but also on shoulders, on the back, and in various other ways. Table 4.4 compiles the results of several studies on the best techniques of carrying weights on the body.

Safe lifting and lowering

Around the middle of the 20th century, information about human lift capabilities concerned largely measurements of static (upward) body strength. Even into the 1980s, the literature contained rather simplistic statements about loads that, supposedly, men, women, and children could lift safely°. With better understanding of the interacting properties of the human body and

Table 4.1   Maximal push forces in N of female and male U.S. workers

			One 2.1 m PUSH every									One 30.5 m PUSH every				
	Height[a]	Percent[b]	6 sec	12	1 min	2	5	30	8 hr	Height[a]	Percent[b]	1 min	2	5	30	8 hr
Initial push forces																
Males	95	90	106	235	255	255	275	275	334	95	90	167	186	216	216	265
		75	275	304	334	334	353	353	432		75	206	235	275	275	343
		50	334	373	422	422	442	442	530		50	265	294	343	343	432
Females	89	90	137	147	167	177	196	206	216	89	90	118	137	147	157	177
		75	167	177	206	216	235	245	265		75	147	157	177	186	206
		50	196	216	245	255	285	294	314		50	177	196	206	226	255
Sustained push forces																
Males	95	90	98	128	159	167	186	186	226	95	90	79	98	118	128	157
		75	137	177	216	216	245	255	304		75	108	128	157	177	206
		50	177	226	275	285	324	334	392		50	147	167	196	226	265
Females	89	90	59	69	88	88	98	108	128	89	90	49	59	59	69	88
		75	79	108	128	128	147	159	186		75	79	79	88	98	128
		50	108	147	177	177	196	206	255		50	98	118	118	128	167

Source: Adapted from Snook, S.H. and Ciriello, V.M. (1991). The design of manual handling tasks: Revised tables of maximum acceptable weights and forces, *Ergonomics*, **34**: 9, 1197–1213.

[a] Vertical distance from floor to hands (*cm*).

[b] Acceptable to 90, 75, or 50% of industrial workers.

Table 4.2 Maximal carry weights in *kg* acceptable to male U.S. workers

Height from floor to hands (cm)	Percent of industrial population	2.1 meter carry One carry every							4.3 meter carry One carry every							8.5 meter carry One carry every						
		6 sec	12 sec	1 min	2 min	5 min	30 min	8 hr	10 sec	16 sec	1 min	2 min	5 min	30 min	8 hr	18 sec	24 sec	1 min	2 min	5 min	30 min	8 hr
111	90	10	14	17	17	19	21	25	9	11	15	15	17	19	22	10	11	13	13	15	17	20
	75	14	19	23	23	26	29	34	13	16	21	21	23	26	30	13	15	18	18	20	23	27
	50	19	25	30	30	33	38	44	17	20	27	27	30	34	39	17	19	23	24	26	29	35
	25	23	30	37	37	41	46	54	20	25	33	33	37	41	48	21	24	29	29	32	36	43
	10	27	35	43	43	48	54	63	24	29	38	39	43	48	57	24	28	34	34	38	42	50
79	90	13	17	21	21	23	26	31	11	14	18	19	21	23	27	13	14	17	18	20	22	26
	75	18	23	28	29	32	36	42	16	19	25	25	28	32	37	17	19	24	24	27	30	35
	50	23	30	37	37	41	46	54	20	25	32	33	36	41	48	22	25	31	31	35	39	46
	25	28	37	45	46	51	57	67	25	30	40	40	45	50	59	27	32	38	38	42	48	56
	10	33	43	53	53	59	66	78	29	35	47	47	52	59	69	32	38	44	45	50	56	65

Table 4.3 Maximal carry weights in *kg* acceptable to female U.S. workers

Height from floor to hands (cm)	Percent of industrial population	2.1 meter carry One carry every							4.3 meter carry One carry every							8.5 meter carry One carry every						
		6 sec	12 sec	1 min	2 min	5 min	30 min	8 hr	10 sec	16 sec	1 min	2 min	5 min	30 min	8 hr	18 sec	24 sec	1 min	2 min	5 min	30 min	8 hr
105	90	11	12	13	13	13	13	18	9	10	13	13	13	13	18	10	11	12	12	12	12	16
	75	13	14	15	15	16	16	21	11	12	15	15	16	16	21	12	13	14	14	14	14	19
	50	15	16	18	18	18	18	25	12	13	18	18	18	18	24	14	15	16	16	16	16	22
	25	17	18	20	20	21	21	28	14	15	20	20	21	21	28	15	17	18	18	19	19	25
	10	19	20	22	22	23	23	31	16	17	22	22	23	23	31	17	19	20	20	21	21	28
72	90	13	14	16	16	16	16	22	10	11	14	14	14	14	20	12	12	14	14	14	14	19
	75	15	17	18	18	19	19	25	11	13	16	16	17	17	23	14	15	16	16	17	17	23
	50	17	19	21	21	22	22	29	13	15	19	19	20	20	26	16	17	19	19	20	20	26
	25	20	22	24	24	25	25	33	15	17	22	22	22	22	30	18	19	21	22	22	22	30
	10	22	24	27	27	28	28	37	17	19	24	24	25	25	33	20	21	24	24	25	25	33

Source: Adapted from Snook, S.H. and Ciriello, V.M. (1991). The design of manual handling tasks: Revised tables of maximum acceptable weights and forces, *Ergonomics*, **34**:9, 1197–1213. Data updates according to Dr. Snook's personal communications of January 1 and 3, 2007.

Table 4.4 Techniques of carrying weights of about 30 kg

	Estimated energy expenditure for carrying 30 kg on a straight flat path	Estimated muscular fatigue	Local pressure and ischemia	Stability of loaded body	Special aspects	
In one hand	?	Very high	Very high	Very poor	Load easily manipulated and released	Suitable for quick pickup and release; for short-term carriage even of heavy loads
In both hands, equal weights	Very high, about 7 kcal/min	High	High	Poor		
Clasped between arms and trunk	?	?	?	?	Compromise between hand and trunk use	
On head, supported with one hand	Fairly low, about 5 kcal/min	High if hand guidance needed	?	Very poor	May free hand(s); strongly limits body mobility; determines posture; pad is needed	If accustomed to this technique, suitable for heavy and bulky loads
On neck, often with sherpa-type strap around forehead	Medium, about 5.5 kcal/min	?	?	Poor	May free hand(s); affects posture	

Continued

Table 4.4 Techniques of carrying weights of about 30 kg (Continued)

	Estimated energy expenditure for carrying 30 kg on a straight flat path	Estimated muscular fatigue	Local pressure and ischemia	Stability of loaded body	Special aspects	
On one shoulder	?	High	Very high	Very poor	May free hand; strongly affects posture	Suitable for short-term transport of heavy and bulky loads
Across both shoulders by yoke, held with one hand	High, about 6.2 kcal/min	?	High	Poor	May free hand(s); affects posture	Suitable for bulky and heavy loads; pads and means of attachment must be provided carefully
On back	Medium, 5.3 kcal/min with backpack; 5.9 kcal/min with bag held in place by hands	Low	?	Poor	Backpack frees hands; forces forward bend of trunk; skin-cooling problem	Suitable for large loads and long-time carriage. Packaging must be done carefully, attachment means shall not generate areas of high pressure on body
On chest	?	Low	?	Poor	Frees hands; easy hand access; reduces trunk mobility; skin-cooling problem	Highly advantageous for several small loads that must be accessible

Distributed on chest and back	Low, 4.8 kcal/min	Lowest	?	Good	Frees hands; may reduce trunk mobility; skin-cooling problem	Highly advantageous for loads that can be divided or distributed; suitable for long-durations
At waist, on buttocks	?	Low	?	Very good	Frees hands; may reduce trunk mobility	Around waist for smaller items, distributed in pockets or by special attachements; superior surface of buttocks often used to partially support backpacks
On hip	?	Low	?	Very good	Frees hands; may affect mobility	Often used to prop up large loads temporarily
On legs	?	High	?	Good	Easily reached with hands; may affect walking	Requires pockets in garments or special attachments
On foot	Highest	Highest	?	Poor	Usually not useful	

Source: Adapted from Kroemer, K.H.E. (1997). *Ergonomic Design of Material Handling Systems*, Boca Raton, FL: CRC Press.

mind, and using improved mathematical modeling of human characteristics, the assessment of lift/lower capabilities can also employ physiological, biomechanical, psychophysical, behavioral, and epidemiological criteria and their combinations°.

Material handling

Physiological criteria relate to metabolic, circulatory, respiratory and musculoskeletal events; see Chapter 10 in this book. Among the biomechanical issues, pressure within the discs of the spinal column is a major concern°. Psychophysical assessments should integrate the physiological and biomechanical criteria while also considering behavioral aspects. With injuries, particularly of the lower back, frequent among material handlers (such as warehouse workers and nurses), epidemiological statistics and accident investigations can pinpoint causes of work-related injuries and, based on such knowledge, indicate ways to avoid risks and to strive for safer conditions.

Lift and lower guidelines

The combination of those disciplinary approaches led to the development of guidelines on lifting and lowering by the U.S. National Institute of Occupational Safety and Health in the 1980s. Notable among the physical criteria used to develop the guidelines was the highest acceptable disc compression in the lumbar spine. The latest 1991 NIOSH° guidelines contain recommended weight limits (RWLs) that 90% of U.S. industrial workers, male or female, may lift or lower.

In the 1991 NIOSH guide, the maximal weight is 23 kg, but the actual recommended load is usually lower, depending on several factors, which include

- The horizontal and vertical paths of the hands, determined by the start and end points
- Whether the action is in front of the body
- The frequency of the lifting and lowering actions
- The quality of coupling between hand and load

These and some other factors become multipliers in an equation by which one calculates the RWL that applies under the given conditions.

At about the same time as the NIOSH developed its equations, Snook and Ciriello conducted psychophysical tests with American workers to determine the efforts that they were willing to exert in lifting and lowering° loads. Some of their results appear in Tables 4.5 and 4.6; these are excerpts from their full recommendations, which contain more complete information.

Table 4.5 Maximal lift weights in *kg* of female and male U.S. workers

Width (a)	Distance (b)	Percent (c)	Floor level to knuckle height — One Lift Every								Knuckle height to shoulder height — One lift every								Shoulder height to overhead reach — One lift every							
			5	9	14	1	2	5	30	8	5	9	14	1	2	5	30	8	5	9	14	1	2	5	30	8
			sec	sec	sec	min	min	min	min	hr	sec	sec	sec	min	min	min	min	hr	sec	sec	sec	min	min	min	min	hr
Males																										
34	51	90	9	10	12	16	18	20	20	24	9	12	14	17	17	18	20	22	8	11	13	16	16	17	18	20
		75	12	18	18	23	26	28	29	34	12	16	18	22	23	23	26	29	11	14	17	21	21	22	24	26
		50	17	20	24	31	35	38	39	46	15	20	23	28	29	30	33	36	14	18	21	26	27	28	31	34
Females																										
34	51	90	7	9	9	11	12	12	13	18	8	8	9	10	11	11	12	14	7	7	8	9	10	10	11	12
		75	9	11	12	14	15	15	16	22	9	10	11	12	13	13	14	17	8	8	9	11	11	11	12	14
		50	11	13	14	16	18	18	20	27	10	11	13	14	15	15	17	19	9	10	11	12	13	13	14	17

Source: Adapted from Snook, S.H. and Ciriello, V.M. (1991). The design of manual handling tasks: Revised tables of maximum acceptable weights and forces, *Ergonomics,* **34**: 9, 1197–1213. Data updates according to Dr. Snook's personal communications of January 1 and 3, 2007. Note that this table is an excerpt from the much more detailed data for North American workers (Snook and Ciriello, 1991), which are regularly updated at www.libertymutual.com.

(a) Handles in front of the operator (cm).
(b) Vertical distance of lifting (cm).
(c) Acceptable to 50, 75, or 90% of industrial workers.
Conversion: 1 kg = 2.2 lb; 1 cm = 0.4 in.

Table 4.6 Maximal lower weights in *kg* of female and male U.S. workers

Width (a)	Distance (b)	Percent (c)	Knuckle height to floor level — One lower every								Shoulder height to knuckle height — One lower every								Overhead reach to shoulder height — One lower every							
			5	9	14	1	2	5	30	8	5	9	14	1	2	5	30	8	5	9	14	1	2	5	30	8
			sec			min				hr	sec			min				hr	sec			min				hr
Males																										
		90	10	13	14	17	20	22	22	29	11	13	15	17	20	20	20	24	9	10	12	14	16	16	16	20
34	51	75	14	18	20	25	28	30	32	40	15	18	21	23	27	27	27	33	12	14	17	19	22	22	22	27
		50	19	24	26	33	37	40	42	53	20	23	27	30	35	35	35	43	16	19	22	24	28	28	28	35
Females																										
		90	7	9	9	11	12	13	14	18	8	9	9	10	11	12	12	15	7	8	8	8	10	11	11	13
34	51	75	9	11	11	13	15	16	17	22	9	11	11	12	14	15	15	19	8	9	10	10	12	13	13	16
		50	10	13	14	16	19	20	20	27	11	13	13	14	16	18	18	22	10	11	11	12	14	15	15	19

Source: Adapted from Snook, S.H. and Ciriello, V.M. (1991). The design of manual handling tasks: Revised tables of maximum acceptable weights and forces, *Ergonomics*, **34**: 9, 1197–1213. Data updates according to Dr. Snook's personal communications of January 1 and 3, 2007. Note that this table is an excerpt from the much more detailed data for North American workers (Snook and Ciriello, 1991), which are regularly updated at www.libertymutual.com.

(a) Handles in front of the operator (cm).
(b) Vertical distance of lowering (cm).
(c) Acceptable to 50, 75, or 90% of industrial workers.
Conversion: 1 kg = 2.2 lb; 1 cm = 0.4 in.

Altogether, there is good overlap among the recommendations. However, the NIOSH values are unisex whereas the Snook and Ciriello data are different for female and male workers. Both sets of recommendations indicate that, as to be expected, the quality of coupling between hand and load determines how much people are willing to exert in lifting and lowering (and carrying, pushing, or pulling). Missing handles, or objects that are so wide that they are difficult to grasp°, reduce the acceptable load values. Also, deep and repeated bending or twisting of the body and far reaches decrease the acceptable efforts. Of course, if several kinds of material handling occur together, the most strenuous task elements establish the activity level limits.

4.5 Designing for easy load handling

Back problems

Lifting, lowering, pushing, pulling, carrying, holding, and dragging loads involve static and dynamic efforts, often heavy work. However, the main problem with load handling is not the strain on muscles but mostly the wear and tear on the back with the risk of lasting injury, especially of the lumbar intervertebral discs. Low back pain° (often simply abbreviated LBP) reduces one's mobility and vitality; it often leads to long absences from work and, as statistics from North America show, is currently one the main causes of early disability. LBP is common even in the younger age groups, with certain occupations particularly likely to suffer from it: nurses, laborers, farmers, baggage handlers, and warehouse workers frequently suffer from back disabilities.

Injuries, especially in the lower back, account for about one quarter of all reported occupational cases in the United States, where some industries report that more than half of all injuries are due to overexertion. Accident and health statistics in the United Kingdom and Germany, for example, show similar figures. Clearly, low back pain is among the most common causes of injury and disability in many industrial populations.

Spinal loading

Many LBP victims cannot pinpoint when their back problems started. In most cases, there is not a certain moment or a specific action when pain appeared, but, rather, the problem developed slowly until it was strong enough to disable. The human spinal column has the shape of an elongated S, as shown earlier in Figure 2.9 of Chapter 2, with intervertebral discs providing elasticity and mobility. As we consider the body weight from the head on down, it becomes obvious that the loading on spinal segments increases from the neck downward; it is greatest at the bottom, in the lumbar area. Weights carried in the hands or on the shoulders (or in the belly of a fat person) increase the load on the spinal column, which is the only solid structure in the human rump that keeps the rib cage from falling into the pelvis.

"Slipped" disc

The spinal discs separate the vertebrae and provide shock absorption and flexibility. Disc degeneration occurs with aging and from repeated motions: when aging and wear come together, a sudden overload easily can lead to an acute injury. Disc degeneration primarily affects the outer layers; tissue change by loss of water can make that fibrous ring brittle and fragile. At first, the degenerative changes mostly make the disc flatter, which negatively affects shock absorption and mobility of the spine. In this case, even small actions such as lifting one's own body or a light load, a slight stumble or similar incident may bring disc injury and severe back pain. Progressive degeneration of the disc or a sudden load can cause disc herniation, where a sudden compression force may squeeze the gel-like nucleus out of the ruptured fibrous ring. Such damage often results in pressure on a spinal nerve, in narrowing the space between vertebrae and in generating tension in muscles and ligaments of the spine. These occurrences can bring about a variety of discomforts, aches, and disabling health problems such as lumbago and sciatica.

Disc strains in the bent back

Since the 1970s, researchers studied the effects of body posture, and of handling loads, on the pressure inside the intervertebral discs. Figure 4.7, based on one of the early studies, illustrates the changes in the loading of the disc between the third and fourth lumbar vertebrae (L3/L4) during various postures and tasks. Compared to no-load upright standing, standing straight with 10-kg weights in each hand naturally increases the disc pressure; a further increase accompanies holding a 20-kg load with bent knees and a slight bend in the back. Yet, the largest pressure occurs when the 20 kg are held with locked legs and a "round back". Table 4.7 lists compression forces in the L3/L4 disc in various postures and with various weights.

Further studies later performed[°] all show essentially the same result: a bent or twisted back causes higher and more dangerous pressure on the intervertebral discs than a straight back. Specifically, the bent back leads to heavy pressure on the front edge of the disc where it increases the risk of rupture and also generates shear, as Figure 4.8 illustrates. Straightening the back eliminates shear, makes the facet joints carry some of the

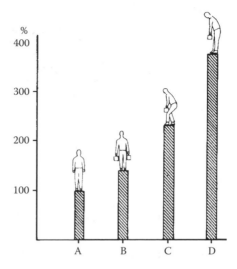

FIGURE 4.7 Strain on the disc between the third and fourth lumbar vertebrae with the body in various straight and bent postures while holding weights of 20 kg. (Adapted from Nachemson, A. and Elfstroem, G. (1970). Intravital dynamic pressure measurements in lumbar discs, *Scandinavian J. Rehabilitation Medicine*, Supplement 1, 1–40.)

Table 4.7   Intervertebral compression force in *N* in the L3/L4 disc

Posture/Activity	Force in *N*
Standing upright	860
Walking slowly	920
Bending trunk sideways 20°	1140
Twisting trunk about 45°	1140
Bending trunk forward 30°	1470
Bending trunk forward 30°, supporting weight of 20 kg	2400
Standing upright, holding 10 kg in each hand	1220
Lifting 20 kg with back straight and knees bent	2100
Lifting 20 kg with bent back and knees straight	3270

Source: Adapted from Nachemson, A. and Elfstroem, G. (1970). Intravital dynamic pressure measurements in lumbar discs, *Scandinavian J. of Rehabilitation Medicine*, Supplement 1, 1–40.

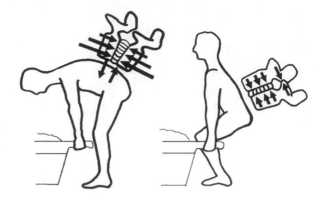

FIGURE 4.8 Spinal loading when lifting with bent and erect back

compressive load, and distributes the compression forces more evenly over the disc surface, all of which reduce the risk of damage.

A simple case of biomechanics

Figure 4.9 illustrates, in much simplified manner, a biomechanical approach° to approximate the pressure on the front edge of a disc: consider l, the length of the lever arm at which a load L pulls downward. (L may represent the weight of the belly, or an external load.) That moment, $L \times l$, is counteracted by a muscle force M which attaches to the spinous process of the vertebra with a lever arm m to the front edge of the disc. There, a force D keeps the biomechanical system in balance. The list of equations that relate these variables to each other appears in Figure 4.9. If we assume the load L to be 100 N, its lever arm 25 cm, and the lever arm of M is 5 cm, then the muscle force M turns out to be 500 N. (Obviously, that rather

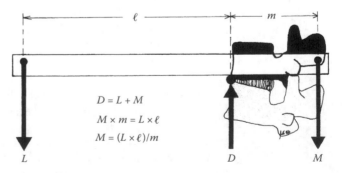

$$D = L + M$$
$$M \times m = L \times \ell$$
$$M = (L \times \ell)/m$$

FIGURE 4.9 A simple biomechanical model to calculate the pressure at the front edge of a vertebra

large value for the muscle force results from the disadvanta-geous ratio of lever arms, 25:5.) A simple addition of the vertical forces shows that the resulting force D at the front edge of the vertebra is 600 N. This force compresses the front portion of the intervertebral disc.

If we increase the load L by assuming an external load of 20 N carried in front, with an accompanying increase of the leverarm from 25 to 30 cm, the counteracting muscle force increases to 720 N, and the compressive force at the front of the disc to 840 N. This simple consideration demonstrates that one should keep a load as close to the front of the body as possible; following the same biomechanical principle, we realize that excess belly protrusion° increases the risk to the spinal column.

Intra-abdominal pressure and lift belts

Load handling, such as lifting, lowering, or carrying, produces a considerable increase of the pressure within the abdominal cavity, naturally accompanied by contraction of the abdominal muscles. The resulting intra-abdominal pressure helps in stabilizing the trunk and in reducing the loading of the spinal column and its supporting structures. (By the way, it is easy to measure that pressure: the subject simply swallows a capsule that contains a pressure-sensitive element.) That observation of increased intra-abdominal pressure led to the idea to put a stiff belt around the trunk, with the purpose of helping material handlers in industry, even competitive weight lifters, to avoid back problems. However, recent investigations have shown that the back belts are often ineffective° and cannot replace proper ergonomic solutions that avert the need for material handling, or at least make it much easier. Incidentally, even weight lifters who wear belts often suffer injuries.

Strong or weak material handlers

For design purposes, the ergonomist must determine single numbers for the strongest and the weakest strength exertions: "average user strength" regularly has no practical importance. (This remark is similar to the one made in Chapter 1 regarding the "average person" phantom.) However, as a *statistical tool* the average (mean) is very useful because it describes, together with the standard deviation, a normal (Gaussian) distribution of data. Multiplying the standard deviation SD of that distribution with, say, 4.25 (from Table 1.6 in Chapter 1), and adding that product to the mean (average), leads to the strength value near the 100th percentile, representing the strongest tabulated person. Looking at the other side of the distribution, subtracting 1.65 SD from the mean leads to the 5th percentile operator, a weak person who is stronger than just 5% of his fellow users. (This is the statistical technique to establish minimal and maximal design limits, discussed in Chapter 1.)

Identifying critical strength values

Figures 3.7 and 3.8 in Chapter 3 illustrate such statistical exercises: the first figure contains mean values with their standard deviations, the second lists 5th percentile strength values. Note that we can calculate an unknown value of SD from a table that lists percentiles. For example, the 5th percentiles in Table 3.8 are 1.65 SDs below the mean. The mean lies, of course, halfway between symmetrical percentiles, such as 5th and 95th, or 10th and 90th.

Nonnormal datasets

Calculations involving average and standard deviation are easy with normally distributed data; but such procedures lead to false results if the dataset is severely skewed or otherwise non-Gaussian. Unfortunately, many collections of strength data show nonnormal distributions. To determine specific points in a nonnormal data distribution, nonparametric procedures are appropriate; they are explained in statistics books°. In some cases, designers might be able to just "eye-ball" or otherwise estimate the percentile values of interest.

Making load handling easy

Machines don't have backs to hurt; therefore, load handling, especially lifting and lowering, is better performed by machinery than people. The second most effective way to avoid lifting and lowering is to convert those activities into carrying, or even better, into pushing and pulling, as shown in Figure 4.10. Carrying is best done on both shoulders, such as with a yoke, depicted in Figure 4.11. Of course, dollies or carts, as in Figure 4.12, can take over the carrying job, converting it into the least risky load-handling category, pushing and pulling. If a person has to generate the needed push or pull, rollers as shown in Figure 4.13, conveyors, or similar ways to facilitate the motion are the technical solutions of choice.

Learning to lift safely

It seems that one should be able to learn how to lift objects safely;° likewise, to do lowering, carrying, pushing, and pulling, and all other "manual material handling"° in safe ways. For example, one should learn not to exert strong force with the body twisted or the trunk severely bent, or in sudden movements. It should be natural to follow such simple advice, but apparently it is human nature as well to not always remember the proper ways. Numerous agencies and commercial outfits have developed systematic and often rather sophisticated ways

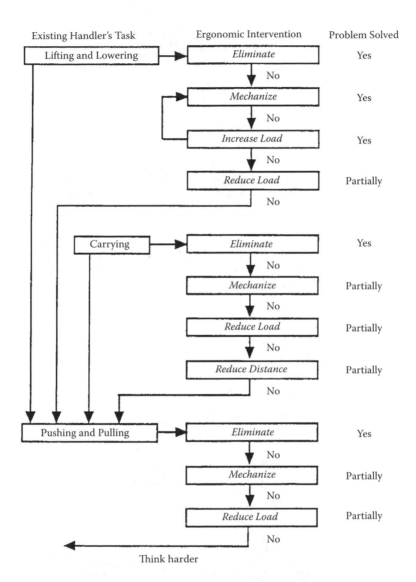

FIGURE 4.10 Converting lifting to carrying or, better, pushing or pulling. (Adapted from Kroemer, K.H.E. (1997). *Ergonomic Design of Material Handling Systems,* Boca Raton, FL: CRC Press.)

FIGURE 4.11 Carrying on both shoulders. (Adapted from Kroemer, K.H.E. (1997). ILO (1988). Maximum weights in load lifting and carrying. Occupational Safety and Health Series #59. Geneva, Switzerland: International Labour Office.)

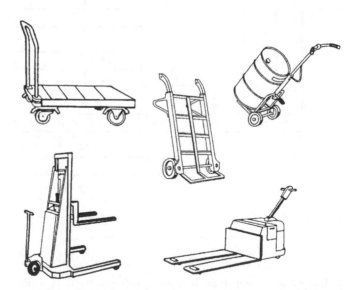

FIGURE 4.12 Using dollies and carts instead of carrying loads. (Adapted from Kroemer, K.H.E. (1997). *Ergonomic Design of Material Handling Systems*, Boca Raton, FL: CRC Press.)

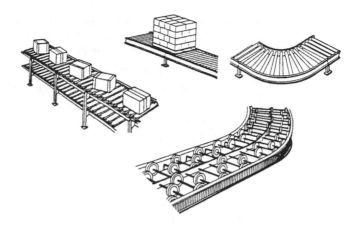

FIGURE 4.13 Rollers and conveyors to move loads. (Adapted from Kroemer, K.H.E. (1997). *Ergonomic Design of Material Handling Systems*, Boca Raton, FL: CRC Press.)

of instructing nurses, warehouse workers, miners, and masons, to mention a few, to perform material handling in safe ways. Unfortunately, many surveys and systematic evaluations of the outcomes of such instructions have shown that a large number of injuries occur even after proper techniques were taught. Hence, it appears that instructions and training are much less effective, if they are successful at all°, than proper engineering of the tasks to eliminate risks of overexertion and injury.

Summary

The engineer or designer wanting to consider operator strength must ask questions and make a series of decisions. These include:

1. Is the *action static or dynamic*? If static, information about isometric capabilities can be used. If dynamic, additional considerations apply, concerning, for example, physical (circulatory, respiratory, metabolic) endurance capabilities of the operator and prevailing environmental conditions. Sections 1, 2, and 5 of this book provide related information. Most body segment strength data are available for static (isometric) exertions. They provide reasonable guidance also for slow motions, although they are probably too high for concentric motions and a bit too low for eccentric motions. Of the little information available for dynamic strength exertions, much is limited to

isokinematic (constant velocity) cases. As a general rule, strength exerted in motion is less than measured in the static positions located on the path of motion.

2. Is the *exertion by hand or by foot*, or with other body segments? For each, specific design information is available as described above in this chapter.

3. Is a *maximal or a minimal exertion* the critical design factor? Maximal user strength determines the structural strength of the object, for example, so that even the strongest operator cannot break a handle or a pedal; therefore, the design value is to be set, with a safety margin, above the highest perceivable strength application. However, normal use depends on the ability of the weakest operator to perform the necessary task.

A general rule for load handling: Lifting and lowering are better performed by machinery than people; we can hurt our backs by simply bending to pick up small objects, because the weight of the body alone, and the biomechanical events within our spinal column, can lead to pain and injury. One way to avoid lifting and lowering is to convert those activities into pushing and pulling.

Fitting steps

Step 1: Determine which specific task must be done. Can a machine do it? If not,

Step 2: Decide on how an operator can do the work best, where in the movement space, with hand or foot.

Step 3: Design, then test. Modify as necessary.

Further reading

CHAFFIN, D.B., ANDERSSON, G.B.J., and MARTIN, B.J. (2006). *Occupational Biomechanics* (4th Edn). New York: Wiley.

KUMAR, S. (ED.) (2008). *Biomechanics in Ergonomics* (2nd Edn). Boca Raton, FL: CRC Press.

LIBERTY MUTUAL (regularly updated since 2004). *Manual Materials Handling Guidelines*. http://www.libertymutual.com/.

Notes

The text contains markers, °, to indicate specific references and comments, which follow.

Individual and situational strength factors: Daams 1993, 2001; Kumar 2004.

4.1 Static and dynamic strength: Astrand et al. 2004; Kroemer 1974, 1999; Marras et al. 1993; Winter 2004.

Special experiments: Hinkelmann et al. 1994, 2005; Williges 2007.

4.2 Maximal or minimal strength, nonparametric statistics: Gibbons 1997, Sprent 2000.

4.3 Hand strength: Adams 2001; DiDomenico et al. 1998, 2003; Kumar 2008.

Shaping the handle: Bullinger et al. 1997, Freivalds 1999, Konz et al. 2000, Kroemer et al. 2003.

4.4 Pulling and pushing:

Static tests: Dempsey 1998, Stanton et al. 2005.

Psychophysical testing: Psychophysics is the branch of psychology that deals with the relationships between physical stimuli and sensory responses. Dempsey 1998; Snook et al. 1991; **45**, 14, 2002 issue of *J. Ergonomics*.

Physiological criteria: these relate to metabolic and circulatory body functions, see Chapter 10 in this book.

Maximal pulling and pushing forces: Liberty Mutual 2004, Kumar 2004, Snook et al. 1991, Snook 2005.

Lifting and lowering:

"Safe" lift loads: ILO 1988.

Combinations of measurement criteria: Dempsey 1998, Kumar 2004, Stanton et al. 2005.

Pressure within the disc: Marras 2006.

1981 and 1991 NIOSH guidelines: NIOSH 1981, Putz-Anderson et al. 1991, Waters et al. 1999.

Maximal lift and lower weights: Liberty Mutual 2004, Snook et al. 1991, Snook 2005.

Objects that are difficult to grasp: Ciriello 2001, 2007; Kroemer 1997.

4.5 Designing for Easy Load Handling:

Low back pain LBP: Deyo et al. 2001, Violante et al. 2000. *Disc strains, further studies:* Chaffin et al. 2006, Kumar 2004, Marras et al. 2006.

Biomechanical approach: Biomechanics is the study of the mechanics of a living body, especially of the forces exerted by muscles and gravity on the skeletal structure. See Zhang et al. 2006, Oezkaja et al. 1991.

Belly protrusion: Marras et al. 2006.

Back belts ineffective: McGill 2006.

Nonparametric statistics: Gibbons 1997, Sprent 2000.

Learning to lift safely: Kroemer 1997, Kroemer et al. 2003. Note the tautology in the often-used term "manual material handling": the Latin word *manus* means hand.

If instructions and training are successful at all: Martimo et al. 2008 reviewed a large number of studies and found no evidence that advice or training in working techniques (with or without lifting equipment) prevents back pain or consequent disability. The findings challenge current widespread practice of advising workers on correct lifting techniques.

The human mind

This second part of this book on "fitting the human" discusses how we receive and perceive information about our environment. Our organ that deals with such information is generally called the mind, consisting physically primarily of our body's network of nerves and the brain. This command and control center has many complex functions, of which the following Chapter 5 treats how and what we see, Chapter 6 deals with our hearing, Chapter 7 concerns our sensing of objects and energy, and Chapter 8 describes how we interact with the climate around us.

CHAPTER FIVE

How we see

The eye senses energy from the outside world in the form of light rays and converts these into nerve impulses, which the brain integrates into a visual picture of the outside world. However, this perceived picture is a subjective modification of what the eye reports. Examples are:

A straight line appears distorted against a background of curved or radiating lines.

A color seems darker when seen against a bright background than when appearing on a dark background.

The perceived hue and intensity of greenish blue are different from person to person, and may change when the beholder's eyes grow old.

People differ greatly in their interpretation of visual data, depending on their age, experience, attitude, and preconceived ideas. People also differ in their abilities to recognize colors and focus clearly on visual targets, and usually significant changes in these abilities occur as one ages. Nevertheless, there is general similarity in how the human visual sense functions, which allows us to develop ergonomic recommendations.

5.1 Our eyes

The eyes continually adjust the amount of light they let in, change their focus on objects far and near, and produce continuous images, which they instantly transmit to the brain.

The eyeball is a roughly spherical organ, about 2.5 cm in diameter, surrounded by a layer of fibrous tissue called sclera. Light enters through the cornea, a transparent dome on the

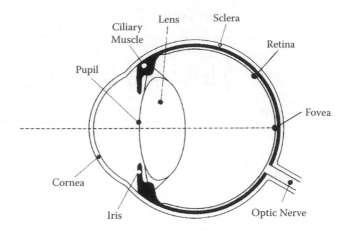

FIGURE 5.1   The right eye as seen from above

front surface of the eye. The cornea serves as both a protective covering and a weak lens that helps to focus light on the retina at the back of the eye; see Figure 5.1.

Pupil in the iris   After passing through the cornea, light enters the pupil, the opening that appears as a small round black area in the middle of the iris. (The pupil looks black because no light emerges from the inside of the eye. However, the pupil can appear red in a photo when a flash lights up the inside, which has many blood vessels.) The iris is the circular colored area of the eye. The pupillary dilator and sphincter muscles open and close the pupil like the aperture of a camera lens that controls the amount of light that enters the eye. It lets more light in when the environment is dim but admits less light when the surroundings are bright.

The lens focuses   The lens is behind the iris. It focuses light onto the retina by changing its shape. When thin, it is a weak lens (in the optical° sense) that focuses on distant objects. The contracting ciliary muscle pulls on the lens and thus makes it thicker and optically stronger so that it may focus on nearby objects. The eyes of a healthy young adult can usually focus on an object as close as 10 cm, but with aging, the lens becomes less flexible and hence less able to thicken, diminishing its ability to deal with nearby objects. As a consequence, the accommodation distance commonly increases. This condition is called farsightedness, presbyopia.

Light focused by the cornea and lens travels through the vitreous humor, a gel-like fluid with refractory properties similar to those of water that fills the interior of the eyeball. Finally, the light reaches the retina, a thin tissue lining about three-quarters of the rear inner surface of the eyeball. Many arteries and veins amply supply the retina with blood.

Rods for white/ grey/black perception

The retina carries about 130 million light sensors. Their arrangement is densest in the center, the fovea, directly behind the lens. The light sensors are of two kinds, named for their shape. The majority, about 120 million, are rods. They contain only one pigment, which responds to even low-intensity light. Rods set off electrical impulses that travel along the optic nerve to the brain for the perception of white, black, and shades of gray. Rods provide us with the most and the most important visual information.

Cones signal color

In addition, the retina has about 10 million cones, located mostly in the fovea. They respond to colored light if it is sufficiently bright. Each cone contains one pigment that is most sensitive to either blue, green, or red wavelengths. An arriving light beam, if intense enough, triggers chemical reactions in one of the three types of pigmented cones, creating electrical signals that pass along the optic nerve to the brain which can compose and distinguish among about 150 color hues.

Optic nerve

The optic nerve exits the eye at its rear, not directly in line with the centers of the cornea and lens but rather offset from the fovea by about 15 degrees toward the inside (medially). Because there are no light sensors in this area, we cannot see an image refracted on this blind spot. However, because both blind spots of the eyes are medially located, they do not overlap in our field of vision and therefore we are usually unaware of their existence.

Visual control system

Proper vision requires continuous actions of a complex control system, sketched in Figure 5.2. It shows, first, the adjustment of the lens (1) behind the cornea to provide a sharply focused image on the retina. The information (2) about that image transmits along the optic nerve (3) to the brain. Various nervous control mechanisms (4) readjust the position of the eyeball, the size of the pupil and the form (refractive property) of the lens continually. The visual perception of the external world takes place in the conscious sphere of the brain (5), and leads to command signals that travel through the spinal cord (6) to instigate appropriate actions of the body.

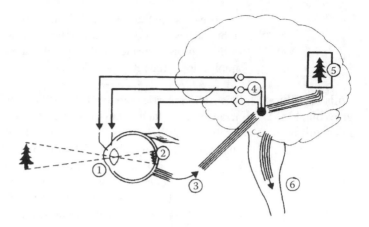

FIGURE 5.2   Diagram of the visual system

5.2 Seeing the environment

The visual field

We can see objects in the visual field°, which is in a roughly conical space in front of the eyes. Yet, within this space, we can discern visual targets with high acuity only if they appear within a very narrow cone. For example, when we focus on written text, only a few adjacent letters appear clearly and the surrounding letters blur, more so the farther away they are.

Fixated eyes

The areas that the two eyes can see do not overlap perfectly, partly because the nose is in the way. When the eye maintains fixation on a single spot straight ahead, researchers measure the size of the visual field by presenting test objects away from this position. Looking to the outside, we can determine the presence of objects within about 90 degrees to the side but only within the inner about 65 degrees can we perceive their color; looking up, we can achieve about 45 degrees yet see color only to about 30 degrees upward; and looking downward, we attain about 70 degrees but the area in which we see color only extends down to about 40 degrees.

Moving the eyes

Rotating the eyeball increases the visual area beyond the field of fixation, adding considerably to the outside of the visual field but nothing in the upward, downward, or inside directions, because the eyebrows, cheeks, and nose stay in place.

Several muscles attach to the outside of the eyeball, working together to move the eye. Rotational movements in pitch, yaw, and roll are most prominent but some forward and back motion of the eyeball also occurs. We can rotate our eyes up to about 50 degrees both in pitch (up and down) and yaw (left and right). However, the eyes seldom rotate to these extreme angles. When we try to look at an object at the periphery of our field of vision, the initial fixation is mostly by eye rotation, but this is quickly replaced by head movement. Given this strong preference to use body movement rather than eye motion, humans commonly use adjustments of body posture to fixate on a peripheral target so that the eyes can operate comfortably and precisely close to their normal resting positions.

The eye can track continuously a visual target that moves left or right at less than 30 degrees per second or cycles at less than 2 hertz. Above these rates, the eye is no longer able to follow continuously but must move in saccades: it lags behind and then jumps to catch up.

Avoid eye fatigue

So-called eye fatigue is often the result of excessive demands on the muscles of the eye; those that move the eyeball, and those that adjust lens and iris. This is particularly a problem for older people whose lenses have become stiff and who often find it difficult to move neck and trunk as easily as they did in their youth. Apparently, many instances of fatigue of which computer users often complain are related to the poor placement of monitor, source documents, or other visual targets (see Section V in this book) or to unsuitable lighting conditions at their workplace, as discussed later in this chapter.

Therefore, it is important to arrange visual targets properly: in front, not to the side; at a distance to which the eyes can easily accommodate; and so low that we can rotate the eyeballs slightly down while keeping trunk, neck, and head in comfortable positions. Makers of eyeglasses and optometrists always knew that looking down on small visual targets, such as printed text, is easier than looking up; therefore, they put the reading section of (bifocal or trifocal) corrective spectacles into the lower parts of the lenses.

Line of sight

When we look at an object in front of us, we unconsciously adjust two pitch angles: that of the eyes within the head and that of the head against the trunk. As we fixate on a target, within the eye the line of sight (LOS) runs from the center of the retina through the midpoints of lens and pupil (and from there to the target). Thus, inside the eyeball the LOS is clearly established. However, a suitable reference is needed to describe the LOS direction from the eye to the visual target.

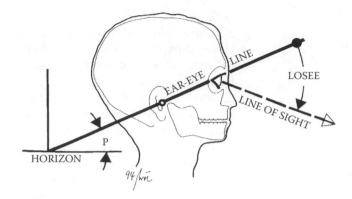

FIGURE 5.3   Line of sight angled against the ear–eye line

Looking down on the job

In the past, the so-called Frankfurt plane served as reference but its anatomical definition by landmarks on the bony skull made it difficult to use. The ear–eye (EE) line is simpler to establish because it runs through the easily discerned earhole and the junction of the eyelids, as shown in Figure 5.3. The angle between the LOS and EE is the pitch angle of the eye, LOSEE, as it looks at the target. LOSEE° is best around 45 degrees when reading a text, on paper or a computer screen. It is natural to look down at close visual targets, such as a written text. LOSEE becomes smaller as the observed object gets farther way; it is about 15 degrees (horizontal) when we look straight ahead.

The position angle P describes how we hold the head compared to the horizon. When the head is upright, P is about 15 degrees.

Size of the visual target

If a visual target is not a point but has a measurable length or breadth perpendicular to the line of sight, then the target size is usually expressed as the subtended visual angle, the angle formed at the pupil. The magnitude of this angle α depends on the size L of the object and on the distance D from the eye, as described in Figure 5.4. The subtended visual angle is usually

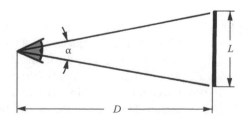

FIGURE 5.4   The subtended visual angle

Table 5.1 Visual angles of familiar objects

Object	Distance L	Visual Angle (arc)
Instrument, 5-cm diameter	0.5 m	5.7 degrees
The sun, the moon		About 30 minutes
Character on computer monitor	0.5 m	About 17 minutes
Book letter at reading distance	About 0.4 m	About 13 minutes
U.S. quarter coin	At arm's length, 0.7 m	2 degrees
	At 82 m (90 yards)	1 minute

Source: Adapted from Kroemer, K.H.E., Kroemer, H.B., and Kroemer-Elbert, K.E. (2003, amended reprint). *Ergonomics: How to Design for Ease and Efficiency* (2nd Edn). Upper Saddle River, NJ: Prentice-Hall/Pearson Education.

given in degrees of arc (with 1 degree = 60 minutes = 60 × 60 seconds of arc). The equation is

$$\alpha \text{ (in degrees)} = 2 \times \arctan (0.5 \times L \times D^{-1})$$

For visual angles not larger than 10 degrees, this can be approximated by α (in degrees) $= 57.3 \times L \times D^{-1}$ or by α (in minutes of arc) $= 60 \times 57.3 \times L \times D^{-1} = 3438 \times L \times D^{-1}$.

The human eye can perceive, at a minimum, a visual angle of approximately 1 minute of arc. Table 5.1 lists the visual angles of some familiar objects. For ease of use, technical products should be designed so that the angle subtends at least 15 minutes of arc, increasing to at least 21 minutes of arc at low light levels.

Diopter

If the target distance D is measured in meters, the reciprocal, $1/D$, is called the diopter. The diopter indicates the optical refraction needed for best focus. Thus, a target at infinity has the diopter value zero, whereas a target at 1 m distance has the diopter value unity (one). Table 5.2 shows values for some typical target distances.

Focusing

Accommodation means the ability of the eye, mostly through shaping its lens, to bring into sharp focus objects at varying distances, from infinity down to the nearest point of distinct

Table 5.2 Focal points, in diopter, and target distances

Target Distance D (m)	Focal Point (Diopter)
Infinity	0
4	0.25
2	0.5
1	1
0.67	1.5
0.5	2
0.33	3
0.25	4
0.2	5
0.1	10

vision. If we hold up a finger in front of the eye, we can sharply focus on it, but the background remains blurred. Alternately, we can concentrate on features of the background, leaving the finger indistinct. We can only see clearly those objects whose images appear focused on the retina.

Incessant changes

If our gaze sweeps to various objects in the near field of vision, the lens must continuously change its curvature to keep adapting its focal length so that sharp images always appear on the retina; even when maintaining its focus on a near target, muscles continuously adjust their contracting forces. For example, when we read a text, the lens does not hold still but oscillates at a rate of about four times per second. Similarly, the iris continuously changes the size of its central opening, the pupil. This keeps its dilator (opening) and sphincter (closing) muscles constantly active to adjust the diaphragm of the eye according to the light conditions in the visual field. During daylight, the pupil size is usually 3 to 5 mm in diameter, which increases at night to more than 8 mm. The aperture of the pupil contracts when we focus on near objects and opens when the lens relaxes. Furthermore, the pupil reacts to emotional states, dilating under such strong emotions such as alarm, joy, pain, or intense mental concentration; it narrows with fatigue and sleepiness. Therefore, changes in pupil size have been used to assess attention and attitudes.

Overcoming ocular problems

A healthy young eye can accommodate from infinity to very close distances, such as 10 cm, meaning that a diopter range from zero to about 10 can be achieved. The minimal distance increases to about 5 diopter (20 cm) at about age 40 and to about 1 diopter (1 m) at age 60, on average. With increasing age, the accommodation capability of the eye decreases, because the lens becomes stiffer by losing water content. The result is difficulty in making light rays from targets at differing distances converge exactly on the retina. If the convergence is in front of the retina, the condition is called myopia; if behind, hyperopia; see Figure 5.5. A nearsighted (myopic) person has no trouble seeing close objects but finds it difficult to focus on far targets. This condition often improves with age, when commonly the lens remains flattened. In contrast, farsightedness (hyperopia) usually becomes more pronounced with age, meaning that it

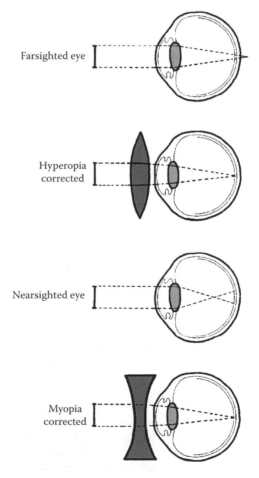

FIGURE 5.5 Convex and concave lenses can correct refractive (focusing) eye deficiencies

gets even more difficult to focus on near objects. Both problems, myopia and hyperopia, can be corrected fairly easily by either contact lenses or eye glasses, as shown in Figure 5.5. (Note that the refractive properties are measured when the eye lens is relaxed.)

More light!

In many people, the pupil shrinks with age. This means that less light strikes the retina, and therefore, many older people need to have increased illumination on visual objects for sufficient visual acuity. Another problem often encountered with increasing age is yellowing of the vitreous humor. The more yellow it becomes, the more energy it absorbs from the light passing through; consequently, again, increasing the illumination of the visual target helps acuity. However, light rays may be refracted within the yellowed vitreous humor, bringing about the perception of a light mist or veil in the visual field. If bright lights are in the visual field, the resulting "veiling glare" can strongly reduce vision. Obviously, artificial lenses cannot correct the yellowing problem.

So-called floaters appear like small flecks in front of the eye. In reality, they consist of small clumps of gel or cells suspended in the vitreous humor. Frequently they are not noticed because the eye adjusts to these imperfections. Floaters are visible to the individual only when they are on the line of sight, casting a shadow on the retina. They are more easily perceived when one is looking at a plain background. Fortunately, occasional floaters are usually harmless.

Another frequent and more serious problem in the aging eye is having cataracts, patterns of cloudiness inside the normally clear lens. When vision is severely impaired, an eye surgeon may remove the cloudy lens and replace it with an artificial implant.

Glaucoma is the leading cause of blindness in the United States, especially among older people. It is a disease of the optic nerve, related to high pressure inside the eye. Regular eye examinations help to detect the beginning of glaucoma and prevent further damage.

There are other vision deficiencies as well, usually not related to aging:

- Astigmatism occurs if the cornea is not uniformly curved, so that, depending on its position within the visual field, an object is not sharply focused on the retina. Often, the astigmatism is a spherical aberration, meaning that light rays from an object located at the side are more strongly refracted than those from an object at the center of the field of view, or vice versa.

- Chromatic aberration is fairly common. An eye may be hyperopic for long waves (red) and myopic for short waves (violet or blue). Placing an artificial lens in front of the eye usually solves problems associated with astigmatism.

- Night blindness is the condition of having less than normal vision in dim light, that is, with low illumination of the visual object.

In North America, color vision deficiencies occur in approximately 8% of the male population whereas only about 0.5% of the female population has that problem. These individuals are not completely color blind, but they have significantly reduced discrimination. A small percentage of the sufferers are missing one of the cone systems and are unable to distinguish some basic colors from one another. Color weakness is more prevalent: here, individuals cannot distinguish as many gradations of colors as people with normal color vision do. Only very few people can see only one color or none at all.

5.3   Dim and bright viewing conditions

The human eye senses and can adapt to increases and decreases in illumination° over a wavelength range of about 380 nm to 720 nm, that is, from violet to red. We can see objects if they are bright enough (either by generating light as the sun does, or by reflecting it) to carry sufficient visible energy onto the retina. The minimal intensity required to trigger the sense of light perception on the retina is 10 photons, causing an illuminance of about 0.01 lux. At such low intensity, the main perception is of dim light, not of color, because only rods are activated. When the illuminance exceeds 0.1 lux, both cones and rods respond, and cones report colors. We experience this at twilight and dawn, when we can see color in the brighter sky, but all dimmer objects appear only in shades of gray as reported by the rods.

Some interesting events can occur in darkness°.

- If one stares at a single light source on a dark background, the light seems to move. This is the so-called autokinetic phenomenon.

- Night vision capabilities deteriorate with decreasing oxygen. Thus, at an altitude of about 1300 m, vision is reduced by about 5%; at 2000 m, the reduction is about

20% yet up to 40% in smokers whose blood has lost some capability to carry oxygen.

- If the horizon shows no visual cues, the lens relaxes and focuses at a distance of about 1 to 2 m, making it difficult for a person to notice faraway objects. This is known as night myopia.

Adaptation to light and dark

The eye can change its sensitivity through a large range of illu-mination and luminance° ranging from dark to bright condi-tions. Figure 5.6 scales the luminance values that we perceive.

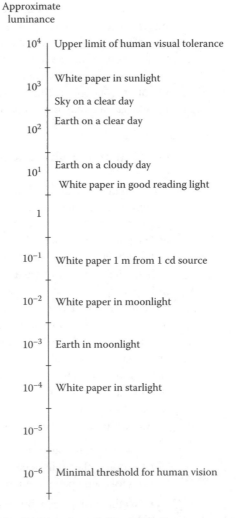

Approximate luminance

10^4	Upper limit of human visual tolerance
10^3	White paper in sunlight
	Sky on a clear day
	Earth on a clear day
10^2	
10^1	Earth on a cloudy day
	White paper in good reading light
1	
10^{-1}	White paper 1 m from 1 cd source
10^{-2}	White paper in moonlight
10^{-3}	Earth in moonlight
10^{-4}	White paper in starlight
10^{-5}	
10^{-6}	Minimal threshold for human vision

FIGURE 5.6 Luminances (in candela per square meter) of objects that humans can see

Adjustments of the pupil, the spatial summation of stimuli, and the stimulation of rods and cones all contribute. The actual change in response thresholds of the eye during adaptation to dark or light depends on the luminance and duration of the previous condition to which the eye had adapted, on the wavelength of the illumination, and on the location of the light stimulus on the retina.

Full adaptation from light to darkness takes up to 30 minutes; during this period, initially the cones are most sensitive, and the rods follow more slowly. After adaptation, the sensitivity at the fovea (with its many cones) is only about one-thousandth of that at the periphery of the retina (where there is a preponderance of rods). Therefore, weak lights can be noticed in the periphery of the field of view, but not if one looks directly at them to make them appear on the fovea.

Adaptation to darkness mostly depends on the change in threshold in the cones. People who suffer from "dark blindness" have nonfunctional rods and can adapt only via cones. Persons who are "color blind" have cone deficiencies (as described above) and adapt via rods only.

In contrast, adaptation to light is quick, fully achieved within a few minutes. During that period, the fovea perceives wavelengths in the yellow region of the visible spectrum most easily. Therefore, it is best to illuminate instruments in vehicles driven at night with reddish or yellowish light, with the yellow color probably more suitable for the older eye. This color arrangement maintains the dark adaptation of the rods so the driver can still observe the mostly black–grey–white events on the road while driving a car. Furthermore, we adjust faster to red and yellow lights than to blue light during adaptation to darkness, after having been in a lighted environment.

Seeing requires light

We cannot see objects without light. Some objects generate light, such as the sun, a lamp, or an electronic display. Other objects reflect light, for example, the moon, the walls of a room, or a page of print. Unless we look directly into a light source, luminance, the light energy reflected from a surface, activates the eye and, hence, is the most important factor for human vision. Because the human eye adapts to lighting conditions, our eyes do not convey reliable information on the absolute lighting level, but rather on the variations in light; they respond to contrasts in luminance and color over time and space.

Visual acuity

Visual acuity can be defined in several ways, usually as the ability to detect small details and discriminate small objects

Snellen Letter   Landolt ring   Grating   Vernier

FIGURE 5.7   Examples of visual targets used in acuity testing

by eyesight. Visual acuity depends on the shape of the object and on the wavelength, illumination, luminance, contrast, and duration of the light stimulus. Acuity is usually measured at viewing distances of 6 m (20 ft) and 0.4 m (1.3 ft), because factors that determine the resolution of an object differ in far and near viewing.

Acuity testing

To assess visual acuity, high-contrast patterns are presented to the observer at a fixed distance. The most common testing procedures use either Snellen letters or Landolt rings; see Figure 5.7. The smallest detail detected or identified is taken as the threshold. These measures of acuity depend primarily on the ability to see edge differences between black and white stimuli at rather high illuminance levels. Such measurement of static edge acuity is simple, but it is neither the only nor the best measure of visual resolution capabilities. For example, people with perfect Snellen acuity may not do well detecting targets on a busy background or observing highway signs at given distances.

Color perception

Sunlight contains all visible wavelengths of the spectrum, but objects onto which the sun shines absorb some of the radiation. Thus, the light that we see on objects is what they transmit or reflect; its energy distribution differs from the light that the objects received. However, a human looking at the object does not have to analyze the spectral composition of the light reaching the eyes; in fact, what appears to be of identical color may have different spectral contents°.

Trichromatic vision

Color-matching experiments show that the human can perceive the same color stemming from various combinations of the three primary colors: red, green, and blue. Therefore, human color vision is called trichromatic°.

Color is an experience

The brain does not measure wavelengths; it simply classifies incoming signals from different groups of wavelengths. We judge colors by comparison and we name them by habit. Some societies do not distinguish between certain colors of the spectrum and lump blue and green together, while others have many words for shades of red. Human color perception is a psychological experience, not a single specific property of the electromagnetic energy we see as light.

Aesthetics and psychology of colors

The physics of color stimuli arriving at the eye can be described well (although often with considerable effort); however, perception, interpretation, and reaction to colors are highly individual and variable. Thus, people find it very difficult to describe colors verbally, given the many possible combinations of individually perceived hue, lightness, and saturation values. People may experience emotional reactions to color stimuli. For example, reds, oranges, and yellows are usually considered warm and stimulating. Violets, blues, and greens are often felt to be cool and to generate sensations of cleanliness and restfulness. However, the attraction to certain colors and their combination varies culturally and regionally, for example, between Asia and Europe.

Designing illumination

The characteristics of human vision provide the bases for engineering procedures to design environments for proper vision. Here are the most important concepts.

- Proper vision requires appropriate luminance of an object, that is, the energy reflected or emitted from it, which meets the eye.
- Luminance of an object is determined by its incident illuminance, and by how much thereof it reflects.
- Carefully select quantity and direction of illumination.
- Special requirements on visibility, including the decreased seeing abilities of the elderly, require particular care in the arrangement of proper illumination.
- Use of colors, if selected properly, can be helpful, but color vision requires sufficient light.

Summary

The eyes provide a huge portion of the information that we need in daily life and at work. In order to see objects and events in detail, they must appear in bright light. When the light dims, objects lose their colors; in the dark, they are invisible. Even a well-lit visual target must be at such a distance from our eyes that we can distinguish particulars. If our eyesight is defective, which usually occurs as we age, it is especially important to have good lighting and we may have to use artificial lenses.

Fitting steps

Step 1: Make sure that the eyes function properly; if in doubt, see an eye physician.

Step 2: Provide a suitable combination of visual targets/tasks and illumination.

Step 3: Adjust conditions as desired.

Further reading

BOYCE, P.R. (2003). *Human Factors in Lighting*. London: Taylor & Francis.

SHEEDY, J. (2006). Vision and work. In Marras, W.S. and Karwowski, K. (Eds.) *The Occupational Ergonomics Handbook* (2nd Edn). *Fundamentals and Assessment Tools for Occupational Ergonomics*, Chapter 18. Boca Raton, FL: CRC Press.

Notes

The text contains markers, °, to indicate specific references and comments, which follow.

5.1 Our Eyes: Optics: The branch of physics that deals with vision and visible light.

5.2 Seeing the Environment:

The visual field: Sheedy 2006.

LOSEE is best: Kroemer 1994; Sheedy 2006.

5.3 Dim and Bright Viewing Conditions:

Illuminance, illumination: light falling upon an object.

Luminance: light reflected/emitted from an object. Definition and measurement of light has been complicated in the past by the use of different procedures and units, as explained by Boyce 2003; Howarth 2005.

Spectral contents of color: For measuring and specifying colors see, for example, Howarth 2005. For ergonomic applications see, for example, Kroemer et al. 2001, 2003.

Trichromatic vision: The data of a color-matching experiment can be displayed as vectors in a three-dimensional (red, green, and blue) space, called a chromaticity diagram, standardized in 1931 and still in common use.

CHAPTER SIX

How we hear

The ear senses energy from the outside world in the form of pressure waves in the air and converts these into nerve impulses, which the brain integrates into a psycho-acoustical perception of the outside sound world. This interpretation is a subjective modification of what the ear reports. For example, the perceived sound and appeal of music differ from person to person, and are likely to change as the person grows older.

People differ greatly in their interpretation of acoustical events, depending on their experience, attitude, and preconceived ideas. People also differ in their ability to recognize sounds and react to them, and aging regularly brings significant changes. Nevertheless, existing similarities in how hearing functions allow us to develop ergonomic recommendations.

6.1 Our ears

Pathways of sound

Sound can reach the inner ear via two different paths. Sound may be transmitted through bony structures, but this requires very high intensities to be effective. Normally, all sound that we perceive is airborne and travels through the ear canal, where it excites the eardrum and the structures behind it, as described below.

The outer ear (auricle or pinna) collects and funnels airborne sound waves into the auditory canal (meatus). At its end, the eardrum (tympanic membrane) closes the canal, separating the air-filled middle ear from the environment. More to the inside, two other membranes separate the middle ear from the inner ear by closing two "windows", one called oval, the other round. A watery fluid (called endolymph or perilymph) fills the inner ear, which carries the organs that sense sound (and body

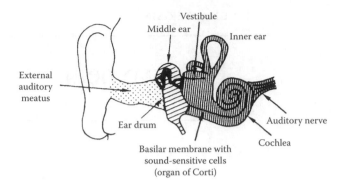

FIGURE 6.1   Main features of the ear anatomy: the air-filled outer and middle ear sections, separated by the eardrum, and the fluid-filled inner ear with the cochlea and the vestibule

position, discussed in Chapter 7). Figure 6.1 shows the main features of the ear anatomy.

Outer ear

An arriving sound wave makes the ear drum vibrate according to the frequency of the sound. Resonance effects within the outer ear and the canal amplify the intensity of the sound by 10 to 15 decibels (dB; see below for an explanation of the units) by the time it reaches the eardrum.

Middle ear

Three bones (ossicles) in the middle ear—the hammer (malleus), anvil (incus), and stirrup (stapes)—mechanically transmit the sound from the eardrum to the oval window.

With the area of the eardrum larger than the surface of the oval window, the sound pressure that enters the inner ear at the oval window's membrane is about 22 times larger than it was at the eardrum.

Inner ear

The inner ear contains the receptors for hearing (and for body position, the vestibulum; see Chapter 7). Fluid shifts in the inner ear propagate sound waves from the oval window to the round window through the cochlea (snail), an opening shaped like a snail shell that has two and one-half turns. The motions of the fluid deform the basilar membrane, which runs the length of the cochlea. The movements stimulate sensors, the feathery hair cells (cilia) and the organs of Corti, located throughout the length of the cochlea. Depending on their structure and location, the Corti organs respond to specific frequencies and generate impulses for transmission along the auditory nerve to the brain for interpretation.

The Eustachian tube connects the middle ear with the pharynx (part of our breathing apparatus). When the tube is open, it allows the air pressure in the middle ear to remain equal to the external air pressure. But when the tube is obstructed, such as in a case of a cold or an ear infection, pressure equalization may not function, and one feels ear pressure, even pain, and cannot hear well. In an airplane, especially during rapid descent, a clogged Eustachian tube can delay the equalization of pressure between the inner ear and the environs. You may try to open the tube by chewing gum or by willful excessive yawning, but "pumping" your outer ear with the hand will not help your middle ear.

6.2 Hearing sounds

Sound

Sound is a vibration that stimulates an auditory sensation. A tone is a single-frequency oscillation, whereas a sound contains a mixture of frequencies. The measurement unit of frequencies is hertz (Hz, oscillations per second). We often describe our personal perception of tone frequencies as "pitch".

Frequencies that we hear

We cannot hear sounds below 16 Hz but we may feel such infrasonic vibrations°; nor can we hear ultrasonics, above 20 kHz, but dogs and other animals do. Infants can hear tones from about 16 Hz to 20,000 Hz (20 kHz), a span of nearly nine octaves. With aging, the human ability to hear high frequencies strongly diminishes. Old people can rarely hear high tones above 10 kHz. Our normal hearing is most sensitive between about 2 and 5 kHz. In human speech, vowel sounds are below 1 kHz, but sibilant consonants can exceed 5 kHz; yet most speech occurs within the 300 to 700 Hz range.

How loud?

"Loudness" describes our sensation of the intensity (sound pressure level, power, amplitude) of an audible vibration. The minimal pressure threshold of hearing is about 20 micropascals (20×10^{-6} N m^{-2}; 1 Pa = 1 N per m^2) in the frequency range of 1000 to 5000 Hz. When the sound pressure exceeds 140 Pa, the ear experiences pain. Figure 6.2 shows the ranges of human hearing.

10 Decibels = 1 Bel

Objectively, loudness is stated in a logarithmic unit° called the decibel (dB). One reason for the use of the logarithmic scale is that the human perceives sound pressure in a roughly logarithmic manner. The sound pressure level is the ratio between two

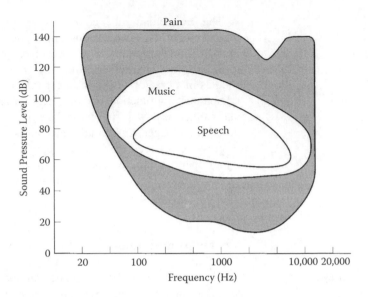

FIGURE 6.2 Typical hearing ranges of adult humans. (Adapted from Kroemer, K.H.E., Kroemer, H.B., and Kroemer-Elbert, K.E. (2003, amended reprint). *Ergonomics: How to Design for Ease and Efficiency* (2nd edn) Upper Saddle River, NJ: Prentice-Hall/Pearson Education.)

sound pressures: P is the present (rms) sound pressure and P_o is the threshold of hearing, which serves as reference. Thus, the definition of sound pressure level is $SPL = 10 \log(P^2/P_o^2)$ or

$$SPL = 20 \log_{10}(P/P_o) \text{ in dB}$$

Thus, in decibels, the dynamic range of human hearing from 20×10^{-6} Pa to 200 Pa is $20 \log_{10}[200/(20 \times 10^{-6})] = 140$ dB; see Figure 6.2.

The sound intensity level is, accordingly,

$$SIL = 10 \log_{10}(I^2/I_o^2) \text{ in dB}$$

where I is the (rms) sound intensity and I_o is 10^{-12} watts per m². Figure 6.3 illustrates the ranges of sound intensity levels.

Emitted sound power (such as from a loudspeaker or a running engine) is measured in Watts; it relates to sound pressure via W/m² = P² / [(air density) × (speed of sound)]. Thus, sound power is proportional to the square of sound pressure.

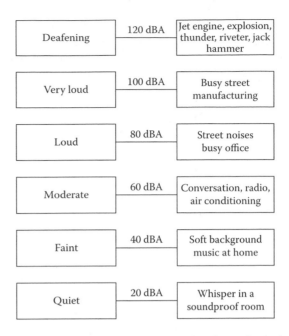

FIGURE 6.3 Typical sound intensity levels. (Adapted from Kroemer, K.H.E., Kroemer, H.B., and Kroemer-Elbert, K.E. (2003, amended reprint). *Ergonomics: How to Design for Ease and Efficiency* (2nd edn) Upper Saddle River, NJ: Prentice-Hall/Pearson Education.)

Sounds occurring together

Because of the use of logarithms, doubling the SPL causes an increase of 6 dB; a 1.41-fold increase in sound pressure causes an increase of 3 dB. If two sounds occur at the same time, one can approximate their combined SPL from the difference between their intensities as follows.

If the difference is 0 db, add 3 db to the louder sound.

If the difference is 1 db, add 2.5 db to the louder sound.

If the difference is 2 db, add 2 db to the louder sound.

If the difference is 4 db, add 1.5 db to the louder sound.

If the difference is 6 db, add 1 db to the louder sound.

If the difference is 8 db, add 0.5 db to the louder sound.

If the difference is 10 dB or more, take the louder sound alone.

Accordingly, one can measure the sound level emitted by a specific source, for example, a loudspeaker or a running machine, even when background sound is present if the source is at least 10 dB louder than the background.

Psychophysics of hearing

Physical measurements describe acoustical events, but people interpret them and react to them in very subjective ways; for example, finding certain sounds either attractive or annoying. The sensation of a tone or complex sound depends not only on its intensity and frequency, but also on how we feel about it.

Loudness

Because of the special sensitivities of human hearing, both frequency and intensity of a sound affect our perception of its loudness in a rather complex manner. Truly earsplitting sound, of 100 dB and more, remains earsplitting at all frequencies, but quieter sounds follow different patterns, especially so at frequencies below 1 kHz. Compared to its intensity at 1000 Hz, at lower frequencies the sound pressure level often must be increased so that it feels equally loud. For example, at 50 Hz a tone must have about 75 dB to sound as loud as a 1000 Hz tone with 50 dB. At high frequencies, however, especially in the range of approximately 2000 to 5000 Hz, the tone intensity can be lowered and it still sounds as loud as at 1000 Hz. (This is where our healthy hearing is most sensitive.) Yet, above about 8000 Hz, the intensity must be increased again above the level at 1000 Hz to sound equally loud. Figure 6.4 shows equal-loudness contours, called phon curves.

These curves of equal loudness are valid only for pure tones; they do not reflect reliably the subjective impression of loudness if we hear different frequencies at the same time, which is the case in most of our daily environments. Nevertheless, the phon curves do indicate effects of different frequency ranges on the human ear; for example, low frequencies at small intensities are difficult to hear, as Figure 6.2 also shows.

Measuring human hearing

Filters on sound-measuring equipment can imitate human sensitivity to tones of different frequencies. These filters, usually in the form of software, correct the physical readings to what the human perceives. The A filter corresponds best to the human hearing response at 40 dB. A-corrected SPL values are identified by the notation dBA or dB(A).

Responses to music

Music is among the oldest human expressions and art forms. Making music has long accompanied activities; examples are singing during fieldwork and rhythmical tunes while marching. Yet even today little is known systematically about the psychophysical consequences of different kinds of music and rhythm and their effects on well-being and productivity.

"Music while you work" is in many respects the opposite of background music°, discussed below. It is not continuous, but

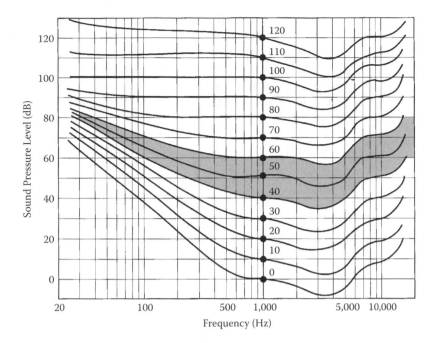

FIGURE 6.4 Phon curves represent combinations of sound pressure and frequency that we sense as equally loud. (Adapted from Kroemer, K.H.E., Kroemer, H.B. and Kroemer-Elbert, K.E. (2003, amended reprint). *Ergonomics: How to Design for Ease and Efficiency* (2nd edn) Upper Saddle River, NJ: Prentice-Hall/Pearson Education.)

programmed to appear at certain times; it usually has varying rhythms with vocals, and it may contain popular hits. Music played while a person works is meant to break up monotony and to generate mild excitement and an emotional impetus toward activity. Although improved morale and productivity have been reported, a clear linking of an underlying arousal theory with the specific components of the music is difficult. Thus, selecting the musical content, the rhythm, the loudness, and the presentation at certain times during work is a matter of guesses rather than of science.

Muzak

Background music known as Muzak, appears like acoustical wallpaper in shops, hotels, waiting rooms, and such. It is meant to create a welcoming atmosphere, to relax customers, to reduce boredom, and to mask disturbing sounds. Its character is subdued, its tempo intermediate, and vocals are avoided. It may produce a monotonous environment for those continuously exposed to it although it may feel pleasant to the transient customer.

In spite of the fact that certain kinds of music do appeal to different individuals under many circumstances, interpreting the relations between listeners and tunes and choosing particular kinds of music for certain activities, environments, and listener populations is an art. Providing music is popular partly because it appears to have beneficial effects and partly because it is a useful market ploy: many people seem to be happier and more productive when music is humming into their ears. And even if not effective, music appears harmless when it is quiet and if the listener can turn it on or off.

Some interesting acoustic events occur°.

- *Directional hearing.* Humans are able to tell where a sound is coming from by using the difference in arrival times (phase difference) or in intensities (as a result of the inverse-square law of energy flux) to determine the direction of the sound. Yet, the ability to use stereophonic cues varies among individuals and is much reduced when earmuffs are worn.

- *Distance hearing.* The ability to determine the distance of a sound source is related to the fact that sound energy diminishes with the square of the distance traveled, but the human perception of energy depends also on the frequency of the sounds, as just discussed. Thus, a source of sound appears more distant when it is low in intensity and frequency and appears closer when it is high in intensity and frequency.

- *Doppler effect.* As the distance between the source of sound and the ear decreases, one hears an increasingly higher frequency; as the distance increases, the sound appears lower. The larger the relative velocity, the more pronounced is the shift in frequency. The Doppler effect can be used to measure the velocity at which source and receiver move against each other.

- *Common difference tone.* When a common frequency interval of 100 Hz or more separates several tones, one hears an additional frequency based on the common difference. This effect explains how one may hear a deep bass tone from a sound system that is physically incapable of emitting such a low tone.

- *Concurrent tones.* When two tones of the same frequency are played in phase, they are heard as a single tone, its loudness being the sum of the two tones. Two identical tones exactly opposite in phase cancel each other completely and cannot be heard. This physical phenomenon (called destructive interference or phase cancellation) can be used to suppress the propagation of acoustical or mechanical vibrations and, with current technology, is particularly effective at frequencies below 1000 Hz.

6.3 Noise and its effects

Noise

Noise is any unwanted and objectionable sound, loud or quiet. Many aspects of noise are psychological and subjective. For instance, dripping of water produces single and short sounds of low intensity, which can appear as noise, just as a neighbor's loud, lasting, complex music may do. Noise surrounds us, day and night: at home, at work, anywhere. There are many sources of noise, such as traffic, barking dogs, construction, industry, spectator events, car races, shooting and hunting, even our own music that we like to hear through loudspeakers or headsets.

What noise can do

Any sound may be annoying and thus felt as noise. The threshold for noise annoyance varies depending on the conditions, including the sensitivity and mental state of the individual. Noise can

- Create negative emotions, feelings of surprise, frustration, anger, and fear.
- Delay the onset of sleep, awaken a person from sleep, or disturb someone's rest.
- Make it difficult to hear desirable sounds.
- Produce temporary or permanent alterations in body chemistry.
- Interfere with human sensory and perceptual capabilities and thereby degrade the performance of a task.
- Temporarily or permanently change one's hearing capability.

PTS, TTS

The noise-effected change in hearing capability of sound is physical, not merely psychological. Exposure to intense sound, such as an explosion, can cause a permanent threshold shift (PTS), which is an irrecoverable loss of hearing; if the exposure is less acute, the result may be a temporary threshold shift (TTS) from which the hearing eventually returns to normal. A PTS may stem from damage to the bones of the middle ear, to the cochlear cilia or the organs of Corti at the basilar membrane in the inner ear, or to the nerves leading to the CNS. Which of these are injured depends on the frequency and, of course, on the intensity of the incident noise. The timing and severity of the loss are dependent upon the duration of exposure, the physical characteristics of the sound (its intensity and frequency), and the nature of the exposure (whether it is continuous or intermittent). Damage may be immediate, such as by an explosion, or may occur over some time with continuous exposure to noise.

High-intensity sound, even when pleasant but loud, can permanently damage hearing. The effect of continuous exposure is usually cumulative. One of the sinister aspects of noise-induced hearing damage is that the victim may not become aware of a developing injury and therefore does not take action to avoid more damage. Noise-induced threshold shift occurs, of course, first at the frequency range of the noise, typically around 4000 hertz, but then it spreads to higher and lower frequencies.

Effects of noise on task performance

Noise can diminish task performance. Simple and repetitive tasks show little impairment in the presence of noise, but noise often degrades the execution of difficult tasks, especially when they require attention, perception, and information processing. Unexpected and irregular noise causes greater degradation than continuous noise does; sudden onset can create a startle response.

Signal-to-noise ratio

One of the most noticeable effects of noise is its interference with spoken communications and the hearing of signals. Workers in loud environments often complain that they must shout to be heard and that they cannot hear and understand others who try to communicate with them. The difference between the intensities of speech (signal) and noise is critical to determine whether speech and signals can be detected in noise. For example, with speech at 80 dBA in noise of 70 dBA, the signal-to-noise relation, $S/N°$, is simply the difference, +10 dB. Detection of a signal is a prerequisite for its discrimination from other signals, for the identification and location of its source, for the recognition of its intended meaning or urgency, and for the judgment of its importance.

Shouting in noise

People have a tendency to raise their voices to speak over noise, and to return to normal when the noise subsides. (This is called the Lombard reflex.) In a quiet environment, males normally produce about 58 dBA, in a loud voice 76 dBA, and, when shouting, 89 dBA. Women normally have a voice intensity that is 2 or 3 dBA less at lower efforts and 5 to 7 dBA less at higher efforts. People can increase the S/N easily by raising their voices at low noise levels, but the ability to compensate lessens as the noise increases. Above about 70 dBA, raising one's voice becomes inefficient, and it is insufficient at 85 dBA or higher. Furthermore, this forced effort of a shouting voice often decreases intelligibility because articulation becomes distorted at the extremes of voice output.

Noise-induced hearing loss

In industrialized countries, noise-induced hearing loss usually occurs around 4000 Hz. From there, it often spreads into lower ranges but mostly it extends into higher frequencies, culminating at about 8000 Hz. Yet, reduced hearing at (and above) 8000 Hz is also brought about by aging. This may make it difficult to distinguish between noise- and age-related causes.

Figure 6.5 shows the hearing capabilities measured on the same person at 20 and 70 years of age. The audiogram of the young person indicates fairly normal hearing with all measurement points close to the zero-loss line. However, at age 70, the hearing has been reduced by about 20 dB in all

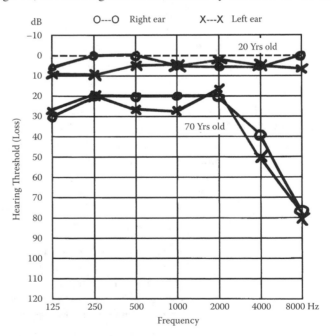

FIGURE 6.5 Audiogram taken on the same person at the ages of 20 and 70 years

frequencies below 2000 Hz. Above 2 kHz, the audiogram shows the typical age- and noise-induced reduction, which is worst near 8 kHz.

Losing some hearing capabilities as one ages is normal. We can expect the following reductions:

- 10 dB at 50 years
- 25 dB at 60 years
- 35 dB at 70 years

Sounds that damage

Sounds that are sufficiently strong or long lasting, and that involve certain frequencies, can damage one's hearing. However, not all persons respond to sound in the same manner. In general, sound levels of about 85 dBA or more are potentially hazardous because of the energy that they contain. The magnitude of the actual hearing loss relates directly to the sound level. In the United States, current regulations allow 16 hours of exposure to 85 dBA, 8 hours to 90 dBA, and 4 hours to 95 dBA, for example.

Simple subjective experiences can indicate whether one is exposed to a hazardous sound. The presence of dangerous sound environments is indicated, for example, by hearing a sound that is appreciably louder than conversational level, by a sound that makes it difficult to communicate, by ringing in the ear (tinnitus) after having been exposed to a noisy environment, and by experiencing sounds that seem muffled to the naked ears after leaving a noisy area.

Three strategies to prevent NIHL

Exposures to excessive sound, in either level or duration or both, generate risks of noise-induced hearing loss, NIHL. Common sources of noise are guns, many occupational environments, power tools, chain saws, airplanes, vehicles, and music (whether heard live or through loudspeakers or headphones). NIHL can occur whether one likes the sound or not.

There are three major approaches to countering the damaging effects of noise.

Avoid generation: The first, fundamental, and most successful strategy is to avoid the generation of excessive sound by properly designing machine parts such as gears or bearings, reducing rotational velocities, changing the flow of air, or replacing a noisy apparatus with a quieter one.

Leave the area: Another successful strategy is to remove humans from noisy places altogether, or do so at least for parts of the work shift.

Impede transmission: The last strategy is to impede the transmission of sound from the source to the listener. In occupational environments, one might install mufflers on the exhaust side of a machine, encapsulate the noise source, physically increase the distance between source and ear, and place sound-absorbing surfaces in the path of the sound.

Planning for "no noise"

Not to generate sound is the one fundamentally successful way to avoid noise. Therefore, planning to avoid noise is most important for engineers and architects. This is done primarily by selecting such technologies and machines that do not produce unacceptable sounds. That principle applies to offices, where machines, keyboards, and conversations should be kept quiet. It applies to assembly facilities where no hand tools should be used that produce a racket, such as pneumatic cutters and mechanical riveters. This approach also applies to manufacturing for which machinery and processes should be selected that are quiet, by design and nature. Noisy engines can be muffled, clinking dies eliminated, and hammering and riveting replaced by welding and gluing. Of course, avoiding the generation of traffic noise in transportation is essential as well: mufflers on trucks and buses, cars, and motorcycles are required, and so are noise-abating take-off and landing procedures for airplanes on airport runways.

To engineer out unnecessary sound is an important task, but certain pieces of machinery and certain jobs are inherently noisy. In this case, the next step is to prevent noise propagation. The laws of physics tell us that sound energy decreases with the square of the distance from the source, so the architect chooses to locate offices and other workplaces apart from the source of noise, such as traffic or machinery. If a factory itself is noisy, the sections where people work should be as far away from the sound sources as possible; intervening spaces, for example, used for storage, can act as buffers.

Noise barriers

Of course, the best way to reduce the propagation of sound is to completely enclose a noise source. However, often we put only a partial barrier between us and the noisemaker. Outdoors, having trees and bushes is helpful, although their usefulness as barriers depends on their density: with all the air spaces among foliage and branches, several rows of trees and bushes may be necessary to achieve sufficient sound abatement.

Another efficient way of reducing noise propagation is to enclose people by buildings. Suitable materials for walls, doors,

Table 6.1   Noise attenuation by building materials

Building Component	Attenuation in dB	Comment
Regular single door	20 to 30	Regular speech clearly understandable*
Regular double doors	30 to 40	Loud speech clearly understandable[a]
Heavy special door	40 to 45	Loud speech audible[a]
Window, single glazing	20 to 25	
Window, double glazing	25 to 30	
Window, soundproofed	30 to 35	
Brick wall, 6 to 12 cm thick	35 to 40	
Brick wall, 25 to 38 cm thick	50 to 55	
Double brick walls, spaced apart, each 12 cm thick	60 to 65	

[a] On the other side of the door.

and windows can easily reduce the inside sound load by 20 to 30 dB. The weak spots are the openings for doors, windows, cables, and plumbing. Table 6.1 lists noise abatements that can be expected from building materials.

Hearing-protec-tion devices

The last resource for protecting human hearing is to wear a hearing-protection device° (HPD) that reduces the harmful or annoying effects of sounds. HPDs are either worn externally (as sound-isolating helmets or earmuffs) or as earplugs inserted into the ear canal. These hearing protectors are variously effective, depending partly on the intensity and the frequency spectrum of the sound arriving at the ear and partly on the fit of the protector to the wearer's ear.

Passive HPDs

Conventional HPDs are passive; they achieve attenuation by making sounds pass through material that absorbs, dissipates, or otherwise impedes energy flow. In spite of these simple features, conventional HPDs can be highly protective—if they are selected properly and worn correctly—by avoiding ambient noise above about 80 dBA that overly strains human hearing. One unfortunate trait of most passive devices is that they attenuate high-frequency sound more than low-frequency sound, which reduces the power of consonants and distorts speech. Persons who already have suffered a hearing loss at higher frequencies therefore experience further difficulties through elevation of their hearing threshold when they use a conventional HPD.

Plugs and muffs Below 500 and above 2000 Hz, as a rule, earplugs provide greater noise attenuation than earmuffs do. However, within those frequency limits, earmuffs typically are more effective in attenuation. For any type of HPD, effectiveness depends strongly on proper fitting and use of the devices. A poor initial fit, loosening of the device during activity, and, of course, failure to wear the equipment cut its effectiveness. The user generally finds muffs easier to fit although, in hot and humid environments, they may be uncomfortable; yet, they can be welcome as ear insulators in a cold environment.

Generally, conventional HPDs cannot differentiate and selectively pass speech versus noise energy. Therefore, the devices do not directly improve the signal-to-noise relation. However, they can occasionally improve clearness in intense noise by lowering the total energy of both speech and noise that reaches the ear, thereby reducing distortion due to overload in the cochlea. An HPD has little or no degrading effect on intelligibility in noise above about 80 dBA, although it can reduce hearing acuity during quiet periods (when worn then even though protection is not needed). Some of the negative effects of the device may be due to the tendency to lower one's voice because the protector amplifies the bone-conductive voice feedback inside the head, mostly at low frequencies. Therefore, we percieive our own voice as louder in relation to the noise than is actually the case, often resulting in a compensatory lowering of the voice by 2 to 4 dB. Thus, one should make a conscientious effort to speak louder when wearing an HPD.

Active HPDs Because conventional hearing protection devices alter all received signals, and may affect the judgment of sound content, direction, and distance, new HPDs have come on the market whose attenuation qualities can be tailored to prevailing noise levels, job demands, and users' hearing abilities. Active noise reduction HPDs use destructive interference by generating sound waves that are of the same nature as the noise but 180 degrees out of phase. Such noise cancellation works particularly well below 1000 Hz. Recent advances in miniature electronic technology and in high-speed signal processing have made such noise abatement and communication headsets practical. They allow, for example, cancellation of all external noise and then introduction of desired signals, such as music or speech.

Voice communications

The ability to understand the meanings of words, phrases, sentences, and entire speeches is called intelligibility. Obviously, this psychological process depends on the present acoustical conditions. For satisfactory communication of most voice messages over noise, at least 75% intelligibility is required. Direct face-to-face communication provides visual cues that enhance intelligibility of speech, even in the presence of background noise. Indirect voice communications lack the visual cues. The distance from speaker to listener, background noise level, and voice level are important considerations. The ambient air pressure and gaseous composition of the air affect the efficiency and frequency of the human voice and, consequently, of speech communication.

Intelligibility

The intensity of a speech signal relative to the level of ambient noise is a fundamental determinant of the intelligibility of speech. The commonly used speech-to-noise ratio S/N is, in fact, not a fraction but a difference, as mentioned above. With an S/N of +10 dB or higher, people with normal hearing should understand at least 80% of spoken words in typical broadband noise. As the S/N falls, intelligibility drops to about 70% at 5 dB, to 50% at 0 dB, and to 25% at −5 dB. People with noise-induced hearing loss often experience larger reductions in intelligibility.

In voice communication, frequencies are from about 200 to 8000 Hz, with the range of about 1 to 3 kHz most important for intelligibility. Men use more low-frequency energy than women do. Filtering or masking frequencies below 600 or above 3000 Hz has little effect on intelligibility, but interfering with voice frequencies between 1000 and 3000 Hz drastically reduces understanding.

In speech (as in written text), consonants are more critical for understanding words than are vowels. Unfortunately, consonants have higher frequencies and, concurrently, generally less speech energy than vowels and therefore are more readily masked by ambient noise; hence, they are more difficult to understand, especially for older persons.

Table 6.2 International spelling alphabet

A: Alpha	O: Oscar
B: Bravo	P: Papa
C: Charlie	Q: Québec
D: Delta	R: Romeo
F: Foxtrot	S: Sierra
G: Golf	T: Tango
H: Hotel	U: Uniform
I: India	V: Victor
J: Juliet	W: Whiskey
K: Kilo	X: X-ray
L: Lima	Y: Yankee
M: Mike	Z: Zulu
N: November	

Components of speech communication

Speech communication has five major components:

1. The message
2. The speaker
3. The transmission of the message
4. The environment
5. The listener

The message becomes clearest if its context is expected, its wording is clear and to the point, and the ensuing actions are familiar to the listener. The speaker should speak slowly and use common and simple vocabulary with only a limited number of terms. Redundancy and use of phonetically discriminable words can be helpful. The International Spelling Alphabet is shown in Table 6.2. Transmission of speech should be by a high-fidelity system that produces little distortion in frequency, amplitude, or time. The listener's natural ability to understand the message is, of course, affected by environmental noise.

Design of warning signals

Warning signals should be designed carefully to penetrate surround sound; one solution is to use frequencies below 500 Hz. A positive side effect of low frequencies is that they diffract easily around barriers. However, in spite of the high-frequency attenuation of conventional HPDs, warning signals should normally be in the range of 1000 to 4000 Hz, unless noise is prevalent in these frequencies. If such masking noise exists, its harmonic frequencies (double, triple, etc.) should not be used for signals. Signals should be in contrast to the masking noise, in tonal qualities such as intensities and frequencies, and in their modulation over time, such as by increases and decreases. A rule of thumb is that the intensity of a signal should be about 15 dB above any masking noise levels. Auditory signals can be combined with other indicators that can be seen or felt, such as lights and vibrators.

Improving defective hearing

Many deficiencies in hearing, natural or caused by noise, can be remedied with modern digital hearing aids. They are more expensive than older analog devices, but they have several major advantages. Many active devices not only amplify sounds but filter out background noise and also make the sounds clearer.

Hearing aids

Hearing aids that fit into the ear canal are popular because they are easily hidden, but they may create a "stuffed up" feeling because they block the ear canal. That blockage may prevent the wearer from benefiting from the existing hearing ability. Therefore, new open-ear hearing aids sit behind the ear, but in contrast to the traditional behind-the-ear fittings, these new devices are smaller and have a barely visible thin tube that extends into the ear canal. In comparison to in-the-ear devices, behind-the-ear models are comfortable to wear and easy to switch on and off.

Adjustments

New hearing aids often have a surround microphone, manual or automatic, which picks up sound no matter from where it comes. Some new devices provide different adjustments for different environments: one setting for a quiet room, one for a concert, and yet another for use in a restaurant, for example. These options are preprogrammed, so all one has to do is to select the proper setting, possibly with a remote control; some devices switch to the suitable setting automatically. Another feature is the automatic adjustment between the aids for the left and right ears, when the person can hear better in one ear than in the other. The technology also can automatically fine-tune the sections in both ears as the listening environment changes.

Surgical implants

Several kinds of surgical implants can help persons with severe hearing loss.

- Bone-anchored hearing aids work for those with single-sided deafness. A transmitter, attached to the skull on the deaf side, picks up sound and then conducts it to the "good" side.

- Middle ear implants can help people with mild to moderate hearing loss. They attach directly to the middle ear bones and amplify sound signals. A component behind the ear houses the microphone, sound processor, and battery.
- Cochlear implants help in the case of severe hearing loss. They convert sound into nerve impulses and send them to the brain. A transmitter is placed under the skin behind the ear, and electrodes are implanted inside the cochlea.

Some techniques can be used in combination, such as cochlear implants with earborne hearing aids.

Summary

Our ears provide information that we need in daily life and at work. The acoustic information is coded in combinations of sound frequencies and intensities that change over time. We interpret that information according to our individual experiences and hearing capabilities.

Noise surrounds us. It can be a mere annoyance, it may interfere with sleep and rest, it can reduce our well-being, affect our performance, and it may damage our ability to hear. To avoid noise-induced damage, various hearing protection devices are available, some of which can actively improve the perception of sound signals.

Fitting steps

Step 1: Make sure that the ears function properly: if in doubt, see an ear physician.

Step 2: Provide a suitable acoustic environment; avoid noise and its effects.

Step 3: Wear appropriate hearing protection as needed.

Further reading

BERGER, E.H., ROYSTER, L.H., ROYSTER, J.D., DRISCOLL, D.P., and LAYNE, M. (2003). *The Noise Manual* (5th edn). Fairfax, VA: American Industrial Hygiene Association.

CASALI, J.G. and ROBINSON, G.S. (2006b). Noise in industry. In Marras, W.S. and Karwowski, K. (Eds.) *The Occupational Ergonomics Handbook (2nd edn) Fundamentals and Assessment Tools for Occupational Ergonomics,* Chapter 31. Boca Raton, FL: CRC Press.

Notes

The text contains markers, °, to indicate specific references and comments, which follow.

6.2 Hearing Sounds:

Infrasound: More on felt mechanical vibrations in Kroemer et al. 2003; Plog 2002; Stanton et al. 2005.

Loudness is stated in a logarithmic unit: The measuring unit decibel, dB, is more convenient to use in acoustics than the basic unit Bell, B. 1 dB = 1/10 B.

Music at work: See Fox 1983; Owen 2006.

Acoustic events: Adapted from Kroemer et al. 2003.

The signal-to-noise relation, S/N: Oddly, the algebraic difference between the intensities of signal and noise has been mislabeled as a ratio.

6.3 Noise and Its Effects:

Hearing-protection devices: See Casali et al. 2006a, b.

CHAPTER SEVEN

How we sense objects and energy

Several kinds of body sensors are usually active at the same time to provide redundant information to us. For example, a bicyclist feels body movement, speed, and terrain through body sensors, sees the path ahead and the objects in the immediate proximity rush by and hears the warning signals coming from other riders or vehicles. A blind person moves the fingertips over Braille to "read" text that the eyes cannot see. We bring our hand cautiously close to an object to find out whether it is hot. We can touch an object to determine whether its surface is smooth or rough. Yet, architects and engineers have made surprisingly little use of the existing information about human sensory capabilities. For example, round doorknobs give no indication, by feel or view, in which direction they must be turned to open the door; emergency bars that one must press to open a hinged door usually provide no cues, neither for touch nor for vision, as to which side of the door will open.

7.1 Sensing body movement

Combined signals

The central nervous system (CNS, see Chapter 9) receives signals from various human senses simultaneously, providing us with specific information and a general picture of the events taking place within and outside the body. For example, as we lift a load, muscle and skin sensors tell us what force we exert on the object. We know how it moves even if we cannot see it because Ruffini joint organs report the locations of our limbs and the angles of the joints, and cutaneous sensors supplement

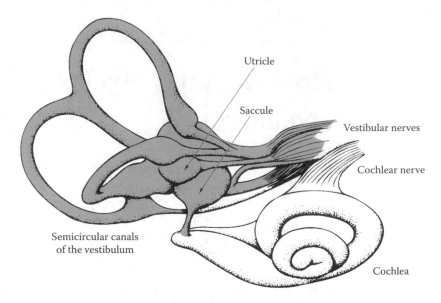

FIGURE 7.1 The three semicircular canals of the vestibulum. (Adapted from Kroemer, K.H.E., Kroemer, H.B., and Kroemer-Elbert, K.E. (2003, amended reprint). *Ergonomics: How to Design for Ease and Efficiency* (2nd Edn), Upper Saddle River, NJ: Prentice-Hall/Pearson Education.)

such information as well because the bending of a joint stretches some skin around the joint and relaxes other sections.

Vestibulum

The vestibulum provides the primary information about the posture and movement of the body. One such pea-sized non-auditory organ is located in each inner ear, next to the cochlea. Each vestibulum has three semicircular canals, shown in Figure 7.1, filled with fluid (endolymph). The arches of the three vestibular canals are at about right angles to each other. Thus, the three canals are sensitive to different rotations of the head. Although the canals share a common cavity in the utricle, each canal functions like an independent fluid circuit. Close to its junction with the utricle, each canal has a widening (called the ampulla) which contains a protruding ridge which carries cilia, sensory hair cells; they respond to displacements of the endolymph. Cilia also are located in both the utricle and saccule. The hair cells generate signals that travel along the vestibular nerves to the CNS.

Sense of body balance

The vestibulum informs us about body balance because it senses the magnitude and direction of accelerations of the head, including the gravitational pull. The system adapts to constant acceleration, and it may not perceive small changes. Placing the head into a new posture requires that the brain compare signals not only with a new spatial reference system, but also with new inputs from the sensors, because the endolymph now loads the cilia in different ways. Given the many other signals, which come from various other sensors simultaneously, that also relate to the body's position and motion, the brain has a complex task of integration and interpretation. Thus, it is not surprising that several vestibular illusions can occur. The best known of these is motion or space sickness, which is probably due to conflicting inputs from the vestibular and other sensors. Others are:

- When the body is aligned with the gravitation vector, for example, of a person in an airplane that turns left or right, one does not perceive the plane's roll.

- Illusionary tilt is the interpretation of linear acceleration as body tilt.

- Elevator illusion occurs when a change in gravitational force produces an apparent rise or lowering of seen objects.

- A person in zero gravity or lying in a prone position may have the illusion of inversion, of being upside-down.

7.2 The feel of objects, energy, and pain

Four different groups of sensory capabilities stem from sensors located in the skin°.

1. Mechanoreceptors, which sense "taction" as contact or touch, tickling, pressure.

2. Thermoreceptors, which sense warmth or cold, relative to each other and to the body's neutral temperature.

3. Electroreceptors°, which respond to electrical stimulation of the skin.

4. Noci(re)ceptors° (from the Latin *nocere*, to damage), which sense pain. Some sensors are deep inside the body.

Taction

The taction sense relates to contact and touch at the skin. One speaks of a tactile sensation when the stimulus solely acts at the skin, whereas a haptic sensation exists when information flows at the same time from sensors in muscles, tendons, and joints (kinesthetic proprioceptors). Most of our everyday perception is actually haptic.

Tactile sensors   The most common type of tactile sensor is a free nerve ending, a proliferation of a nerve that dwindles in size and then
disappears. Thousands of such tiny fibers extend through the
layers of skin. They respond particularly to mechanical displacements and are particularly sensitive near hair follicles. In
smooth and hairless skin, encapsulated receptors are also common, shown in Figure 7.2. Among these are Meissner and the
similar Merkel corpuscles, which respond to pressure. They are
especially numerous in the ridges of the fingertips. The Pacinian
corpuscle is an encapsulated nerve ending of a single, dedicated
nerve fiber. These highly responsive tactile receptors are located
densely in the palmar sides of the hand and fingers and in distal joints. They are also prevalent near blood vessels, at lymph
nodes, and in joint capsules. Krause end bulbs are particularly
sensitive to cold, but probably respond to other stimuli as well.

In 1826, E. H. Weber[o] demonstrated that skin sensors react
to the location of a stimulus, and to certain kinds of stimuli,

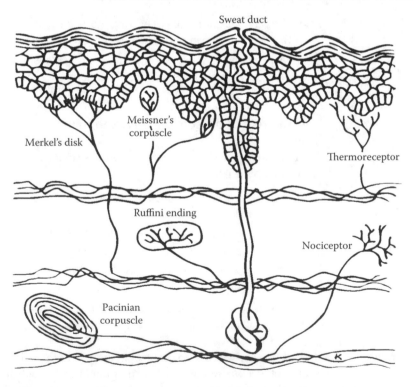

FIGURE 7.2 The common skin sensors. (Adapted from Kroemer, K.H.E., Kroemer,
H.B., and Kroemer-Elbert, K.E. (2003, amended reprint). *Ergonomics: How to
Design for Ease and Efficiency* (2nd Edn) Upper Saddle River, NJ: Prentice-Hall/
Pearson Education.)

especially force or pressure, warmth, and cold. Yet, in spite of nearly 200 years of research, the sense of taction still is not fully understood. Taction stimuli often are not well defined. What is the relation among pressure, force, and touch? Which sensors respond to each stimulus in what way? Do the sensors respond singly or in groups? If in groups, in what patterns do they respond? Does one sensor respond to only one stimulus or to several stimuli? A single Meissner corpuscle may connect with up to nine separate nerves, which may also branch to other corpuscles. It is not clear how such simultaneous convergence and diversion of the neural pathways can reliably generate and code neural signals.

Temperature

Our temperature sensations are relative and adaptive. Objects at skin temperature appear as neutral or indifferent, at physiological zero. We call a temperature below this level *cool* and a temperature *warm* if above. Slowly warming or cooling the skin near physiological zero may not elicit a change in sensation. This range of temperature neutrality differs in parts of the body; for the forearm, the neutrality zone is from approximately 31 to 36 degrees Celsius.

Feeling cold or warm

Some nerve sensors respond specifically to falling and cold temperatures, whereas others react to increasing and hot temperatures. The two scales may overlap, which can lead to paradoxical or contradictory information. For example, spots on the skin, which consistently register cold when stimulated, may also report cold to a warm object of about 45°C. An opposite paradoxical sensation of warmth also exists. We sense more easily changes toward warm temperatures from physiological neutral than changes toward cold temperatures. Warm sensations adapt within a short period of time, except at rather high temperature levels. Adaptation to cold is slower and does not seem to occur completely. Rapid cooling often causes an "overshoot" phenomenon; that is, for a short time one feels colder than one physically is. The following Chapter 8 discusses more interactions of the body with the environment.

Sensing electricity

Although one feels electrical currents on the skin, we know currently of no receptor specialized to sense electrical energy. Electricity apparently can arouse almost any sensory channel of the peripheral nervous system. The threshold for electrical stimulation depends heavily on the configuration and location of the electrode used, on the waveform of the electric stimulus, the rate of stimulus repetition, and on the individual subject.

Generally, the threshold is about 0.5 to 2 mA if the pulse lasts 1 ms.

Pain

Tissue injury, hard touch or pressure, electricity, warmth, and cold can arouse unpleasant, burning, itching, or painful sensations. Some related research results are difficult to interpret because of the various levels and categories labeled "pain". Current understanding is that nocireceptors and other nerve cells possess specialized molecules for detecting pain-causing stimuli. The cells transmit signals to nerve cells in the spinal cord and on to the brain, which identifies the messages as painful. We may feel sharp or piercing pain, usually associated with events on the surface or in teeth or the head, and dull or numbing pain, often coming from deep in the body.

Pain can range from barely felt to unbearable. The threshold for pain is highly variable, probably because pain is so difficult to separate from other sensory and emotional components. Under certain circumstances and to certain stimuli, one can even adapt to pain. Some people have experienced so-called second pain, which is a new and different pain wave following a primary pain after about two seconds. Referred pain indicates the displacement of the location of the pain, usually from its visceral origin to a more cutaneous location; an example is the appearance of cardiac anginal pain in the left arm.

Not all pain is worrisome. For example, the acute kind that accompanies a minor tissue injury is protective because it encourages us to avoid further damage. This pain tends to be temporary and to lessen over time.

7.3 Designing for tactile perception

In modern daily life, we often overstrain our vision and audition for information input while other senses remain underused, for instance, touch and smell. Some of that disparity of use has to do with what is technically practical, but other reasons stem from the lack of specific and quantitative knowledge about how human senses work. Apparently, many past and current engineering applications rely mostly on everyday experience because very limited experimental data are at hand. Yet, new technology leads easily to novel uses such as making a portable phone produce a sound and, at the same time, also vibrate to alert us that somebody is trying to call.

Research needs

Indeed, fairly little is still known about using our sense of smell (olfaction) for engineering purposes, but we know that our reactions to odors are slow, similar to our (gustation) reactions to tastes in food and drink. Surprisingly, after about two centuries of studying the human senses, even our knowledge about the cutaneous sensitivities is still spotty. First, stimuli were often not well defined in older research. Second, sensors are located in different densities all over the body. Third, the functioning of sensors is not understood exactly: many sensors react to two or more distinct stimulations simultaneously and produce similar outputs, and it is often not clear whether or which specific sensors respond to a given stimulus. Fourth, the pathways of signal conduction to the central nervous system are complex; they may be a composite of several afferent paths from different regions of the body. Finally, the CNS interprets arriving signals in unknown manners. Given all these uncertainties, much more research needs to be performed to provide complete, reliable, and relevant information on the cutaneous senses for use in human-engineered design°.

Taction sensitivity

Given the lack of reliable experimental information, the following guidelines for design applications of the taction sense rely mostly on extrapolations of experience and educated guesses.

Mechanoreceptors can differentiate touch information by

- The strength of the stimulus
- The temporal rate of change
- The size of the area of the skin (i.e., the number of receptors) stimulated
- The location of the skin stimulated

Sensitivity to taction stimuli is highest in the facial area and at the fingertips, and fair at the forearm and lower leg. Other body areas, such as the eyes, are most sensitive, but obviously out of bounds.

Tactile sensitivity also depends on temperature: for example, if a stimulus of low intensity appears slowly on cold skin, it may not be noticed.

Using temperature signals

For several reasons, our sense of temperature is difficult to use for communication purposes: it has a relatively slow response time and it is poor in identifying location. Furthermore, it adapts to a stimulus over time and it may integrate into one signal several stimuli distributed over an area of skin. Finally, interactions exist between mechanical and temperature sensation; for example, a cold weight feels heavier on the skin than a warm weight.

The strength of thermal sensation depends on the location and size of the sensing body surface. The temperature sensation is made stronger by increasing

- The absolute temperature of the stimulus, and its difference from physiological zero
- The rate of change in temperature
- The exposed surface (an example is immersion of the whole body in a bath as compared to immersion of a hand or foot)

Assuming a neutral skin temperature of about 33°C, the following rules of thumb apply for naked human skin.

- A skin temperature of 10°C appears painfully cold; 18°C feels cold; and 30°C still feels cool.
- The highest sensitivity to changes in coolness exists between 18 and 30°C.
- Heat sensors respond well throughout the range of about 20 to 47°C.
- Thermal adaptation—that is, physiological zero—can be attained in the range of approximately 18 to 42°C, meaning that changes are not felt when the temperature difference is less than 2°C.
- Distinctly cold temperatures (near or below freezing) and hot temperatures above 50°C provoke sensations of pain.
- Compared to warmth, cold is sensed more quickly, particularly in the face, chest, and abdominal areas. The body's ability to feel warmth is less distinct, but it is best in hairy parts of the skin, around the kneecaps, and at the fingers and elbows.

Using the smell sense

Olfactory information is seldom used by engineers, because few research results are available, because people react quite differently to olfactory stimuli, because smells can be easily masked, and because olfactory stimuli are difficult to arrange. Among the few industrial applications are adding odorous methylmercaptan to natural gas and pyridin to argon to allow people to smell leaking gas.

Using electric signals

Electricity is seldom used as an information carrier, although it has great potential for transmitting signals to the human.

Attaching electrodes is convenient. The energies that are transmitted are low, requiring only about 30 microwatts at the electrode–skin junction, up to a tolerable limit of about 300 milliwatts. Coding can be via placement, intensity, duration, and pulsation. Electrical stimulation can provide a clear, attention-demanding signal that is resistant to masking; however, responses to weak electrocutaneous signals usually have long response latencies, many misses, and false alarms. In complex environments that require a heightened level of vigilance, electrical signals are suitable to provide redundancy.

Do not use pain as signal stimulus

Pain does not lend itself to engineering applications, primarily because one is ethically bound not to cause pain, but also because the sensation of pain follows any damage already done too slowly to prevent more damage.

Reaction and response

Reaction time° is the period from the appearance of a stimulus (such as an electrical signal, a touch, or a sound) to the beginning of a responding effector action (e.g., movement of a hand):

$$\text{Reaction time} + \text{motion time} = \text{response time}$$

Simple reactions

Table 7.1 lists approximate shortest possible simple reaction times. This is a selective compilation of data from many experiments on reaction time which were conducted before the 1960s and have appeared in engineering handbooks. Today, the origins of most underlying data, the experimental condi-

Table 7.1 Shortest reported simple reaction times

Sensory stimulus	Approx. shortest reaction (ms)
Electric shock	130
Touch and sound	140
Sight and temperature	180
Smell	300
Taste	500
Pain	700

Source: Adapted from Kroemer, K.H.E. (2006). *"Extra-Ordinary" Ergonomics: How to Accommodate Small and Big Persons, the Disabled and Elderly, Expectant Mothers and Children.* Boca Raton, FL: CRC Press.

tions under which they were measured, and the accuracy of the measurements are no longer known.

This list shows small time differences in reactions to electrical, tactile, and sound stimuli. The slightly longer reaction times for sight and temperature stimuli may lie well within the range of measuring accuracy or of variations among persons. However, the time following a smell stimulus is distinctly longer and that for taste yet longer; it takes by far the longest to react to the infliction of pain.

Complex reactions

Table 7.1 contains so-called *simple reaction* times; they apply if a person knows that a particular stimulus will occur, is prepared for it, and knows how to react to it. If conditions become more complex, such as when there is uncertainty about the appearance of the signal, the reaction takes longer. If a person has to choose among several actions that can be taken, so-called *choice reaction* is involved, which takes longer than a simple reaction. Time expands even further if it is difficult to distinguish between several stimuli of which only one should trigger the response. This indicates that one should expect, in real-life situations, reaction times that are considerably longer than listed in the table.

Motion time

Motion time follows reaction time. Movements may be simple, such as lifting a finger in response to a stimulus, or complex, such as swerving a car to avoid a collision. Swerving the car involves not only more intricate movement elements than lifting a finger but also larger body and vehicle masses that must be moved, which takes time.

Response time

Minimizing response time—the sum of the reaction and motion lags—is often a goal of human factors engineering. This involves three intertwined choices:

1. The stimulus, most appropriate to the situation, that initiates the shortest possible reaction time
2. The body part that is best suited to do the task of reacting
3. The equipment that allows the fastest execution of the task

Chapter 9 expands the foregoing discussion of actions and reactions in feedforward and feedback loops.

Summary

Much of our daily life relies on receiving information from our environment and appropriately interpreting it to guide our actions. Obviously, this is one of the most ancient and natural features that we use instinctively. Research done during the last century or so has explained the overall sensory processes and many of their intricacies, but many details are still unknown or uncertain. The lack of technically useful information offers many opportunities and challenges for future investigations; examples are the perception of smell, of temperature, and electrical signals.

Fitting steps

Step 1: Select the primary stimulus to provide critical information.

Step 2: Add a secondary stimulus of a different type as supplement or backup.

Step 3: Establish and fine-tune the parameters of the stimuli.

Further reading

BOFF, K.R., KAUFMAN, L., and THOMAS, J.P. (EDS.) (1986). *Handbook of Perception and Human Performance*. New York: Wiley.

Notes

The text contains markers, °, to indicate specific references and comments, which follow.

7.2 The feel of objects, energy, and pain:

Skin: Sensors located in the skin are called cutaneous (from *cutis*, Latin for skin) or somesthetic (from soma, Greek for body).

Electroreceptors: It is has been disputed whether such specific sensors exist, or whether the related sensations stem from the stimulation of other taction sensors. A similar discussion concerns *nocireceptors*: some researchers hold that there are no specific pain sensors, but that other sense organs transmit pain. For more information, read Basbaum et al. 2006; Boff et al. 1988.

Ernst Heinrich Weber, 1795–1878, founder of experimental psychology.

7.4 Designing for tactile perception:

Research needs on human sensory capabilities for engineering applications: Check Chapter 2, Assessment Methods and Techniques, in Kroemer 2006. Read Gawande 2008.

Reaction and motion times: For more details, see Boff et al. 1988, Kroemer et al. 2003.

CHAPTER EIGHT

How we experience indoor and outside climates

The human body generates heat: its production fluctuates with the physical work done. The body must dissipate heat to its surroundings; this is easy in a cold climate but difficult in a hot environment.

Two overriding facts govern the heat flow within the body, and the exchange of heat with the environment:

1. Body core temperature must remain close to 37°C.

2. Heat flows from the warmer to the colder matter.

The physiological requirement of keeping body core temperature the same establishes the need for proper human engineering of the environmental climate. The physics principle of heat energy flow from warm to cold provides the engineering tools to create suitable environmental climates.

8.1 Human thermoregulation

The human body has a complex control system to maintain its internal body temperature near 37°C (about 99°F). Deviation in deep body temperature of only a few degrees from that value affects cellular structures, enzyme systems, and other body functions and it impairs physical and mental work capacities. There are minor fluctuations in temperature throughout the day and from day to day, mostly due to circadian rhythms (see Chapter 15) and larger changes with physical activities. However, the main impact upon the human thermal regulatory system results from the interaction between the (metabolic) heat generated within the body and external energy gained in hot surroundings or lost in cool environments.

The body produces heat in its metabolically active tissues: primarily at skeletal muscles, but also in internal organs, fat, bone, and connective and nerve tissue. Blood carries that heat energy throughout the body. The circulatory system modulates the flow of blood by constriction, dilation, and shunting of its vessels, mostly its superficial arteries and veins. This generates and controls heat dissipation from warmer to colder parts of the body. Heat exchange with the environment takes place at the body's respiratory surfaces of the lungs and, of course, through the skin.

The body regulates its heat content to prevent undercooling or overheating. The temperature of core tissues (such as brain, heart, lungs, and abdominal organs) must be kept constant at about 37°C. So, all temperature control takes place in the external tissues. Accordingly, there is often a large temperature difference between core and shell. Under normal conditions, the gradient between skin and deep body is about 4°C at rest, but in the cold the temperature difference may be 20° or more.

Hot skin in a hot environment

Usually, the body must dissipate heat energy in order to avoid overheating of its core. Dissipation is not a problem if the environment is colder than lung and skin surfaces because of the natural flow of heat from warm to cold. However, because of the same physics principle, heat dissipation becomes a problem in an environment that is hotter than body surfaces are. Nature's solution is trying to increase the skin temperature above the environment's temperature and especially to evaporate water, sweat, on the skin. Evaporation of 1 cm³ of water requires energy of about 2440 J (580 calories) mostly extracted from the skin; that cools the body. When it cannot disperse enough energy, the body must reduce its internal heat production, meaning that it abandons physical activities as much as possible.

Cold skin in a cold environment

If an environment is very cold, the body may have to struggle to conserve its heat; it primarily does this by the reducing blood flow to the skin, which lowers its temperature. Obviously, increasing insulation by wearing proper clothing is a prudent action. When the body loses too much heat, it can increase its heat production by contracting muscles, done voluntarily or by reflex shivering; both generate heat energy.

Heat exchanges with the environment

The body's heat control systems (muscular, vascular, and sudomotor, discussed above) must interact with the physical components of the environment. The interactions are by convection, conduction, radiation, and evaporation.

Heat exchange by convection or conductance

Traditionally, one calls energy exchange with air or water *convection* but *conduction*° when with a solid matter. However, both rely on the same principle: energy flows from the warmer body to the colder one; as the temperatures of the contact surfaces become equal, the energy exchange ceases. Heat exchange by convection or conductance is a function of

- The body surface participating in the heat exchange
- The temperature of the body surface
- The temperature of the medium that is in contact with the body
- The heat conduction coefficient between the body and the medium

Cork or wood that we touch feel warm because their heat conduction coefficients are below that of human tissue, but cool metal accepts body heat easily and conducts it away. Heat exchange is facilitated if the medium moves quickly along the skin surface, which helps in maintaining a temperature differential. As long as there is a temperature gradient between the skin and the medium, there persists some natural movement of air or fluid: this is called free convection. A forced action (by an air fan, or while swimming in water rather than floating motionless) can produce more movement: this is called induced convection, shown in Figure 8.1.

Heat exchange by radiation

Heat exchange through radiation° depends primarily on the temperature difference between two opposing surfaces, for example, between a windowpane and a person's skin. Heat always radiates from the warmer to the colder surface, for example, from the body to the cold window in winter or from a sun-heated pane to the body in summer see Figures 8.2 and 8.3. This radiative heat exchange does not depend on the temperature of the air between the two opposing surfaces. It depends on

- The body surface participating in the energy exchange
- The temperature of the emitting surface
- The emission coefficient of the emitting surface
- The absorption coefficient of the receiving surface
- The temperature of the receiving surface

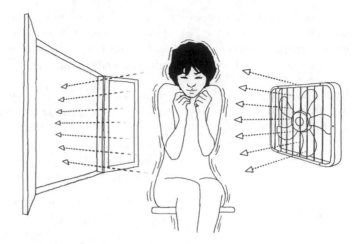

FIGURE 8.1 Air flow helps heat exchange by convection. (Adapted from Kroemer, K.H.E. and Kroemer, A.D. (2005). *Office Ergonomics*, authorized translation into Korean. Seoul: Kukje.)

FIGURE 8.2 Heat radiated from the body to a cold windowpane. (Adapted from Kroemer K.H.E. and Kroemer, A.D. (2005). *Office Ergonomics*, authorized translation into Korean. Seoul: Kukje.)

FIGURE 8.3   Heat radiated by the sun to the body. (Adapted from Kroemer, K.H.E. and Kroemer, A.D. (2005) *Office Ergonomics*, authorized translation into Korean. Seoul: Kukje.)

The wavelengths of radiation from the human body are in the infrared range. Hence, it radiates like a black body, that is, with an emission coefficient close to 1 (unit), independent of the color of the radiating human skin. However, its absorption coefficient depends on skin color.

Heat exchange by evaporation   Heat exchange by evaporation° is in only one direction. The human body loses heat by evaporation; there is no condensation of water on the skin, which would add heat. Evaporation of water (sweat) on the skin requires an energy of about 2440 J (580 cal) for every cubic centimeter of water. The heat lost by evaporation depends on

- The volume of sweat evaporated (which is partly a function of the wet body surface participating in evaporation)
- The vapor pressure at the skin
- The relative humidity of the surrounding air
- The vapor pressure in the surrounding air

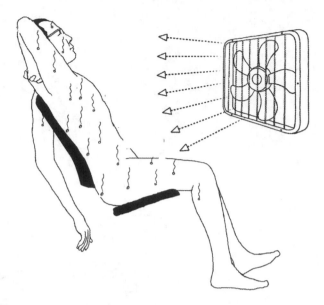

FIGURE 8.4 Air flow helps evaporation. (Adapted from Kroemer, K.H.E. and Kroemer, A.D. (2005). *Office Ergonomics*, authorized translation into Korean. Seoul: Kukje.)

Of course, heat loss by evaporation can only occur if the surrounding air is less humid than the air directly at the skin. Therefore, movement of the air layer at the skin, which enhances convection, also increases the heat loss through evaporation because drier air replaces humidified air, as sketched in Figure 8.4. Evaporative heat loss occurs even in a cold environment because secretion of some sweat onto the skin surface continues during physical work but, more importantly, there is always evaporation of water in the warm lungs.

Of all pathways that can generate heat loss from the body when the temperature of the environment is warmer than about 30°C, evaporation is by far more effective than convection and radiation are, as Figure 8.5 illustrates.

8.2 Climate factors: temperatures, humidity, drafts

Four physical factors define the thermal environment:

1. Air (or water) temperature
2. Air humidity
3. Air (or water) movement
4. Temperatures of surfaces

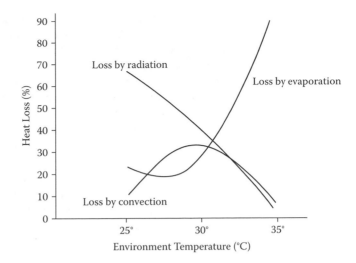

FIGURE 8.5 Cooling the body in a warm environment. (Adapted from Kroemer, K.H.E., Kroemer, H.B., and Kroemer-Elbert, K.E. (2003, amended reprint). *Ergonomics: How to Design for Ease and Efficiency* (2nd Edn) Upper Saddle River, NJ: Prentice-Hall/Pearson Education.)

The combination of these four factors determines the physical conditions of the climate, its effects on us, and how we perceive it.

Measuring temperature

Measurement of temperature is commonly performed with thermometers, now usually filled with alcohol instead of mercury; or with thermistors or thermocouples. Whichever technique is used, it must be ensured that the ambient air temperature is not affected by humidity and air movement. To measure the so-called dry temperature of air, one keeps the sensor dry and shields it to reflect radiated energy. Hence, proper measurement of (dry) air temperature is done with a so-called dry bulb (DB) thermometer.

Measuring humidity

Air humidity may be measured with a hygrometer, originally a human or horse hair that changed its length with wetness, now an instrument whose electrical conductivity (resistance/capacitance) alters with the existing humidity. Another instrument is the psychrometer, which uses one DB and one wet bulb (WB) thermometer, with the wet more cooled by evaporation (that is proportional to the humidity of the air): higher vapor pressure reduces evaporative cooling. A psychrometer is called *natural* if there is no artificial air movement and *forced* if there is.

Air humidity may be stated either in absolute or in relative terms. The highest absolute content of vapor in the air exists when any further increase would lead to the development of water droplets falling out of the gas. This "dewpoint" depends on air temperature and barometric pressure: higher temperature and pressure allow more water content than lower conditions. One usually speaks of relative humidity, which indicates the actual vapor content in relation to the possible maximal content at the given air temperature and air pressure.

Measuring air flow

Air movement was measured with various types of anemometers, which were basically windmills turned by the flow of air; various other electric and electronic techniques have largely replaced the measuring mechanisms.

Measuring radiant temperature

Radiant temperature is determined by measuring the temperature via a thermometer placed inside a black globe, which absorbs radiated energy. That globe temperature (GT) reading needs a correction for the existing air temperature.

Interacting climate factors

Air temperature is the climate factor that most obviously affects body temperature; in fact, the labels that we use commonly to describe our perception of a climate are "hot" or "cold". Air humidity determines our ability to sweat, which is the body's last resort to avoid overheating. Movement of air around our skin affects our ability to exchange energy by convection and to lose heat by evaporation. The energy exchange by radiation depends on the temperatures of the emitting and receiving surfaces.

These examples demonstrate that most climate factors can affect the measurement values of other components; furthermore, the manifold combinations of the factors can lead to very different effects on the human by that composite, called climate, and on how humans perceive it. Various techniques exist to express the combined effects of the environmental factors in one model, chart, or index. One of these was the effective temperature chart, used widely in the late 20th century, which relied on measurements of air temperature, air velocity, and relative humidity.

Combining instruments

A convenient way to assess the existent climate is to combine a set of special measurement instruments in one device, which weighs the separate measurement results according to their effects on the human body and from there calculates a single index. The currently used version is the so-called wet bulb globe temperature, abbreviated WBGT. It applies primarily to warm environments and weighs the effects of several climate parameters:

- WB is the wet bulb temperature of a sensor in a wet wick exposed to natural air current.
- GT is the globe temperature at the center of a black sphere of 150 mm diameter.
- DB is the dry bulb air temperature.

For outdoors, with solar load:

$$WBGT \text{ (outdoors)} = 0.7 \text{ WB} + 0.2 \text{ GT} + 0.1 \text{ DB.}$$

The WBGT is simpler for indoors (or outdoors without sun):

$$WBGT \text{ (indoors)} = 0.7 \text{ WB} + 0.3 \text{ GT.}$$

8.3 Our personal climate

What is of importance to the individual is not the climate in general, the so-called macroclimate, but the climatic conditions with which one interacts directly. Every person prefers a microclimate that feels comfortable under given conditions of work, adaptation, and clothing. The suitable microclimate is highly individual and variable. It depends somewhat on age, where with increasing years muscle tonus is likely to decrease; older persons tend to be less active and to have weaker muscles, to have a reduced caloric intake, and to start sweating at higher skin temperatures. The suitable microclimate also depends on the surface-to-volume ratio which, for example, in children is much larger than in adults, and on the fat-to-lean body mass ratio.

Personal thermal comfort

The work performed, its type and intensity, are major determinants of a person's thermal comfort. Physical work in the cold may lead to increased heat production and hence to less sensitivity to the cold environment, whereas in the heat hard physical work could make it very difficult to achieve an energy balance. The effects of the microclimate on mental work are rather unclear; the only sure statement is that extreme climates hinder mental work.

Clothing

Clothing can have decisive effects on the thermal balance of the body and comfort. Clothing has three major traits: insulation, permeability, and ventilation.

Table 8.1　Typical clothing insulation values

Type of clothing	Insulation in CLO[a]
None	0 (zero)
Light informal summer clothing	About 0.3
Light work outfit	About 0.6
Light business suit	About 1
Heavy business suit	About 1.5
Snow suit	About 2
Arctic quilted thermo parka outfit	About 2.5

Source: Adapted from Havenith, G. (2005). Thermal conditions measurement. In Stanton, N., Hedge, A., Brookhuis, K., Salas, E., and Hendrick, H. (Eds) *Handbook of Human Factors and Ergonomics Methods,* Chapter 60. Boca Raton, FL: CRC Press.

[a]　$1\ \text{CLO} = 0.155\ \text{m}^2\ {}^\circ\text{C}\ \text{W}^{-1}$.

Insulation is a measure of the resistance to heat exchange by convection, conductance, and radiation. In hot or cold environments, it impedes energy exchange by convection and radiation in either direction. Light-colored clothing minimizes heat gain by radiation from the sun on a sunny day, whereas dark-colored clothes absorb the sun's radiated heat. In the cold, clothing helps to reduce body heat loss by convection and radiation. Clothing also reduces the risk of injury by conduction when touching a hot or a very cold object.

The insulating value of clothing is defined in CLO units, the reciprocal of thermal conductivity: $1\ \text{CLO} = 0.115\ \text{m}^2\ {}^\circ\text{C}\ \text{W}^{-1}$ is the insulating value of normal clothing worn by a sitting subject at rest in a room at about 21°C and 50% relative humidity; see Table 8.1. Air bubbles contained in the clothing material or between clothing layers provide increased insulation, both against hot and cold environments.

Permeability is a measure of how clothing permits movement of water vapor through the fabric. This is important in hot environments, where good permeability allows evaporation from the skin and subsequent cooling of the body. However, permitting water vapor to escape is also important in a cold climate to avoid the clammy feeling of water trapped in and under clothes. Permeability may also play a role in protecting the body against chemical exposure.

Ventilation is a measure of the ability of ambient air to move through the clothing. Such airflow is advantageous in a hot environment to enhance evaporative and convective cooling; in contrast, it is usually undesirable in a cold environment.

Clothing also determines the surface area of exposed skin. More exposed surface areas allow better dissipation of heat in a hot environment but can lead to excessive cooling in the cold. Fingers and toes need special protection in cold conditions because they are far away from the warm body core. Head and neck are close to the core and have warm surfaces, which is advantageous in the heat but not in the cold.

Wind chill

Clothing can affect the individual microclimate strongly by preventing so-called wind chill, which is a special kind of convective heat loss. If the air moves more swiftly along exposed skin surfaces, their cooling becomes more pronounced. Figure 8.6 shows the energy loss, expressed as equivalent wind chill temperatures, depending on wind speed and actual DB air temperature, but not considering humidity, on the naked human face as determined in 2001 by the Canadian and U.S. National Weather Services°. Naturally, with stronger wind and colder air, the cooling of exposed skin and hence the danger of frostbite increases, as both Table 8.2 and Figure 8.6 show. Obviously, simply covering the skin, for example with a facemask increases insulation against heat loss by wind chill.

Table 8.2 Wind chill index conditions

Wind chill index (WCI[a])	Effects on the body
200	Pleasant
400	Cool
1000	Cold
1200	Bitterly cold
1400	Exposed flesh freezes
2500	Intolerable

Source: Adapted from Parsons, K. (2005). Ergonomics assessment of thermal environments. In Wilson, J.R. and Corlett, N. (Eds) *Evaluation of Human Work* (3rd Edn), Chapter 23. London: Taylor & Francis.

[a] $WCI = (10\sqrt{v} + 10.45 - v)(33 - DB)$ with v the air velocity and DB the dry bulb air temperature.

Wind Velocity in km/h	Actual Dry Air Temperature in deg C										
0	5	0	−5	−10	−15	−20	−25	−30	−35	−40	−45
5	4	−2	−7	−13	−19	−24	−30	−36	−41	−47	−53
10	3	−3	−9	−15	−21	−27	−33	−39	−45	−51	−57
15	2	−4	−11	−17	−23	−29	−35	−41	−48	−54	−60
20	1	−5	−12	−18	−24	−30	−37	−43	−49	−56	−62
25	1	−6	−12	−19	−25	−32	−38	−44	−51	−57	−64
30	0	−6	−13	−20	−26	−33	−39	−46	−52	−59	−65
35	0	−7	−14	−20	−27	−33	−40	−47	−53	−60	−66
40	−1	−7	−14	−21	−27	−34	−41	−48	−54	−61	−68
45	−1	−8	−15	−21	−28	−35	−42	−48	−55	−62	−69
50	−1	−8	−15	−22	−29	−35	−42	−49	−56	−63	−69
55	−2	−8	−15	−22	−29	−36	−43	−50	−57	−63	−70
60	−2	−9	−16	−23	−30	−36	−43	−50	−57	−64	−71
65	−2	−9	−16	−23	−30	−37	−44	−51	−58	−65	−72
70	−2	−9	−16	−23	−30	−37	−44	−51	−58	−65	−72
75	−3	−10	−17	−24	−31	−38	−45	−52	−59	−66	−73
100	−3	−11	−18	−25	−32	−40	−47	−54	−61	−69	−76
Equivalent Windchill Temperatures in deg C											

FIGURE 8.6 Wind chill temperature is only defined for temperatures at or below 5°C and for wind speeds above 5 km/h. Air humidity is not considered. Bright sunshine may increase the wind chill temperature by 5 to 10°C. The table above relies on the Wind Chill Calculator by the U.S. National Weather Service, updated on 27 November 2006, downloaded from www.nws.noaa.gov/om/windchill on 4 October 2007.

Acclimatization   Individual thermocomfort is also a function of acclimatization, the adjustment of an individual's body (and mind) to changed environmental conditions. Acclimatization of the body to a hot environment mostly improves the control of blood flow to the skin, facilitates sweating, and increases stroke volume of the heart without increase in heart rate. Healthy persons achieve heat acclimatization in one or two weeks but can lose it just as quickly. Whether any truly physiological acclimation of the whole body to a moderate cold environment takes place is questionable, because most of the adjustments made concern proper clothing, beneath which the body performs at its usual microclimate. However, blood flow to exposed surfaces of the face and neck and to the hands and feet can adapt to cold conditions.

A climate that was uncomfortable and restricted ability to perform physical work during the first exposure may be quite agreeable after a couple of weeks. Seasonal changes in climate, in type of work, of clothes worn, and personal attitude play major roles in determining what feels acceptable. In the summer, most people are willing to accept as comfortable warmer, more humid, and draftier conditions in a room than they would tolerate in the winter.

8.4 Working in hot environments

Blood distribution

The human body produces heat, much of which it must release to the surroundings to prevent overheating. As already discussed, the physical means to disperse heat are convection, conduction, radiation, and evaporation. These heat dissipations work best when the skin temperature is well above the temperature of the immediate environment. To achieve this, the body's circulatory system redistributes its blood flow to facilitate heat transport to the skin: skin vessels are dilated and superficial veins fully opened, actions contrary to the ones taken in the cold ("red skin in the heat"). This may bring about a four-fold increase in blood flow above the resting level, increasing the conductance of the tissue.

Sweating

If heat transfer is still not sufficient, the body activates its sweat glands, and the evaporation of the produced sweat cools the skin. Recruitment of sweat glands from different areas of the body varies among individuals. Some persons have few sweat glands, whereas most have at least two million sweat glands in the skin— large differences in the ability to sweat exist among individuals. The activity of each sweat gland is cyclic. The overall amount of sweat developed and evaporated depends very much on clothing, environment, work requirements, and on the individual's acclimatization. Sweating reduces the water content of the body, which must be replenished by drinking water.

Reducing physical effort

If heat transfer by blood distribution and sweat evaporation remains insufficient, the body reduces its muscular activities to lower the amount of energy generated through metabolic processes. In fact, this is the final and necessary action of the body if the core temperature would otherwise exceed a tolerable limit. If the body has to choose between unacceptable overheating and continuing to perform physical work, the choice will be in favor of core temperature maintenance, which means reduction or cessation of work activities.

Signs of heat strain

There are several signs of severe heat strain on the body. One is an increase of circulatory activities: cardiac output enlarges, mostly brought about by higher heart rate. This may be associated with a reduction in systolic blood pressure.

A high sweat rate is another sign of excessive heat strain. Above the so-called insensible perspiration (in the neighborhood of about 50 cm³/hr), sweat production increases depending on the heat that must be dissipated. During normal work, about one liter is produced in one hour, but total sweat losses up to 12 L in 24 hours have been reported in extremely strenuous efforts in hot climates. Sweat begins to drip off the skin when the sweat generation has reached about 1/3 of the maximal evaporative capacity. Of course, sweat running down the skin contributes very little to heat transfer.

Drink water

The water balance within the body provides another sign of heat strain. Dehydration of only 1 or 2% of body weight can critically affect the ability of the body to control its functions. Therefore, we must maintain a sufficient fluid level, best by frequently drinking small amounts of water. Sweating extracts not only water from the plasma but also carries some salt from the blood onto the skin. Normally, it is not necessary to add salt to drinking water because in western diets the salt in the food is more than sufficient to resupply the salt lost by sweating.

Heat distress

Among the first reactions to heavy exercise in excessive heat are sensations of discomfort and perhaps skin eruptions ("prickly heat") associated with clogged sweat ducts. As a result of sweating, so-called heat cramps may develop, which are muscle spasms related to local lack of salt. They may also occur after quickly drinking large amounts of fluid. Heat exhaustion is a combined function of dehydration and overloading the circulatory system. Associated effects are fatigue, headache, nausea, and dizziness, often accompanied by giddy behavior. Heat syncope indicates a failure of the circulatory system, demonstrated by fainting. Heat stroke indicates an overloading of both the circulatory and sweating systems and is associated with hot dry skin, increased core temperature, and mental confusion.

Working hard in the heat

The ability for short exertion of maximal muscle strength is not affected by heat or the body's water loss. However, the ability to perform high-intensity and endurance-type physical work is severely reduced during acclimatization to heat. Even after one is acclimatized, the demands on the cardiovascular system for heat dissipation and for blood supply to the muscles compete with each other. The body prefers heat dissipation, with a proportional reduction in a person's capability to perform.

8.5 Working in cold environments

The human body has few natural defenses against a cold environment. Most of the actions that we can take are behavioral in nature, such as putting on suitably heavy clothing covering the skin, seeking shelter, or using external sources of warmth. The body has only two major ways to regulate its temperature: redistribution of the blood flow and increase in metabolic rates.

Redistribute blood

To conserve heat, the body lowers the temperature of the skin to reduce the temperature difference against the outside. This is done by constricting blood vessels near the body surface, thus displacing the circulating blood toward the core, away from the skin ("pale skin in the cold"). The effects can be rather dramatic; for example, the blood flow in the fingers may be only 1% of what exists in a moderate climate.

The hunting reflex, a cold-induced automatic vasodilation, is an interesting phenomenon: after initial constriction has taken place, a sudden dilation of blood vessels occurs which allows warm blood to return to the skin, such as of the hands, which rewarms that body section. Then vasoconstriction returns and this sequence may be repeated several times.

Wear gloves, caps

Displacement of the blood volume from the skin to the central circulation is an efficient way to keep the body core warm and its surfaces cold. The associated danger is that the temperature in the peripheral tissues may approach that of the environment. Thus, very cold fingers and toes may result, with possible damage to the tissue if the temperatures get close to freezing. The blood vessels of the head do not undergo as much vasoconstriction so the head stays warm even in cold environments, with less danger to the tissues; however, the resulting large difference in temperature to the environment brings about a large heat loss, preventable by wearing a cap and a collar or scarf to create an insulating layer.

Increase metabolism

The other major reaction of the body to a cold environment is to increase its metabolic heat generation. This may occur involuntarily: shivering usually begins in the neck, apparently to warm the most important flow of blood, to the brain. Shivering is caused by muscle units firing at different frequencies of repetition and out of phase with each other. Because they do no mechanical work to the outside, the total activity is transformed into heat production, allowing an increase in the metabolic rate to up to four times the resting rate. If the body does not become warm, shivering may become rather violent when motor unit innervations become synchronized so that large muscle units are contracted.

Of course, muscular activities also can be done voluntarily, such as by either increasing the dynamic muscular work performed, or by moving body segments, contracting muscles, flexing the fingers, and the like. Because the energy efficiency of the body is very low, dynamic muscular work may easily increase the generation of metabolic heat to ten times or more than that at rest.

"Goose bumps"   Incidentally, the development of goose bumps of the skin helps to retain a layer of stationary air close to the skin, which is relatively warm and has the effect of an insulating envelope, reducing energy loss at the skin.

How cold does it feel?   An individual's decision to stay in the cold or to seek shelter depends on the subjective assessment of how cold body surfaces or the body core actually are. Dangerous situations can develop either when a person fails to perceive and to react to the body's signals that it is becoming dangerously cold, or if the body temperature becomes so low that further cooling is below the threshold of perception.

The perception of the body getting cold depends upon signals received from surface thermal receptors, from sensors in the body core, and from a combination of these signals. As skin temperatures decrease below 35.5°C, the intensity of the cold sensation increases; cold sensation is strongest near 20°C; yet, at even lower temperatures, sensitivity decreases. It is often difficult to separate feelings of cold from pain and discomfort.

The conditions of cold exposure may greatly influence the perceived coldness. Obviously, it makes quite a difference whether or not one wears suitable protective clothing and what one is actually doing when exposed to cold air or water. Experiments have shown that subjects find it very difficult to assess how cold they actually are: neither core nor surface temperatures reliably correlate with cold sensations. When the temperature plunges, each downward step can generate an "overshoot" sensation of cold sensors, which react not only to the difference in temperature, but also to the rate of change. Exposure to very cold water accentuates the overshoot phenomenon observed in cold air. This may be due to the fact that the thermal conductivity of water is about a thousand times greater than that of cold air at the same temperature.

Signs of cold strains

The subjective sensation of cold is an unreliable, possibly dangerous, indicator of core and surface temperature of the body. Measuring ambient temperature, humidity, air movement, and exposure time and rationally acting on these physical measures is a better strategy than relying on subjective assessments.

If vasoconstriction and metabolic rate regulation cannot prevent serious energy loss through the body surfaces, the body will suffer some effects of cold stress. As just discussed, the skin is first to suffer from cold damage while the body core is protected as long as possible. If skin temperature approaches freezing, ice crystals develop in the cells and destroy them, a result known as frostbite.

In the hands, joint temperatures below 24°C and nerve temperatures below 20°C reduce the ability to perform fine motor tasks. Manual dexterity drops as finger skin temperatures fall below 15°C. Tactile sensitivity diminishes below 10°C and becomes severely reduced as the skin temperature falls below 8°C. At about 5°C, skin receptors for pressure and touch cease to function and the skin feels numb.

As nerve temperature falls to 8°C, peripheral motor nerve velocity may decrease to near zero; this generates a nervous block, which explains the rapid onset of physical impairment by local cooling. Deep cooling goes along with increasing inability to perform activities, even if they could save the person ("cannot light a match") leading to apathy ("let me sleep") and final hypothermia.

Reduction of core temperature is even more serious. Vigilance begins to drop at temperatures below 36°C. At core temperatures of 35°C, central nervous system coordination suffers so that one may not be able to perform even simple activities. When the core temperature drops even lower, apathy sets in and the mind becomes confused. Loss of consciousness occurs around 32°C. At core temperatures of about 26°C, heart failure may occur. At very low core temperatures, such as 20°C, vital signs disappear, but the oxygen supply to the brain may still be sufficient to allow revival of the body from hypothermia.

Working hard in the cold

With appropriate clothing, a cold environment does not restrict one's ability to perform medium, even heavy, work. However, bulky clothing may hinder the exertion of certain tasks.

Climate effects on mental tasks

It is difficult to evaluate the effects of moderate heat or cold on mental and intellectual performance because of large subjective variations and a lack of practical yet objective testing methods. However, as a rule, mental performance deteriorates with rising room temperatures, starting at about 25°C for the unacclimatized person; that threshold increases to 30 or even 35°C if the individual is used to heat. Brain functions are particularly vulnerable to heat; keeping the head cool improves the tolerance to elevated deep body temperature. A high level of motivation may also counteract some of the detrimental effects of heat. Thus, in laboratory tests of perceptual motor tasks, onset of performance decrement can occur in the low 30°C WBGT range whereas very simple mental performance is often not significantly affected by heat as high as 40°C WBGT.

8.6 Designing comfortable climates

Comfortable climates

In confined spaces, such as buildings or the cabins of vehicles, many techniques are available to generate a warm environment° that suits persons who perform very light physical work, and wear appropriate clothing. Generally, comfortable climate conditions are in the ranges of about 21 to 27°C in a warm climate or during the summer, and from 18 to 24°C in a cool climate or during the winter. Preferred ranges of relative humidity are between 30 and 70%. In rooms, air temperatures at floor level and at head level should differ by less than 6°C, best about 3°C, keeping the head cool, the feet warm. Differences in temperatures between body surfaces and windows, walls, and other surfaces should not exceed approximately 10°C. Velocity of cool air, especially if irregular, should not exceed 1.5 m/s.

The technical means to influence the climate must be considered in interaction with the work to be performed, with the acclimatization conditions of the individuals, with their clothing, and their psychological inclination either to accept given conditions or to consider them uncomfortable. The preferred personal climate varies from person to person, and within a person. For example, selecting clothing and changing the intensity of physical effort can make major differences on the perceived suitability of the physical conditions of the climate, temperatures, humidity, and air movement. Various combinations of these climate factors can subjectively appear as similar. However, with that many variables, it is not surprising that some people may consider the same combination of climate factors as too warm and others find that condition too cold, as sketched in Figure 8.7. Selecting appropriate clothing is often a simple and effective way to create a comfortable personal climate: adding layers of clothing when one feels cold, and changing into lighter clothes when one feels hot.

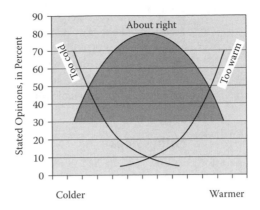

FIGURE 8.7 Different opinions about climates

Summary

The body must maintain a core temperature near 37°C with little variation even in the presence of major changes in internally developed energy (heat and physical work), in heat energy received from a hot environment, or in heat energy lost to a cold environment.

Heat energy may be gained from or lost to the environment by the following physical exchanges:

- Convection
- Conduction
- Radiation
- Evaporation (which can only transfer energy from the body to the environment)

In a hot environment, the body tries to keep the skin hot to prevent heat gain and to achieve heat loss. Sweating is the ultimate means to cool the body surface. In a cold environment, the body tries to keep the skin cold to avoid heat loss. Proper clothing strongly affects the energy exchanges in both hot and cold environments.

The thermal environment is determined by combinations of

- Air humidity (mostly affecting evaporation)
- Air temperature (affecting convection and evaporation)
- Air movement (affecting convection and evaporation)
- Temperature of solids in touch with the body (affecting conduction)
- Temperature of surfaces adjacent to the body (affecting radiation)

The combined effects of all or some of these physical climate factors can be expressed in the form of a climate index. Various scales are in use, with the WBGT index currently favored. It identifies certain ranges of air and surface temperatures, and of air velocity and humidity, which are physically suitable and perceived as comfortable for given tasks and clothing.

Fitting steps

Step 1: Determine what and how work is to be done.

Step 2: If indoors, establish a comfortable climate. For outdoors, set up a suitable work regimen and provide appropriate clothing and shelter.

Step 3: Encourage individual adjustments, especially in clothing.

Further reading

BERNARD, T.E. (2002). Thermal stress. In Plog, B.A. (Ed) *Fundamentals of Industrial Hygiene* (5th Edn), Chapter 12. Itasca, IL: National Safety Council.

PARSONS, K.C. (2003). *Human Thermal Environments* (2nd Edn). London: Taylor & Francis.

Notes

The text contains markers, °, to indicate specific references and comments, which follow.

8.1 Human thermoregulation:

Heat exchanges by convection or conductance: Both follow "Newton's law of cooling," see Bernard 2002.

Radiative heat exchange follows the "Stefan-Boltzmann Law of Radiative Heat Transfer," see Bernard 2002.

"Wind chill temperatures": go to the Web site of the U.S. Weather Service www.weather.gov.

8.2 Climate factors: temperatures, humidity, drafts:

Evaporation: For equations for heat exchange by convection or conductance, radiation and evaporation, see Bernard 2002; Havenith 2005.

8.3 Our personal climate:

Wind chill: The latest version of the Wind Chill Index is available at http://www.nws.noaa.gov/om/windchill/. This table was developed in the USA jointly with the Canadian weather service and is used in both countries with no regionalization. Other countries have their own versions, which one can obtain through their meteorological agencies, or through the World Meteorological Organization.

8.4 Designing comfortable climates:

Further information for the built environment, such as in offices in Europe and North America, is contained in the ANSI-ASHRAE Standard 55, latest edition, and more generally, in ISO standards such as listed by Parsons 2003.

Body and mind working together

As we try to understand the human, we often consider certain characteristics separately, as if they functioned independently of each other. Of course, in most cases, there is strong interaction. The first part of this book treated specific physical aspects of the human body and, in the second part, the main issue was information obtained through the senses. In this section, such physiological and sensory aspects are brought together, to better represent the human where, indeed, body and mind work together.

Chapter 9 treats mental activities, Chapter 10 is concerned with hard labor, Chapter 11 addresses work that is less physically demanding, and Chapter 12 deals with workload and stress.

Mental activities

Two internal systems control the human's functioning: one is the endocrine system, which consists of a group of organs, the glands of internal secretion, whose main function is to produce hormones and secrete them into the blood stream. Hormones are substances that affect activities in body cells at another site. Although the biochemistry of hormones and of their effects is known, the endocrine system does not lend itself to ergonomic exploitation. Therefore, that control system of the human body is not a topic in this book.

The nervous system, the other controller of the human body, is well understood. This knowledge provides ample opportunities to appreciate how the human body and brain function, and hence indicates to the human factors engineer how to design work and work equipment to take advantage of the intricacies of the nervous control system.

9.1 The brain–nerve network

Brain

The brain is the control center for the entire body. It generates thoughts and beliefs, memories, behaviors, and moods. The brain coordinates the abilities to move, touch, smell, hear, and see. The brain requires the continuous flow of oxygen and blood, about 20% of the total output of the heart.

Parts of the brain

The brain has several major anatomic components. As Figure 9.1 shows, they include the cerebrum, the cerebellum, and the brain stem.

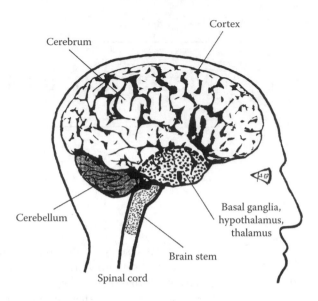

FIGURE 9.1   Major components of the brain

Cerebrum

The *cerebrum* consists of dense convoluted masses of tissue divided into two halves, the left and right cerebral hemispheres, connected in the middle by nerve fibers. The cerebrum is further divided into the frontal, parietal, occipital, and temporal lobes. The frontal lobes control skilled motor behavior, including speech, mood, thought, and planning. The parietal lobes interpret sensory information from the rest of the body and control body movement. The occipital lobes deal with vision. The temporal lobes generate memory and motions, process and retrieve long-term memories, and initiate communication or action.

Cerebellum

The *cerebellum*, which lies beneath the cerebrum just above the brain stem, coordinates the body's movements.

A special collection of nerve cells is at the base of the cerebrum: the basal ganglia, hypothalamus, and the thalamus. The basal ganglia help to smooth out movements. The hypothalamus coordinates automatic functions of the body, such as sleep and wakefulness; it maintains body temperature and regulates the body's water balance. The thalamus organizes sensory messages to and from the highest level of the brain, the cerebral cortex.

Brain stem

The *brain stem* connects the brain with the spinal cord. The brain stem automatically regulates body functions such as the rate at which the body uses food, and it controls body posture,

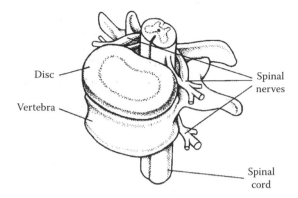

Disc

Vertebra

Spinal nerves

Spinal cord

FIGURE 9.2 Typical segment of the vertebral column with the spinal cord and emanating nerve extensions. (Modified from Kapandji, I.A. (1988). *The Physiology of the Joints.* Edinburgh: Churchill Livingstone; *The Merck Manual of Medical Information,* 1997 Home Edition. Whitehouse Station, NJ: Merck & Co., Inc.)

breathing, swallowing, and heartbeat. It increases alertness when needed. Severe damage to the brain stem destroys these functions and death follows.

Spinal cord

The *spinal cord* is an extension of the brain. It makes reflex decisions and it carries pathways for nervous signals. The spinal cord begins below the brain stem and continues down the vertebral column, which provides protection in its bony ring that encircles the foramen, already shown in Figures 2.10 and 2.11 of Chapter 2. This is the opening through which the spinal cord passes, a long fragile structure that can be damaged by displacements of the vertebral bones, often in an automobile accident or by a fall. Figures 9.2 and 9.3 show details of a vertebral segment with the spinal cord and its nerve extensions. Nerves at the front (anterior) of the spinal cord carry information from the brain to the muscles. These are the motor nerves. Nerves at the rear (posterior) of the spinal cord carry information from sensors of the body toward the brain. These are the sensory nerves.

Nerves

The spinal cord provides the main pathway for nervous signals, which relay both incoming (afferent) sensory messages and outgoing (efferent) motor commands.

Nerve impingements

Between pairs of vertebrae, two nerve bundles emerge like roots from the spinal nerve, as Figure 9.3 shows. Displacements

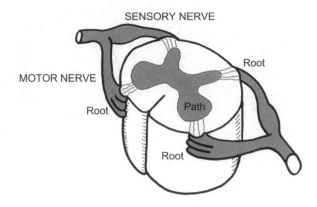

FIGURE 9.3 Segment of the spinal cord. (Modified from Kapandji, I.A. (1988). *The Physiology of the Joints*. Edinburgh: Churchill Livingstone; *The Merck Manual of Medical Information*, 1997 Home Edition. Whitehouse Station, NJ: Merck & Co., Inc.)

of vertebrae and material of a ruptured disc can impinge on the spinal nerve extensions with dire consequences for efferent or afferent signals. Wounding a motor nerve may cause weakness, even paralysis of the muscle it serves. Damage to a sensory nerve can lead to painful feedback that, in certain cases, may mislead the brain. For example, pain often seems to come from body sections such as the buttocks and thighs, which are innervated by the sciatic nerve; but, in most cases, the problem is not there but at the point of impingement of the sciatic nerve at the spinal column.

Neurons

The basic functional unit of the human nervous transmission system is a nerve cell, the *neuron*. The nervous system contains about one billion such cells that extend like strings throughout the body. A neuron has a cell body, soma, which is a few thousands of a millimeter thick, and short extensions, dendrites, where connections from other neurons arrive across synapses. Figure 9.4 sketches a motor nerve cell with its single long extension, called the axon, which can be more than a meter long. Nerves transmit their messages electrically in one direction: from the axon of one neuron to dendrites of the next neurons. At the synapses, the contact points between neurons, the axons secrete chemical neurotransmitters, which trigger the receptors on the next neurons to generate a new electrical current.

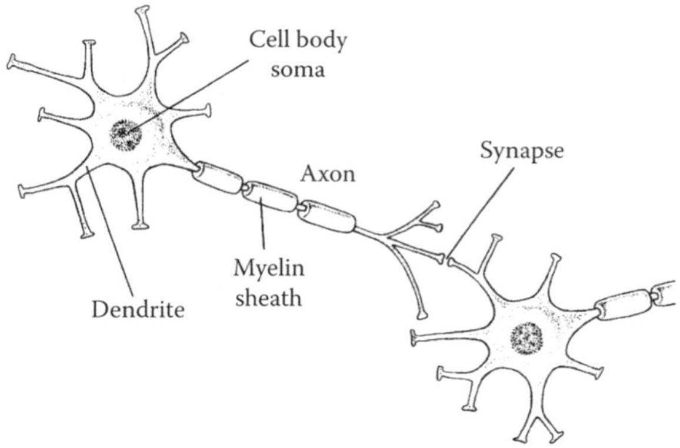

FIGURE 9.4 Scheme of a motor neuron. (Adapted from Kroemer, K.H.E., Kroemer, H.J., and Kroemer-Elbert, K.E. (1997). *Engineering Physiology. Bases of Human Factors/ Ergonomics* (3rd Edn). New York: Wiley.)

Signal transmission

Synapses are not just simple transmitters, but also serve as filters or switches, because they can inhibit the transmission of incoming signals that are infrequent and weak. Neurons come in several types and sizes, some of which are able to transmit as many as 1000 impulses per second, whereas others may not convey more than 25 signals per second. The velocity of impulse travel is a constant for each nerve fiber, ranging from about 0.5 to about 150 meters per second. The speed is slow in thin fibers such as the ones transmitting pain. Thick axon fibers with myelin sheaths serve skeletal muscles; this setup allows the largest speed of conduction.

Action potential

Incoming signals, arriving via synapses, may be too weak to stimulate the nerve cell. If strong enough, the cell generates an electrical spike, shown in Figure 9.5. The peak is some 100 mV above the resting potential; this impulse, called the action potential, travels to the next neuron. In a motor neuron, the impulse passes from the axon to the so-called motor endplates in the muscle where it triggers a contraction.

Motor unit

At the muscle, a motor nerve divides into several fibers, each controlling several muscle fibers. A *motor unit* consists of all the muscle fibers that one motor neuron innervates. In muscles that carry out precise movements, there are only 3 to 6 muscle fibers in a motor unit whereas muscles set to perform heavy work may have 100 or more muscle fibers controlled by a single neuron.

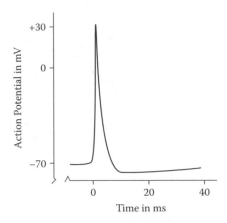

FIGURE 9.5   Action spike of about 100 mV above the resting potential of –70 mV

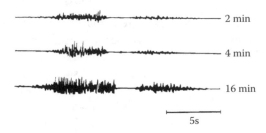

FIGURE 9.6   Sample electromyograms taken after 2, 4, and 16 minutes of contractions of equal strength

EMGs

Electromyograms (EMGs) allow us to observe the strength and frequency of activation commands sent from the central nervous system. These electrical signals, picked up by surface or indwelling electrodes, provide information not only about the frequency of activations of muscle units, but also about their status of fatigue. Figure 9.6 shows sample electromyograms of the extensor muscle of the upper arm. The muscle executed a series of contractions of equal strength. After 4 minutes of repetitions, the amplitude of the EMG is already enlarged, whereas after 16 minutes the large signal amplitude and shows extensive fatigue.

Feedforward/ feedback loop

Figure 9.7 sketches the innervation of a muscle by its motor nerve and the sensory feedback loop to the spinal cord and brain. The incoming action impulses transmit contraction commands to the muscle at the motor endplates. Feedback information about the ensuing events travels in sensory nerves toward

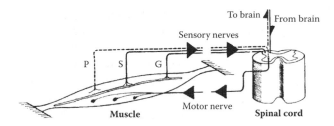

FIGURE 9.7 Muscle innervation and feedback control (G: from Golgi organ, S: from muscle spindle, P: pain signal)

the spinal cord. Important signals originate at Golgi organs (G) which react to tension in tendons; they also start at muscle spindles (S) which sense muscle stretch; and they can include messages of pain (P). Such feedback partly goes to the spinal cord for reflex-type automatic adjustments of the muscular effort; other signals arrive at the brain for continuous evaluation and control of task performance.

**Design
for simple
movement
control**

Motor nerves carry movement orders from the brain to muscles where they cause contractions and, via sensory nervous feedback, control the muscular activities.

The body can perform some motions without the involvement of the brain proper; yet, most muscular activities need fine regulation. This requires various degrees of involvement of higher brain centers, such as the cerebral cortex, the basal ganglia, and the cerebellum. To channel information along lengthy transmission paths through numerous neurons and decision processes is laborious and time consuming. Thus, learning to execute complex movements is difficult and slow, as we know from the experience of acquiring a new complicated skill, especially as we get older. For the human factors engineer this indicates that job activities should be performed in the simplest possible way and sequence, using the least decision making and the fastest path of information transmission, and employing the smallest possible body mass. Clearly, controlling the speed of an automobile by complicated movements of the feet to and from invisible pedals is not a good example of modern human factors engineering.

**The brain–nerve
network**

Brain and nerves are structures of a complex communication system. Normally, the system can send and receive copious amounts of information simultaneously. The brain is the control center. It communicates with the body through the nerves that run up and down the spinal cord. The sensory nerves carry information to the brain about pressure, pain, heat, cold, vibration, and the feel and shape of objects. They also carry messages from internal body sensors, for example, about positions and motions of body parts and about the related tension in muscles and tendons. This information about events outside and inside the body enables the brain to make decisions about what to do; the information loop closes when the brain then receives feedback about the results of the taken actions.

Reflexes

To a lesser extent than the brain, the spinal cord is also a source for coordination of certain actions, such as reflexes. A reflex usually begins with the stimulation of a sensory receptor, for example, one located just below the front of the kneecap. This sends a signal to the spinal cord, which evokes an immediate response that travels rapidly to the appropriate muscle and commands a knee jerk. In this way, a muscle reaction is possible within a few milliseconds after the stimulus occurred because no time-consuming higher brain functions are involved.

CNS and PNS

It is customary to partition the human nervous system into divisions. One approach is to divide by anatomy, as done above: accordingly, the *central nervous system* (CNS) consists of the brain and the spinal cord. The CNS controls the body by gathering information, making decisions, and initiating actions. The *peripheral nervous system* (PNS) runs from sensors, located all over the body, to the CNS, and from there to organs, especially muscles. Peripheral nerves are bundles of single nerve fibers, some of which are very thin, less than half a millimeter in diameter, whereas others are thicker than 5 mm. The PNS does not control, but delivers signals.

Somatic system

Another way to distinguish is by function. The *somatic nervous system* comprises the sensory and motor nerves of the PNS, together with their associated parts in the CNS, all needed to control conscious actions and mental activities. The somatic system links the body with the outside world through perception, awareness, and actions, especially by muscle activation.

Autonomic system

Other functional components form the *autonomic nervous system*. It governs the internal organs that are essential to the life of the body such as blood circulation, breathing, and digestion. This system (also called *visceral*) is responsible for automatic functioning of the body, emergency responses, and emotions.

Sensory receptors

The nervous system monitors all the sensations that come from inside the body and its surfaces as well as through the eyes and ears. If the signals are strong enough they are transmitted to the CNS where the information is perceived and a judgment is made regarding whether an action is needed. If that is the case, the CNS activates muscles and monitors the results of the actions taken.

External receptors

How we perceive the world around us by seeing, hearing, and touch, Section II of this book discusses in some detail. Body sensors also provide information via the senses of taste (gustation) and smell (olfaction), and we can feel temperature, electricity, pressure, and pain. Different kinds of nerve sensors are embedded in the layers of the skin in varying concentrations. Figure 9.8 shows divisions of skin surface areas (dermatomes) which convey information to specific sensory nerve extensions of the spinal cord. Certain parts of our bodies, for example, our lips and fingertips, have dense concentrations of sensors whereas others are fairly numb, such as our backs.

9.2 Taking up information

Processing information

Much of our mental work consists of sensing information, then processing it, and finally acting on it. Of course, most of these activities intermingle with other mental tasks that occur simultaneously; nevertheless, it is convenient to consider these mental activities as occurring in stages that follow each other in series. Figure 9.9 illustrates the concept of the human as a linear processor of signals. A sensor detects information and sends related impulses along the sensory pathways of the PNS to the central processor in the CNS. It interprets the information and chooses an action (which may include no action). Then the CNS generates appropriate feed-forward impulses and sends these along the motor pathways of the PNS to the effectors, usually the voice or muscles in the hand or foot. A sensory feedback loop serves to compare the output of the system with the desired performance. If the comparison between actual and intended performance shows a difference, that disparity gives rise to revised action.

Sensors inside the body

Interoceptors are sensors inside the body; they include the Golgi organs and spindles, associated with muscle, already mentioned. Ruffini organs are receptors in the body joints where they inform on their angulation. Other sensors are in

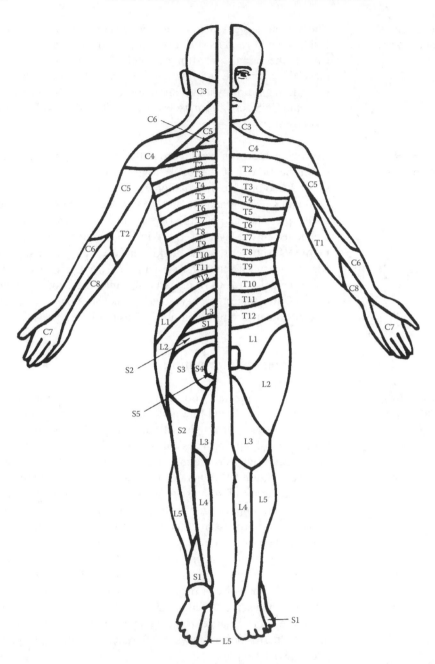

FIGURE 9.8 Sensory dermatomes identified by their nerve roots (C: cervical, T: thoracic, L: lumbar) at the spinal column. (Modified from Jarrett, A. (Ed.) (1973). *The Physiology and Pathology of the Skin*. London: Academic Press; Kroemer, K.H.E., Kroemer, H.J., and Kroemer-Elbert, K.E. (1997). *Engineering Physiology. Bases of Human Factors/Ergonomics* (3rd Edn). New York: Wiley.)

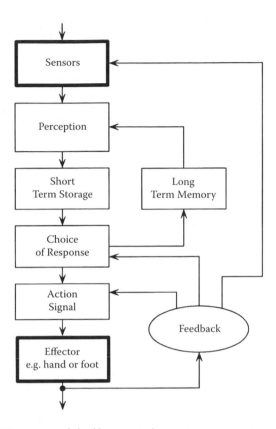

FIGURE 9.9 Model of human information processing. (From Kroemer, K.H.E. (2006). *"Extra-Ordinary" Ergonomics: How to Accommodate Small and Big Persons, the Disabled and Elderly, Expectant Mothers and Children.* Boca Raton, FL: CRC Press.)

the semicircular canals of the ears: these vestibular receptors detect and report the position of the head in space. Other intero-ceptors respond to events within the internal organs, such as in the abdomen, chest, and the head. They relay such sensations as pain and pressure.

Sensors near the surface

Exteroceptors provide information about the exterior of the body. They are involved in sight, sound, taste, smell, and touch. Some of these senses are of particular importance for the control of body activities: the sensations of touch, pressure, and pain provide feedback to the body regarding the direction and intensity of muscular activities transmitted to an outside object. The receptors are located throughout the skin of the body, although in different densities; see the dermatomes shown in Figure 9.8. The sensors are mostly free nerve endings, Meissner's and Pacinian corpuscles. Because the PNS nerve pathways leading to the CNS interconnect extensively in the PNS, the reported sensations are not always specific to a given input; for example, very hot or very cold sensations can be associated with pain that may be caused also by hard pressure on the skin.

Adaptation and speed

Almost all sensors respond vigorously to a change in the stimulus but will report less data when the load stays constant. This adaptation makes it possible to live with continuously present but unimportant stimuli, such as the pressure of clothing. The speed of adaptation varies with the sensors. Furthermore, the velocities of transmitting sensations to the CNS are quite different for diverse sense pathways: light and sound, for example, are reported quickly but pain appears rather slowly.

Modifying input signals

The human has a wide range of sensations, yet, there are important signals that we cannot naturally perceive. For example, we cannot hear ultra- and infrasounds, frequencies that are of great importance to many animals such as dogs and elephants. Another case is the presence of x-rays, for which our body has no sense organs. One of the major tasks of the ergonomist is to modify the physical propensities of external signals, which we are not able to sense but that are important to us, so that we become aware of them. If the signal is not within the bandwidth of the human senses, the engineer must change the nature of the signal. Examples of technical solutions are generating a buzzing sound or flashing light to make the presence of radiating energy known.

In other cases, we adjust signal or change the environment so that the signal can penetrate the clutter of the surroundings. Examples are on/off blinking of signals, changes in the light colors of an emergency vehicle, modulating the sound of its sirens, or adjusting the qualities of a loudspeaker system. Computer offices provide more examples: here, the overall illumination level must be low to avoid glare on the computer screen or washing out the image on it; yet, in spots we increase the illumination by separate task lights to make reading of paper documents easier, especially for older persons, which in turn might generate glare. Providing proper input signals can be a daunting ergonomic task. Figure 9.10 presents a schematic that shows how we can change external signals so that the human senses may perceive them.

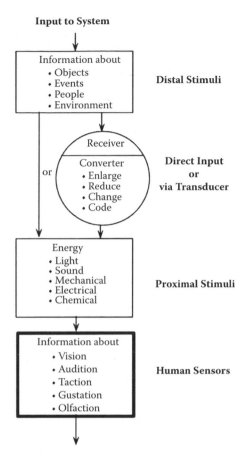

FIGURE 9.10 Transforming distal signals into proximal stimuli that can serve as inputs to the human processor. (From Kroemer, K.H.E. (2006) "Extra-Ordinary" Ergonomics: How to Accommodate Small and Big Persons, the Disabled and Elderly, Expectant Mothers and Children. Boca Raton, FL: CRC Press.)

9.3 Making decisions

Models of information processing

Traditional models of information processing (IP) have a sequence of stages through which information passes, like flowcharts used in industrial engineering. The simplest "model of mental work" includes two distinct phases: first, evaluation of the events occurring, and then bringing about changes, execution. The resulting new environment is then evaluated again, and so begins a new sequence of phases. A refinement of this concept is the recognition that multiple sources of information are present simultaneously, which we evaluate using multiple processing facilities. In each stage, our mind filters and transforms information from previous stages, then integrates and compares it with information from memory° before it passes to the next stage. By this model, the overall cognitive performance of a person depends on (a) the number and type of stages required, and (b) the efficiency with which the operations proceed at each stage. Accordingly, some IP models emphasize limits in the amount of information that can be processed, often expressed as channel capacity. For the human factors engineer, awareness of channel capacity leads to the design of systems that pose the least possible number of inputs to a human and the lowest demands on processing.

Perception of sensory signals

In the first stage of IP, we select from the overall onslaught certain kinds of sensory information for further processing. The brain does this by comparing the new inputs with information in our memory°, searching for and recognizing familiar features. This means that our past experiences generate expectations, which we then use to guide perception and selection of information. New, surprising incoming sensory information may either be rejected or, in contrast, it may easily penetrate our expectation-based barriers, but then we may not know what to make of the unfamiliar inputs.

Short-term memory

Apparently, humans have short-term and long-term memory capacities. Short-term (or working) memory keeps registered information for a brief duration, such as a second, in mind to freeze a picture of the rapidly changing world around us that can serve as a comparison and filtering frame. Another example is looking up a phone number and keeping the number active in memory until dialing it. The working memory seems to have a very limited capacity, both in duration and channel capacity. One theory is that the boundaries of the short-term memory may be the limiting capacities in information processing.

Long-term memory

In contrast, long-term memory has much larger capacity and no obvious duration constraints. Common distinctions describing memory are those that involve general knowledge, called semantic memory, or specific events, referred to as events memory. Quite often, these are linked, such as by an episode in our past, for example, an accident. The ability to remember (to find and call up) essential information in our long-term memory is important for many tasks. Forgetting information may be due to never having successfully encoded and engraved it in the memory bank, or one may be unable to retrieve it from the buffer because one lacks memory queues, or present irrelevant information may clutter it.

To interpret information shown on a display and then to choose an appropriate response requires us to think about task-relevant information. Activation of the knowledge stored in long-term memory may be triggered by recalling an action taken in the past in similar conditions. Two major factors influence the availability of information and its reactivation from long-term memory. One is the strength of the information trace, determined by its initial importance, the number of times it has been activated, and how recently that occurred. The other factor is its association with related items and events, such as the illumination of the workplace when an accident occurred. Such associated links mean that we store much information in chunks, not in isolated pieces.

Making decisions

Our brain must integrate incoming sensory stimuli with other relevant information and then process that knowledge in order to select a suitable response; this part of our mental work is often called central processing°, consciousness, or thinking. Of course, it requires keeping the overall goal of the task in mind in order to select the proper response from all possible ones. Thus, decision making and response selection include the essential necessity of understanding the outcomes of several possible responses, and recognizing which activities are needed to execute the responses.

The thinking process must integrate many pieces of information, some newly acquired, most already engraved in the long-term memory. Many of our daily tasks involve this kind of central processing. Shrinking the IP requirements, using limited inputs and integration needs, makes the task easier. One example is driving an automobile in dense traffic where one might decide to follow a preceding car, or to turn into another lane to overtake it. Following the car by accelerating and braking is usually easier and less dangerous than overtaking; that action requires checking whether a neighboring lane is open, turning the wheel one way and the other way, and at the same time carefully regulating the speed of one's own car.

Fitting the human to the job

A traditional model of human information processing appeared in Figure 9.9. It shows that, in the CNS, the signal transmitted from sensors is perceived, decoded, interpreted, and processed, using sequential stages. Little is known specifically about how this complicated procedure actually develops, yet, obviously it involves a comparison with previous experiences by pairing the current information with related knowledge stored in the long-term memory. Such experience comes naturally with aging and intentionally with teaching and training. That need for teaching and training of novices in decision making is an example of fitting the human to the job as opposed to fitting the job to the human.

New models of IP

The computer metaphor still provides the basis for most current concepts of human mental work: they rely on the assumption that our brain functions in ways similar to how computers process information, although, naturally, a comparison between a living organism and a technical product is always imperfect. For example, computers have specialized input and output devices; this is somewhat similar to the specialized sensory and response modality of humans. They and we have different types of memory stores that can be accessed or searched in different ways; searches, evaluations, and decisions follow hierarchies of importance, and capacities are often limited in amount and function time.

However, advanced concepts° are finding their ways into ergonomic thinking. These include models based on new findings in neurophysiology, for example, using multiple and parallel feed-forward and feedback paths together with distributed processing neural networks. Such models would overcome the limitations of the serial-stage models with discrete processing in each stage; they should provide a more detailed and realistic description of how information storage and processing actually function. Nevertheless, for the time being, the traditional stage-based concept still provides a useful framework and supplies guidance for the human factors engineer.

9.4 Actions and reactions

After the central nervous system has chosen a response to the received signals, the CNS generates action signals, usually to an effector such as the hand or foot, to activate muscles in order to fight or run. With today's technology that usually means that the body applies force to a handle or pedal or other input control element of machinery, be it a simple hand tool or a motorcycle or an airplane. If the effector tool is a hammer or a screwdriver, the task is direct. If there is a more complex piece of machinery, such as an airplane or ship, so-called transducers are involved, which modify the action of the human body, as Figure 9.11

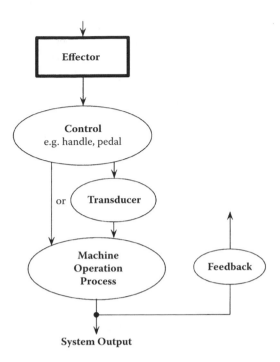

FIGURE 9.11 Transforming effector output for machine operation. (From Kroemer, K.H.E. (2006) *"Extra-Ordinary" Ergonomics: How to Accommodate Small and Big Persons, the Disabled and Elderly, Expectant Mothers and Children.* Boca Raton, FL: CRC Press.)

illustrates. For example, in a small car or boat, turning the steering wheel creates immediate conversion of the human movement into vehicle motion by means of mechanical gears, hydraulic transmissions, or electric/electronic commands. In complex human-operated systems, such as in a large ship or a space vehicle, the response to the action of the human master is slow and extends into the future. In this case, a computerized system can predict the future results of the current action.

The design of the transducers and their feedback signals is a challenge to the human factors engineer. No particular effort is needed to assess the effect of swinging a hammer the next time; the result of the first blow is immediately apparent and serves to redirect the next stroke in order to achieve the goal of hammering a nail deeply into a wooden beam while avoiding hitting one's other hand. However, for more complicated human–machine interactions, it is useful to distinguish among specific aspects of human actions and reactions.

Response time   The time from the appearance of a stimulus (such as light) to
the beginning of an effector action (for example, movement of
a foot) is the *reaction time*. Usually, *motion time* follows, and
combined, they result in *response time*°:

Response time = reaction time + motion time

Reaction time   Time passes between the appearance of a signal on the sensory
input side of the nervous system and the emergence of a (re)
action on the output side. Time delays occur at the sensor, in
afferent signal transmission along the PNS, in processing by
the CNS, in efferent signal transmission in the PNS, and in
muscle activation. Estimates of the time delays are:

- At the receptor: 1 to 38 ms
- Along the afferent path: 2 to 100 ms
- In CNS processing: 70 to 100 ms
- Along the efferent path: 10 to 20 ms
- Muscle latency and contraction: 30 to 70 ms

Simply adding the shortest time delays leads to the theoretically shortest possible reaction times. In reality, there is little
reason to expect that all the conditions are ideal and, hence, the
times are shortest.

If a person knows that a particular stimulus will occur, is
prepared for it, and knows how to react to it, the result is called
simple reaction time. Its duration depends on the kind of stimulus and its intensity. Many different tables of reaction times
have appeared in engineering handbooks, but how the original
data were measured usually is no longer known. The following list of shortest possible reaction times (in milliseconds, see
Table 7.1) is typical of generally used, although somewhat dubi-
ous, information.

- electric shock, touch and sound: at least 130 ms
- sight and temperature: at least 180 ms
- smell: at least 300 ms
- taste: at least 500 ms
- pain: at least 700 ms

This listing shows practically no time differences in reactions to electrical, tactile, and sound stimuli. The slightly longer reaction times for sight and temperature stimuli may lie
well within the range of measuring accuracy. However, the

time following a smell stimulus appears distinctly longer and that for taste even longer, whereas it takes by far the longest to react to the infliction of pain. The simple reaction times change little with age from about 15 to 60 years, but reactions are substantially slower at younger ages and slow again as one grows older.

Choosing between reactions

If a person has to choose among several feasible actions, *choice reaction time* is involved. It is longer than the simple reaction time and expands further if it is difficult to distinguish among several similar stimuli when only one should trigger the response. The length of a choice reaction time is a logarithmic function of the number of alternative stimuli and responses. The mathematical formula is

$$\text{Reaction time} = a + b \log_2 N$$

where a and b are empirical constants and N is the number of choices. N may be replaced by the probability of any particular alternative; this means $p = 1/N$, and the preceding equation changes to

$$\text{Reaction time} = a + b \log_2 (1/p)$$

Motion time

Motion time follows reaction time. Movements may be simple, such as lifting a finger in response to a stimulus, or complex, such as swinging a tennis racket. Swinging the racket involves more intricate movement elements than lifting a finger and it also engages larger body and object masses to be moved; all of this requires more time.

Motion time depends on the distance of the movement and its required precision. This relationship, called Fitts' law, is:

$$MT = a + b \log_2 (2D/W)$$

where D is the distance covered by the movement and W the width of the target. The constants a and b depend on the particulars of the situation (such as the body parts involved, the masses moved, and the tools or equipment used), the number of repetitive movements, and skill (training and experience).

Response time

Minimizing response time—the sum of the reaction and motion lags—often is a goal of human factors engineering. Optimizing the stimulus and selecting the body part that is best suited to the task are the obvious choices. To do so usually requires both careful task selection (and of associated equipment and procedures) and assessment of related capabilities of the prospective user. This is often an iterative process until the best match becomes apparent.

Summary

At the center of the human nervous system is the brain. It has the abilities to perceive and interpret signals sent from body sensors, then to process that information by integrating it with knowledge culled from memory and, finally, to make decisions about actions to be taken. Then, signals are sent along the efferent pathways of the peripheral nervous system to body organs, especially muscles, to take action. Body sensors pick up the results of these actions, together with independent new information, and corresponding signals are sent along the afferent pathways of the peripheral nervous system to the brain for new decision making.

Current concepts of the brain functions presume similarities between the ways the human brain and computers operate. Hence, the models depict information processing in sequential stages. We can look forward to the use of more realistic models and their practical applications. For the time being, however, we must make do with serial stage models. These provide guidance to the ergonomist for the design of human-controlled systems.

Fitting steps

Step 1: Select external signals that human sensors accept easily and quickly.

Step 2: Plan for decisions that depend on the sensory inputs, which are familiar, secure, and fast to make.

Step 3: Design for appropriate quick reactions. Test the system and revise as needed.

Further reading

BAILEY, R.W. (1996). *Human Performance Engineering*. Upper Saddle River, NJ: Prentice Hall.

WICKENS, C.D., LEE, J., LIU, Y., and GORDON-BECKER, S. (2004). *An Introduction to Human Factors Engineering.* (2nd Edn) Upper Saddle River, NJ: Prentice-Hall/Pearson Education.

Notes

The text contains markers, °, to indicate specific references and comments, which follow.

9.1 Taking up information:

Memory: Arswell et al. 2001, Stanton et al. 2005.

Central processing: Bailey 1996, Chapanis 1996, Wickens et al. 2004.

Advanced concepts: Arswell et al. 2001, Vicente 2002.

9.4 Actions and reactions:

Response times: Kroemer 2006, Kroemer et al. 2003, Swink 1996; Wargo 1967.

CHAPTER TEN

Hard physical work

In many industrialized areas, mechanization has reduced the demands for human labor. Nevertheless, heavy physical work still prevails in sections of agriculture and forestry; demanding labor remains frequent in mining, in building construction, and in some industries; it exists even in modern jobs, for example, in baggage handling by airline personnel. Periods of heavy demands may alternate with times of light duty work.

10.1 Physiological principles

Hard labor is an activity that intensely employs skeletal muscles that can convert chemical energy into work (physical energy) by moving body segments against internal and external resistances. From resting, muscle can increase its energy generation up to 50-fold. Such enormous variation in metabolic rate not only requires quickly adapting supplies of nutrients and oxygen to the muscle but also generates large amounts of waste products that need removal. The bloodstream, powered by the heart, provides the transport means. The lungs absorb the needed oxygen and feed it into the blood. Thus, while performing physical work, the body's ability to maintain an internal equilibrium largely depends on the proper functioning of the circulatory and respiratory systems to serve the involved muscles. The services include the supply of energy carriers and oxygen and the removal of wastes and heat. Among these functions, the control of body temperature is of importance, especially in hot and humid environments (see Chapter 8).

Assessing labor demands and the worker's capacities

Heavy work calls for great physical exertion with high energy consumption; thus, it poses high demands on the worker's metabolic functions with consequent strains on the circulatory and respiratory functions of the body. Usually, energy consumption and cardiac effort set limits to the performance capability of an individual. Therefore, the demands on metabolic and heart functions often serve to assess the severity of a physical task. Measurements of the worker's capacities in metabolic, cardiovascular, and respiratory functions establish his or her ability to execute heavy physical work.

10.2 Energy consumption

The skeletal muscles make the body work by moving body segments against internal and external resistances. Muscles need

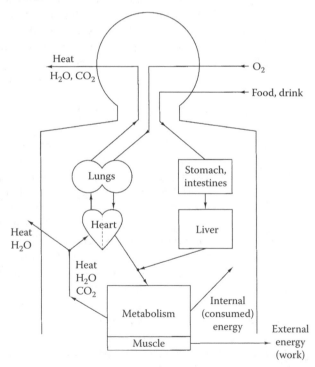

FIGURE 10.1 Diagram of the energy flow in and out of the human body. (From Kroemer, K.H.E., Kroemer, H.J., and Kroemer-Elbert, K.E. (1997). *Engineering Physiology. Bases of Human Factors/Ergonomics* (3rd Edn). New York: VNR–Wiley.)

energy for contraction. The mitochondria inside muscles can convert chemical energy into physical energy (see Chapter 3). Running this "human energy machine" involves a complex metabolic process that has similarities to combustion of fuel in an engine: it

- Yields energy that moves parts
- Needs fuel and oxygen to proceed
- Produces heat and other byproducts

Figure 10.1 provides a simple diagram that illustrates the sequence of events in the human body associated with the conversion of nutrients into mechanical energy and heat.

Energy units

The measuring units for energy (work) are joules (J) or calories (cal) with 4.19 J = 1 cal. (1 J = 1 Nm = 0.2389 cal = 107 ergs = 0.948×10^{-3} BTU = 0.7376 ft lb.) The units for power are watts, 1 W = 1 J/s, or kcal/hr = 1.163 W.

Metabolism

Metabolism is a fundamental biological process: the body takes in food and drink, which contain chemically stored energy and converts it into mechanical energy. As Figure 10.2 illustrates, food passes from the mouth to the stomach, where it is

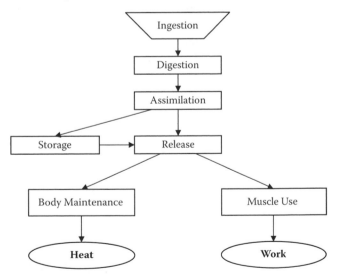

FIGURE 10.2 Diagram of the energy conversions from ingestion to heat and work. (Modified from Kroemer, K.H.E., Kroemer, H.J., and Kroemer-Elbert, K.E. (1997). *Engineering Physiology. Bases of Human Factors/Ergonomics* (3rd Edn). New York: VNR–Wiley.)

liquefied; alcohol, when present, is absorbed in the stomach and passed from there into the bloodstream. Carbohydrates, fats, and proteins, the common energy carriers in food and drink, are digested in the intestines. Digestion is the chemical conversion of large complex molecules into smaller ones that can pass through the membranes of the intestine cells and then become absorbed into blood and lymph. The liver largely controls what happens to the absorbed energy carriers: they are assimilated (reassembled) into new molecules that can be

- Stored as energy reserves (mostly as fat; one can "get fat" without ever eating any°)
- Employed for body growth and repair with the rest converted into heat
- Degraded for use of their energy content (glucose and glycogen)

When needed for doing work, first glucose and then glycogen serve as energy sources; stored fat provides the body's largest energy resource but is the last to be used.

Metabolic byproducts

Human metabolism is a complicated process. Only part of the chemically stored energy is actually converted into mechanical work performed by muscles; most serves to build and maintain structures of the human body and finally converts to heat. Because the human body core temperature must stay at 37°C (99°F), the excess heat must be dissipated to the environment, a task quite difficult to achieve in a hot climate (see Chapter 8). The heat is transported within the body by the bloodstream: some to the lungs, where it is dispersed into the air to be exhaled, and some to the skin where it is dispelled to the outside air, often with the help of evaporating sweat. Another byproduct is water, which the blood also carries to lungs and skin. A third byproduct is carbon dioxide, which the blood transports to the lungs for dissipation. Proper management of the metabolic byproducts is a major prerequisite for the body to be able to maintain energy generation, and hence to be able to continue hard physical labor.

Comparing the combustion engine with the human energy machine

In the cylinder of the engine, an explosive combustion of a fuel–air mixture transforms chemically stored energy into physical kinetic energy and heat. The energy moves the pistons of the engine, and gears transfer their motion to the wheels of the car. Cooling the engine is necessary to prevent overheating, and waste products need removal.

In the human machine, muscle fibers are both cylinders and pistons; bones and joints are the gears. The fuel, mostly derivatives of carbohydrates and fats in the nutrients, needs oxygen to yield energy in a slow combustion. When the muscles work, they produce metabolic byproducts, including heat, that need removal.

Energy content of food and drink

The kilojoule (1 kJ = 1000 J) or the kilocalorie (1 Cal = 1 kcal = 1000 cal) commonly serve as measures of the energy content of food and drink. Their nutritionally usable energy contents per gram are, on average (4.19 J = 1 cal):

- Alcohol: 30 kJ (7 Cal)
- Carbohydrate: 18 kJ (4.2 Cal)
- Protein: 19 kJ (4.5 Cal)
- Fat: 40 kJ (9.5 Cal)

Prepackaged food and drink usually carry labels that list the number of kilocalories or kilojoules of its content and per serving size, which often is in rather fanciful amounts. The label also breaks down the energy content in terms of carbohydrates, fats, and proteins.

Basal metabolism

A minimal amount of energy is necessary to keep the body functioning even if a person performs no activity at all. Under strict conditions (complete physical rest in a neutral ambient temperature, after fasting for 12 hours, with protein intake restricted for at least 2 days) one can measure the basic metabolism. The results show that the basal metabolic values depend primarily on age, gender, height, and weight—the last two variables are occasionally expressed as body surface area. Among healthy adults, there is little variation, hence a commonly used value is 1 kcal (4.2 kJ) per kg per hour, or 4.9 kJ/min for a person weighing 70 kg.

Resting metabolism

The highly controlled conditions needed to measure basal metabolism are rather difficult to accomplish for practical applications. Instead, one often measures the metabolism before the working day, with the subject as much at rest as possible. Depending on the given conditions, resting metabolism is around 10 to 15% higher than basal metabolism.

Work metabolism

The increase in metabolism from resting to working is called work metabolism. This increase above resting level represents the amount of energy needed to perform the work.

Measuring the energy above resting level is one way to assess the demands of work; another often employed measure is to determine the total amount of energy used by the body including the resting or basal levels, respectively.

Measuring heaviness of work

One way to assess the heaviness of work is, simply, to ask the working person to describe how hard the effort feels. For this, often some sort of a rating scale is used, such as the Borg scale described in Chapter 12. However, in many cases, objective measurements are desired, and for this purpose three different procedures are in common use. One solution is to observe the energy supplied to the body over a given time, the second approach is to take into account the heart rate during work, and the third technique measures the volume of oxygen consumed during work°.

Energy supply to the body

Taking into account the energy input by observing what a person eats and drinks and weighs is one of the oldest techniques. The underlying assumption is that, after deducting what is needed to maintain the body, all surplus energy is used for doing work. Naturally, this approach requires long observation periods, days or even weeks, during which the observed person performs various kinds of the physical activities, interspersed with resting periods. Therefore, the method is notoriously inaccurate unless performed under strictly controlled conditions.

Oxygen consumption at work

When the body performs work, oxygen consumption (and carbon dioxide release as well) is a measure of the associated metabolic energy production. A variety of measurement techniques is at hand; they all rely on the principle that differences in O_2 content between the exhaled and inhaled air indicate the oxygen absorbed in the lungs. Assuming an overall "average energy value" of oxygen of 5 kcal (21 kJ) per liter of O_2, the volume of oxygen consumed allows calculation of the energy that the body converts to do an activity during the observation period.

RQ

The *respiratory exchange quotient*, RQ, provides a more detailed assessment of the nutrients actually metabolized. The RQ compares the volumes of carbon dioxide expired to oxygen consumed. Metabolizing 1 g of carbohydrate requires 0.83 L of oxygen and releases the same volume of carbon dioxide. Hence, for carbohydrate, the RQ is one (unit). The energy released is 18 kJ per gram, 21.2 kJ per liter O_2. The RQ for protein conversion is 0.8, and for fat and alcohol conversion about 0.7. Measuring the volumes of CO_2 and O_2 during work assesses which energy carrier is metabolized.

10.3 Heart rate as a measure of work demands

Heart rate during work

Counting the heart rate during work is also a time-honored and widely used method. It relies on the knowledge that the body's bloodstream must transport nutrients and oxygen, and metabolic byproducts as well. The higher the energy requirements, the more blood flow is needed. To attain more blood flow, the heart must produce higher outputs, which it primarily achieves by increasing the number of heartbeats per minute. Thus, the heart's pulse rate varies in accordance with work demands.

Relations between heart rate and O_2 measurements

During dynamic work, a close relationship exists between metabolic processes and their support systems because the metabolizing muscle must be supplied with nutrients and oxygen and metabolic byproducts be removed from it for proper functioning. Therefore, heart rate (as indicator of circulatory functions) and oxygen consumption (representing the metabolic conversion) have a linear and reliable relationship. (However, this relationship differs among persons and may change with a person's training or deconditioning.) Therefore, one often can simply substitute heart rate countings for measurement of metabolic processes, particularly O_2 assessment. This is a very attractive shortcut because heart rate responds more quickly to changes in work demands. Furthermore, pulse counting is easier than oxygen measurements.

Reactions of heart rate and O_2 uptake to work

As Figure 10.3 shows schematically, at the start of physical work there is a sudden need for oxygen; the actual oxygen uptake lags sluggishly behind the demand. After its slow onset, oxygen intake rises rapidly and finally approaches the level at which it meets the requirements of the work. Thus, during the first minutes of labor, the body incurs a deficit in available oxygen. After work stops, the oxygen demand falls again to resting level, quickly at first and then leveling off. During this time, the still elevated cardiac, respiratory, and biochemical functions require that, as a rule, "the body pays 100% interest on the oxygen borrowed from the anaerobic bank": the oxygen debt repaid is approximately twice as large as the oxygen deficit initially incurred. Of course, given the close interaction between the circulatory and the metabolic systems, heart rate reacts similarly;

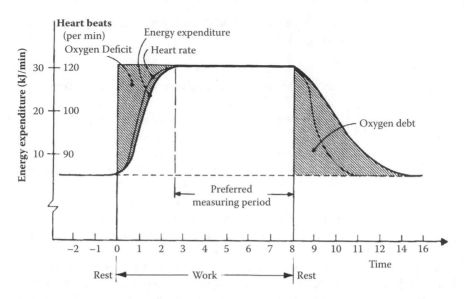

FIGURE 10.3 Schematic of energy liberation, energy expenditure, and heart rate before, during and after steady state work. (Modified from Kroemer, K.H.E., Kroemer, H.J., and Kroemer-Elbert, K.E. (1997). *Engineering Physiology. Bases of Human Factors/Ergonomics* (3rd Edn). New York: VNR–Wiley.)

yet, it increases faster at the start of work than oxygen uptake and it also falls back to resting level more quickly.

Steady-state work

If a required work effort stays below a person's maximal capacity, then blood flow, oxygen supply, and respiration can achieve and maintain their required levels. This condition of stabilized functions at work is called "steady state"; it is shown in Figure 10.3. Obviously, a physically fit person can attain this equilibrium between demand and supply at a relatively high workload, whereas an untrained person would be able to achieve this only at a lower demand level.

Classifying work demands

Obviously, expenditure of energy and the number of heart pulses during extended work are objective indicators of the heaviness of the demands of work. When asked for descriptive words, the workers might call some demands light or easy, and others heavy or hard. However, such descriptions can vary with circumstances and experiences: some grandparents, for example, accustomed to physical labor, might call a certain level of work demands moderate which their grandchildren might find heavy. Table 10.1 classifies work demands. The energy values listed

Table 10.1 Classifications of work demands

Classification in words	By energy expenditure		By heart rate in beats/min
	in kJ/min	in kcal/min	
Light, easy	10	2.5	90 or less
Medium, moderate	20	5	100
Heavy, hard	30	7.5	120
Very heavy	40	10	140
Extremely heavy	50	12.5	160 or more

Source: Modified from Kroemer, K.H.E., Kroemer, H.J., and Kroemer-Elbert, K.E. (1997). *Engineering Physiology. Bases of Human Factors/Ergonomics* (3rd Edn). New York: VNR–Wiley.

contain basal and resting metabolism; all values are unisex so many men will find the work a bit easier and many women harder than the labels imply.

10.4 Limits of human labor capacity

A maximal effort raises many body functions considerably, as Table 10.2 shows. As long as the body can meet the work demands by staying at steady state, the work can continue, as Figure 10.3 showed. However, if work demands exceed the body's capabilities, the supply functions cannot achieve a steady state but increase until they reach their limits. This forces work stoppage; then heart rate and other functions slowly fall back to their resting levels, as Figure 10.4 illustrates. The level of demand that an individual can follow depends on the person's physical fitness and skill.

Measuring people's fitness to do heavy work

Most medical and physiological assessments of human energetic capabilities rely primarily on the measurement of oxygen consumption, with heart rate as a secondary indicator. Standardized tests allow comparisons of work capacities. These tests employ normalized external work, mostly using bicycle ergometers, treadmills, or steps. Their use stresses primarily leg muscles. Because leg mass and musculature are substantial components of the body, their extensive exercising in a bicycle test also strains pulmonary, circulatory, and metabolic functions of the body. The treadmill also stresses primarily lower body capabilities, but in contrast to bicycling, the legs must support

Table 10.2 Changes in physiological functions from rest to maximal effort

Energy consumption	From 1 to 20 kcal/min	× 20
Oxygen uptake	From 0.2 to 4 L	× 20
Cardiac action	Heart rate from 60 to 180 beats/min	× 3
	Stroke volume from 50 to 150 mL	× 3
	Cardiac output = minute volume × heart rate, from 5 to 35 L/min	× 7
	Blood pressure, systolic, from 90 to 270 mmHg	× 3
Respiration	Breathing rate from 10 to 50 breaths/min	× 5
	Minute volume = tidal volume × breathing rate, from 5 to 100 L/min	× 20

Source: Modified from Kroemer, K.H.E., Kroemer, H.J., and Kroemer-Elbert, K.E. (1997). *Engineering Physiology. Bases of Human Factors/Ergonomics* (3rd Edn). New York: VNR–Wiley.

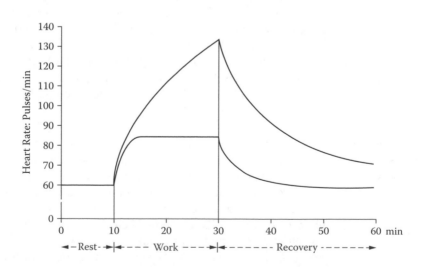

FIGURE 10.4 Heart rate increase at exhausting work versus heart rate at steady state during light physical work

and propel the weight of the whole body. Hence, the treadmill test strains the body in a more complete manner than bicycling but both omit trunk and arm capabilities from consideration. These examples show that selection of the test equipment and procedure can lead to different evaluations of physical fitness; for instance, the outcomes of testing well-trained bicyclists and of long-distance runners would differ when done on either bicycles or treadmills. Neither of these tests resembles actual work conditions.

Selecting persons fit for heavy work

Health and fitness tests are important means to make sure that only able persons are employed to perform heavy physical work. However, in ergonomic terms, it is better to design work tasks and equipment so that that they impose relatively small demands on human physical capabilities. This not only ensures that people are not overtaxed on the job, but also that more persons, even those with less than superb physical capabilities, can do the job.

Static work

One work demand is maintaining the same posture over some time: this requires continued muscular contraction. If such isometric (static) muscular contraction° exceeds about 15% of the muscular strength, blood flow through the muscle becomes reduced because the muscle begins to compress its own arteries and veins that transverse the muscle tissue. Stronger compression, which comes with increased muscle tension, further reduces the blood flow, and may even cut it off completely even though the heart tries to increase blood pressure to overcome the flow obstruction. This leads to fatigue, which finally makes us abandon the tiring posture to relax and recover. The ability to endure a static contraction depends on the magnitude of contraction, as Figure 10.5 demonstrates. In suitable dynamic work, in contrast, rhythmic alterations between tension and relaxation in a muscle facilitate blood flow through its tissues—this has been called a "muscle pump"—as illustrated in Figure 10.6.

Static effort increases the pulse rate as the heart strives to increase blood pressure in order to overcome the flow resistance in compressed blood vessels within the straining muscle. Increased heart rate associated with a sustained static muscular effort is indeed an indicator of physical strain; however, because blood flow is diminished, metabolism is reduced as well. Therefore, in the case of static effort at work, the linear relation between heart rate and energy consumption, mentioned above, falls apart.

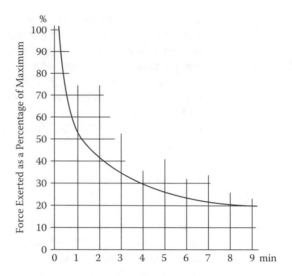

FIGURE 10.5 Maximal duration of static muscle contractions

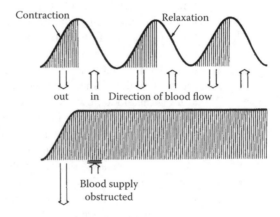

FIGURE 10.6 Blood flow through a muscle doing dynamic and static work

10.5 Designing heavy human work

Human energy efficiency at work

Engineers and economists like to express work efficiency as the ratio of gain versus effort. If we assume that the energy storage in the body does not change (meaning that the observed person does not put on or lose weight) and that the body neither gathers heat from the environment nor loses any, we can state a simple energy balance equation as

$$I = H + W,$$

where I is the energy input, H is the heat developed, and W is the performed work.

Human energy efficiency (work efficiency) is the ratio between work performed and energy input:

$$e \text{ (in percent)} = 100 \ W/I.$$

In everyday activities, only about 5% or less of the energy input converts into work, that is, energy usefully transmitted to outside objects; highly trained athletes may attain, under favorable circumstances, perhaps 25%. The remainder (that is, most) of the input converts into heat, usually at the end of a long chain of internal metabolic processes. Because humans are so inefficient, in energy terms, their capabilities are better used to think and manage, to control machinery and processes rather than to serve as physical movers.

Design work to fit the human
The engineer determines the work required and how it is to be done. To arrange for a suitable match between capabilities and demands, the engineer needs to adjust the work to be performed (and the work environment) to the body's energetic capabilities. These human capabilities are determined by the individual's capacity for energy output (physique, training, health), by the neuromuscular function characteristics (such as coordination of motion, muscle strength, etc.), and by psychological factors (such as motivation).

Avoid exhausting work
Coal miners and lumberjacks, who do their work with hand tools, are among the male workers with the highest energy consumption: about 19,000 kJ per day, measured in the 1960s. Such extreme efforts are probably becoming rare in many countries because mechanized tools and modern machinery can make work less demanding. The daily energy consumption for moderately demanding work is around 12,000 to 15,000 kJ for men and 10,000 to 12,000 kJ for women.

Provide rest breaks
Breaks in physical work provide recovery and rest periods. Their provision is essential in hard physical labor, and desirable even at lighter work for physiological and psychological reasons. Providing many breaks of short duration is more beneficial than allowing a few longer interruptions. The reason is that recovery is steepest at the beginning of a break in work, as the curves of energy and heart rate in Figures 10.3 and 10.4 illustrate.

No "static work"

As already mentioned, the traditional understanding of physical work is that it consists of dynamic actions. Under these conditions, heart rate and energy consumption are closely related. However, many actual work tasks include static efforts, where parts of the body must be kept in a frozen position for a while. Such static effort strains the cardiovascular system even though it does not constitute work in physics terms. Therefore, heart rate shows an increase and energy consumption does not. Static muscle efforts are often tiresome but not productive; therefore, they should be designed out by changing the process or replacing human effort by use of mechanical solutions.

Summary

Figure 10.7 provides an overview of the human traits and the conditions at work that determine, by interacting with each other, how much hard work a person can do. The ergonomist exerts no power over individual characteristics but, in contrast, has latitude and responsibility for the layout of work task and schedule and, relatedly, for the equipment and tools used.

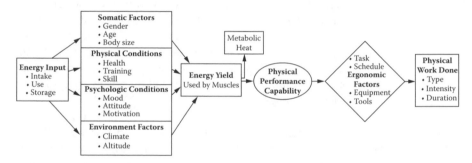

FIGURE 10.7   Interactions between the human operator and the design of work that determine the person's ability to perform heavy physical work. (Adapted from Kroemer, K.H.E., Kroemer, H.B., and Kroemer-Elbert, K.E. (2003, amended reprint). *Ergonomics: How to Design for Ease and Efficiency* (2nd Edn). Upper Saddle River, NJ: Prentice-Hall/Pearson Education.)

Fitting steps

Step 1: Determine whether hard physical labor is indeed necessarily done by humans.

Step 2: Try to alleviate the workload.

Step 3: Provide rest breaks, encourage taking time off as needed.

Further reading

ASTRAND, P.O., RODAHL, K., DAHL, H.A., and STROMME, S.B. (2004). *Textbook of Work Physiology. Physiological Bases of Exercise* (4th Edn). Champaign, IL: Human Kinetics.

RODAHL, K. (1989). *The Physiology of Work*. London: Taylor & Francis.

Notes

The text contains markers, °, to indicate specific references and comments, which follow.

10.2. Energy Consumption:

Getting fat without eating fat: Flier et al. 2007.

Measure heaviness of work: Kroemer et al. 1997.

Static muscular contraction: Because there is no displacement in an isometric (static) muscle contraction, this muscular effort does not produce work in the definition of physics: work is the integral of force and displacement.

CHAPTER ELEVEN

Light and moderate work

Even when we are at rest or when our work requires only light effort, the basic metabolic functions of our bodies run the same as when supporting hard labor (see Chapter 10), but at a lower level: energy stored in chemical compounds in our food and drink is digested, then assimilated into simpler molecules. These serve to build and maintain our organs and to provide the energy that muscles need to function. The byproducts of the metabolism must be dissipated through the lungs and the skin. This loads the circulatory and respiratory body functions, but naturally less than during heavy physical work.

Therefore, in contrast to hard labor, during light and moderate work neither energy use nor blood flow pose the limiting factors for our capabilities to perform our tasks. The chief demands are on our abilities to toil diligently with endurance and attention to detail during long shifts at home and at the work site, in commerce and transportation, in manufacture and assembly, and in agriculture and trade.

The environment in which we work can contribute strongly to our perception of the work as being easy or demanding. The psychosocial environment is an important aspect; it determines, for example, how we get along with our coworkers and supervisors. In physics terms, the indoor work environment primarily depends on air temperature and humidity, on the surrounding sound and noise and on lighting, already discussed in Section II of this book. Our feeling well at work also depends on the arrangement of our working hours, including shiftwork, treated in Chapter 16 of this book.

11.1 Physiological and psychological principles

Most jobs in modern offices or driving vehicles are sedentary and require little physical effort, as the examples in Table 11.1 show. Therefore, many persons with sit-down jobs try to bring up their daily expenditures by doing demanding leisure activities, work-outs, and sports. Figure 11.1 shows energy expenditures in leisure activities. Such exercise activities rev up otherwise idle body functions and help to compensate for unneeded intake of energy by food and drink which otherwise might lead to excess weight.

Table 11.1   Energy expenditures at sample activities

Sedentary office work	About 2 kJ/min
General house work	5 to 20 kJ/min
Walking, on level ground at 4 km/h	About 14 kJ/min
Walking with 30 kg load on the back, on level ground at 4 km/h	About 23 kJ/min
Running, on level ground at 10 km/h	About 45 kJ/min
Climbing stairs, 30 deg incline, 100 steps/min gaining height at 17 m/min	About 60 kJ/min

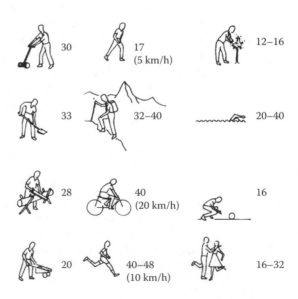

FIGURE 11.1   Energy consumption in leisure activities

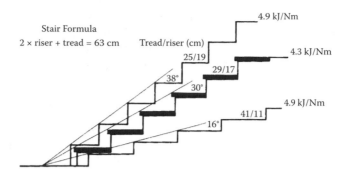

FIGURE 11.2 Energy (in kJ) per body weight (in N, with 9.81 N equivalent to 1 kg) raised (in m)

Stairs and ramps

One simple way to "work the body" for people who have little exercise at work is to walk stairs instead of taking the elevator in tall buildings. Stair climbing demands heavy muscle use to lift the body and hence keeps the metabolic and circulatory systems fit. Figure 11.2 provides information on energy use depending on step designs and steepness of stairs. However, this information can also help in devising stairs that persons can use who, by injury, illness, or age, are physically handicapped. Of course, ramps without steps are even easier to walk and allow wheelchairs to go up and down as well.

Measuring work load effects

When the energy requirements of work are low, measuring the worker's oxygen consumption, for example, provides little useful information. However, heart rate (HR)—which is easier to obtain—may be a useful indicator because it also responds to static muscle loading, such as when holding the body or its segments in one given position. This sensitivity of HR may be an advantage over energy assessment, but the responsiveness extends also to mental reactions such as excitement or frustration; so, reactions of HR may reflect not just loading by physical work. (Yet, that very responsiveness of HR may provide means for nontraditional assessments.) Other, often specialized approaches to measure workload effects° rely on electrical events in the body, for example, related to muscle activation (electromyograms, EMGs), brain activities (electroencephalograms, EEGs), and eyelid closure frequencies. Simply observing the general work output, especially as it fluctuates with changing working conditions, is also a realistic and nonintrusive yet nonspecific measure.

Scaled judgments

Another way to assess satisfaction or discontent with the conditions of work is to ask people for their opinions and judgments. This can be done in either an informal manner or, more frequently, by using prepared questionnaires. One example of such a questionnaire might use the following pairings of descriptors.

Fresh ——— Weary

Sleepy ——— Wide-awake

Vigorous ——— Exhausted

Weak ——— Strong

Energetic ——— Apathetic

Dull, indifferent ——— Ready for action

Interested ——— Bored

Attentive ——— Absent-minded

The person may choose between one of the paired descriptors: this is called a forced choice. Or the person may put a mark on the line between two opposing statements to indicate how interested, or how bored, for example, he or she actually is. Of course, many other, often complicated questionnaires may be of value to assess the conditions of comfort or discomfort at work.

Often, responses are collected using scales, of which three different kinds are in common use. Ordinal scales arrange judgments in rank order; they allow only statements of higher or lower. Interval scales require judgments set in steps that are at equal distances from each other. Ratio scales are interval scales that are anchored by an absolute zero; therefore, numbered responses in ratio scales allow mathematical calculations. The Borg RPE Scale in Table 12.2 is an interval scale whereas the Borg CR 10 Scale is a ratio scale—see Chapter 12.

Because of the inherent difficulties related to obtaining objective numbers, many studies use subjective judgments (by the worker or by another expert) that assess the suitability of an existing work situation. Although the procedures vary widely among researchers, most rely on templates that ask for discomfort or pain statements. The manner of administering the survey may affect the outcome. Being involved in such a study may generate a heightened awareness of problems at work that should be in everybody's interest, although management sometimes views this issue with apprehension.

An often-used, well-standardized inquiry tool is the Nordic Questionnaire. This survey requires forced binary or multiple-choice answers. It consists of two parts, one asking for general information, with the other specifically focusing on regions

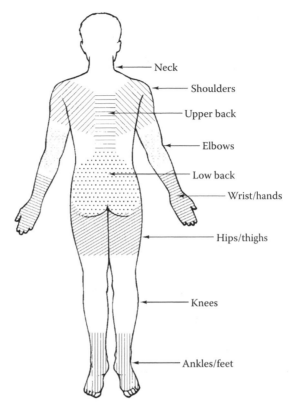

Neck
Shoulders
Upper back
Elbows
Low back
Wrist/hands
Hips/thighs
Knees
Ankles/feet

FIGURE 11.3 Body sketch in the Nordic Questionnaire. (Adapted from Kroemer, K.H.E., Kroemer, H.B., and Kroemer-Elbert, K.E. (2003 amended reprint). *Ergonomics: How to Design for Ease and Efficiency* (2nd Edn). Upper Saddle River, NJ: Prentice-Hall/Pearson Education.)

of the body. This section uses a sketch of the body, divided into nine regions, shown in Figure 11.3. The interviewed person indicates whether she or he experiences any musculoskeletal problems in these areas. A number of modifications to the Nordic Questionnaire have been proposed, however, it is widely administered and, therefore, provides internationally standardized information.

11.2 Tiredness, boredom, and alertness at work

Feelings of tiredness and boredom are familiar to many from everyday life. They describe states in which we feel unable, or at least disinclined, to continue what we were doing, that we take little interest in our task. Truly, tiredness relates to fatigue,

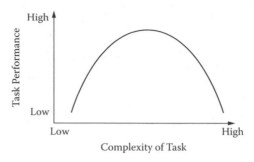

FIGURE 11.4   Conjectured relation between task complexity and performance

the physiological effect of having spent our energy at work, of muscles being overstrained, so that our body needs rest to recuperate from the preceding effort, as discussed in Chapter 10. Strictly speaking, boredom is a psychological or emotional condition in which a lack of events lulls us into a state of indolence. Yet, in everyday parlance, we feel "tired" as a result of either (physical) fatigue or (psychological, emotional) boredom.

Diversity versus monotony

Many individuals enjoy jobs that have diverse parts, changing tasks that challenge the mental and physical capabilities. The variety of demands keeps the workers interested and they find satisfaction in successful solutions. However, to others there is appeal in a job that predictably presents the same tasks, or at least similar things to do. They may find satisfaction in skillful repetition, such as on an assembly line, while thinking, daydreaming, or conversing. So, monotony, the lack of unusual stimuli, is something to loathe or to like, depending on one's inclinations. For most people, however, performance—and job satisfaction—is best with a job that is neither overly complex nor too simple, as depicted in Figure 11.4. However, any person's preferred spot on that U-shaped curve may change according to skill, health, and mood.

Vigilance and event frequency

Many jobs require alert watchfulness, vigilance, over long periods of time. The ability to sustain concentration turned out to be of special importance during World War II. Then it was noticed that the frequency at which radar observers reported submarines to appear on their screens diminished with the length of their watch. Half of all occurrences were reported during the first 30 minutes on duty; during each of the following 30-minute periods, the reports of sightings fell to 26, then to 16, and finally to 10%. Obviously, alertness° deteriorated with time on watch.

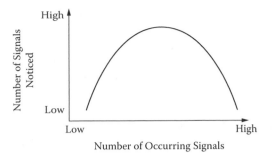

FIGURE 11.5 Inverted U function depicting the number of signals occurring and reported

Other experiences showed that the frequency at which signals occur greatly influences the number of reported signals. If so few signals appear that the observer gets tired of waiting for them and therefore pays little attention anymore, a signal may not even be recognized when it finally shows. On the other side, at very high frequencies of occurring signals, the observer may no longer be able to follow and report them. Between these extremes of too few and of too many signals is a broad range of signal frequencies in which the signal appearances are well reported. Figure 11.5 depicts these relations: it suggests that the observer's performance follows the shape of an inverted U. This finding led to the hypothesis that, at very few events, the observer is underloaded, and so the mind drowsily wanders; conversely, with too many events, the observer is overloaded and hence fails in recognizing and reporting all of them.

These wartime observations opened an important field of research that generated extensive knowledge and many theories about vigilance and arousal. In addition to the experience that performance decreases with time, it also shows that performance worsens if the intervals between signals vary much; if the observer is fatigued or under mental stress; or if the observer must perform in environments that are obnoxious, for instance, because of heat, noise, or vibration. On the other hand, strong signals at suitable frequencies, meaningful to the task, improve performance.

Monotonous jobs

In the 1920s, industrial managers thought it appropriate to split jobs, when possible, into a small number of identical tasks, which the operator was to repeat over and over. This approach, often traced to the ideas of Frederick Taylor°, created monotonous and repetitive tasks, each with the same demands on the operator's dexterity. Therefore, after practice, the worker became highly skilled and could perform the tasks at high speed. There are, however, serious problems with this Tayloristic approach: although some people like to do repetitive work, others are bored by drudgery and monotony; furthermore, excess repetition can lead to injury of the human musculoskeletal system, as discussed below.

Satisfaction at work

Individuals' preferences and aspirations differ, yet, almost every person wishes to have the satisfying experience of mastering the work task and exercising power over how to do the job, especially how it could be done more easily and the outcome improved. So-called detail (or instrumental) control is at the task level, including decisions regarding workplace arrangements, tools and procedures, and the immediate environment. Conceptual (managerial) control is at a higher level: it may include decisions about general design, work organization, or even administrative policies. Having control over the job provides feelings of power and importance whereas lack of control can be a work deterrent and a social stressor. Feedforward and feedback among workers and management is a vital part of participatory and team work, as discussed in some detail in Section IV of this book. Active communication can greatly improve job satisfaction and create a social climate conducive to high productivity. Recognition, rewards, and incentives, directly related to successful work, create personal interest and ego involvement°, which improve performance.

11.3 Suitable postures at work

No static work

In light and moderate work, fatigue often results from the requirement to maintain positions of the limbs, or the posture of the whole body, over considerable time. Maintaining a posture often requires the involved muscles to keep up a constant contraction, which compresses its own tissues. That pressure hinders the flow of blood through muscle (discussed in Chapter 10), and the obstructed bloodstream cannot remove all metabolic byproducts. With accumulating metabolic byproducts, the muscle "fatigues" and must relax to recover. During that time, the blood flow gets restored and the metabolic waste removed. To avoid such fatiguing and debilitating postures, various—and often simple—ergonomic solutions are at hand.

Avoiding fatiguing body postures

Many tasks still contain requirements for tiring body postures°. Standing, stooping, and kneeling increase the metabolic cost

Sitting	Standing	Stooping	Kneeling
3–5%	8–10%	50–60%	30–40%

FIGURE 11.6 Metabolic cost increases associated with body postures; the reference is resting while lying down

FIGURE 11.7 Avoid trunk twisting

over sitting, as Figure 11.6 shows. Other examples of fatiguing conditions are twisting the trunk; bending close to the ground or reaching up; working with extended arms; at sales counters, standing for hours; in computerized offices, long sitting with eyes on the display and hands on keys. Figure 11.7 shows that severe twisting of the trunk is necessary in order to move a heavy packet onto a shelf. Figure 11.8 illustrates that thoughtlessly putting a large box on the floor requires a person to "dive" into it in order to sort and take out the content. Simply putting the box on its side, atop a support, allows the worker to work in a more upright posture.

Sitting at work   Sitting is less tiring than standing. However, there should be space for the legs and a provision for effortless sitting; not every operator likes to fold his legs under his body, as the

FIGURE 11.8 Putting a large box on its side, elevated, replaces "diving" into the container to take out its content. (Adapted from Kroemer, K.H.E. (1997). *Ergonomic Design of Material Handling Systems.* Boca Raton, FL: CRC Press.)

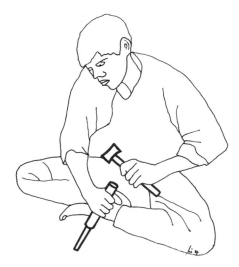

FIGURE 11.9   Wood carver sitting on his work piece

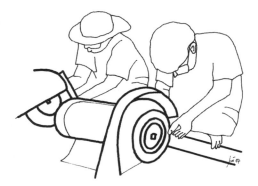

FIGURE 11.10   Grinding in miserable work postures

wood carver does in Figure 11.9. One could argue that carving in this way, where the toes actually serve to hold the work piece, is the traditional way to which local craftsmen have long been accustomed. However, no such excuse applies to the grinder workplaces in Figure 11.10. This drawing shows that the grinders must tuck their feet under, bend forward, and work with extended arms; we can just hope that at least proper safety provisions exist at these miserable workplaces. The operator shown in Figure 11.11 has better legroom, but an edge of the machine seems to press against her knees; furthermore, she must do much of her work with arms stretched upward and forward.

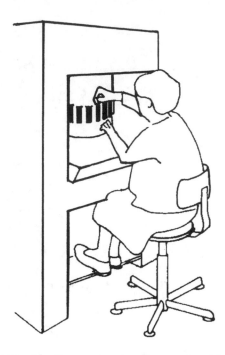

FIGURE 11.11   Loading and unloading parts with extended arms

**Too much
sitting**

Cases of nearly intolerably long and immobile sitting occur in long-distance automobile driving and airplane flying. A truck driver who needs to cover a long distance may have to remain seated with hands on the wheel, feet on the pedals, and eyes on the road, for many hours before he or she can get up and stretch the body. A military pilot who has to fly a long mission is even worse off, because there is no way of getting up and out. The solution tried in World War II airplanes was to have separate, air-filled sections in the pilot seat that would automatically inflate and deflate. Such pulsating body support can provide some relief from constant pressure on the body. Seriously ill persons who must remain in bed for days and cannot move often suffer bedsores on those body parts that transfer most weight. These examples show that moving about instead of remaining still is essential for well-being and comfort.

11.4 Accurate, fast, skillful activities

**Exact
manipulations**

In light work, especially if it requires frequent and exact movements, the work area of the hands should be at about waist

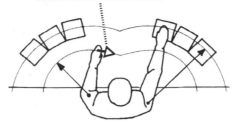

FIGURE 11.12 Work area for fast and accurate light handwork

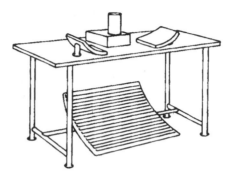

FIGURE 11.13 Workplace with supports for the elbows and feet

height, in front of the trunk. This preferred work area is shown in Figure 11.12, which also indicates the fastest and most accurate motions of hand and forearm, accomplished by inward rotation about the elbow and shoulder joints. If the work requires particularly exact manipulation, armrests as in Figure 11.13 might be helpful to stabilize the upper body.

Seeing what we are doing   Being forced to keep the neck bent or twisted can cause a real headache; such an unbecoming posture is often enforced by having to look at a work object that is not within easy view. Work with fixed loupes or microscopes has alway been difficult because, in order to keep the eyes in the exactly prescribed position, neck and trunk muscles must maintain strict static contractions, as Figure 11.14 sketches. Electronic devices can eliminate this problem because they project the image on a monitor, which the operator can view from various positions of the eyes the head.

Describing the head and neck posture used to be a chore with the so-called Frankfurt Plane (still found in older textbooks),

FIGURE 11.14   Work with loupes and microscopes

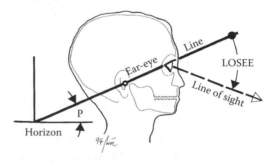

FIGURE 11.15   The Ear–Eye Line runs through the ear hole and the junction of the eyelids. It describes the posture of the head and serves as reference for the angle of the line of sight.

but is easy using the Ear–Eye Line, shown in Figure 11.15. The EE Line runs through two simple markers on the head, the ear hole and the meeting point of the eyelids (described earlier with Figure 5.3).

The line of sight connects the eye with the visual targets. If the visual target is far ahead, the favored angle of the sight line against the EE Line, LOSEE in Figure 11.15, is about 15 degrees; but the LOSEE angle becomes larger as the object, upon which one must focus, gets closer. At reading distance, at about forearm length (around half a meter) from the eyes, the preferred angle is about 45 degrees. However, there are large

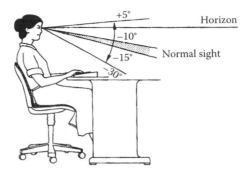

FIGURE 11.16 The "cone of easy sight"

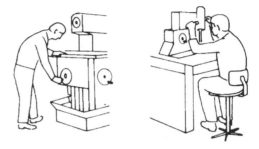

FIGURE 11.17 Improper machine tool design causes bent body postures

differences in the preferred angles and distances from person to person, and wearing eye corrections can play a role as well. Eye movements within about 15 degrees above and below, and to the sides, of the average line of sight angle are still comfortable; that means that the common viewing task should be within a 30-degree cone around the normal line of sight, as shown in Figure 11.16.

Better tools

Forcing the body into contorted, especially bent, positions is frequently the result of improper design, such as mispositioning the controls on machinery. Figure 11.17 illustrates such design flaws on machine tools, which are simple to avoid by considering body sizes and reaches. One can improve even such traditional tasks as gardening and field work by simple means; Figure 11.18 shows an example for seeding and fertilizing. Variations in basic fieldworking tools such as hoes can make work easier or harder: in loose soil, an open hoe needs less energy to use than a conventional blade, and it also works well in compacted earth; see Figure 11.19.

FIGURE 11.18   Hand seeding/fertilizing

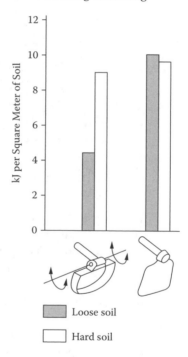

FIGURE 11.19   Energy use with two kinds of hoes

Repetitive work   Much skillful work consists of often-practiced tasks. Repetition-caused injuries to muscle and other connective tissues, and to the joints of the human body, have become a major problem in occupational safety and health. This is surprising because, already in 1713, Ramazzzini[o] described occupation-related diseases and injuries that occurred in jobs with physical labor; he also mentioned that even "secretaries" (scribes) suffered from muscle pains and cramps stemming from their repetitive tasks[o].

Doing the same activity over and over, using the same muscles, tendons, and joints, often harms them. Apparently, the human body is not designed for overly repetitive actions, whether they occur at work or at leisure, as Table 11.2 illustrates. The underlying reasons for the listed health problems are plainly obvious from the descriptive names of the injuries. Low back, wrists,

Table 11.2 Repetition-related disorders

- Writer's cramp or scribe's palsy
- Telegraphist's or pianist's wrist
- Stitcher's or tobacco primer's or meat cutter's or washerwoman's wrist
- Goalkeeper's, seamstress,' or tailor's finger
- Bowler's or gamekeeper's or jeweler's thumb
- Bricklayer's hand
- Jackhammerer's or carpenter's arm
- Carpenter's or jailer's or student's elbow
- Porter's neck
- Shoveller's hip
- Weaver's bottom
- Housemaid's or nun's knee

20th and 21st Century Disorders:
- Knitters or crocheter's wrist
- Ballet dancer's or nurse's foot
- Baseball catcher's hand
- Letter sorter's and yoga wrist
- Golfer's or tennis elbow
- Letter carrier's shoulder
- Nurses' back
- Carpet layer's knee
- Typist's and cashier's wrist
- Typist's myalgia
- Typist's tenosynovitis
- Keyboarder's mouse elbow
- Keyboarder's carpal tunnel syndrome
- Texters' thumb

Source: Adapted from Kroemer, K.H.E. (2001). Keyboards and keying. An annotated bibliography of the literature from 1878 to 1999. *International Journal Universal Access in the Information Society UAIS,* **1**: 2, 99–160.

elbows, and shoulders are frequent locations for repetition-caused injuries.

Rest breaks

Breaks in physical work, which provide time for rest and recovery, are beneficial for both physiological and psychological reasons. Providing many breaks of short duration is more helpful than allowing a few longer interruptions. The reason is that recovery is steepest at the beginning of a break in work. Interrupting the flow of work, especially if it is repetitive, monotonous, or otherwise demanding in either the mechanical or psychological sense, helps to prevent overload. Furthermore, it gives opportunity to engage in social interaction with others, often while the energy supplies of the body are replenished by food and drink. Thus, generous provision of frequent rest breaks, freely selected by the worker if at all possible, can be an important means to improve feeling well, job-related attitude, and hence performance in all kinds of work, mental and physical, heavy and light.

Summary

Light or moderate work does not impose heavy burdens on our physical capabilities; instead, the chief demands are to toil diligently with endurance and attention to detail during long shifts at home and at the worksite. Therefore, many persons with sit-down jobs choose to bring up their daily expenditures by doing demanding leisure activities and sports.

Measuring the heart rate—which is easy to obtain—can be a useful indicator of a person's work strain because it responds to dynamic work and to static muscle loading, such as when holding the body or its segments in one given position. This sensitivity of HR extends also to mental reactions such as excitement or frustration: so, reactions of HR may reflect more than the loading by physical work. Another way to assess the conditions of work is to ask people for their opinions and judgments. This can be done in either an informal manner or, more frequently, by using prepared questionnaires and scales.

Many individuals enjoy jobs that have diverse tasks that challenge the mental and physical capabilities. However, to others there is appeal in a job that predictably presents the same tasks, or at least similar things to do. For most people, performance, and job satisfaction, are best with a job that is neither overly complex nor too simple and that allows them some control over task execution, workplace arrangements, and the immediate environment.

Physical fatigue can result from the requirement to maintain positions of limbs, or the posture of the whole body, over considerable time. Although sitting is, as a rule, less tiring than standing, cases of nearly intolerably long and immobile sitting occur in long-distance automobile driving and airplane flying. These examples show that moving about instead of remaining still is essential for well-being and well-feeling.

Work that requires frequent and exact movements of the hands should be done at about waist height, in front of the trunk. Being forced to keep the neck bent or twisted can cause a real headache; such an unbecoming posture is often enforced by having to look at a work object that is not within easy view. Commonly, a viewing task should be within a narrow cone around the normal line of sight, which is inclined forward–downward. However, there are large differences in the preferred angles and distances from person to person.

Some skillful work consists of often-practiced tasks. However, doing the same activity over and over, using the same muscles, tendons, and joints, often harms them. Apparently, the human body is not suited for overly repetitive actions, whether they occur at work or at leisure.

Breaks in physical work, which provide time for rest and recovery, are beneficial for both physiological and psychological reasons. Thus, generous provision of frequent rest breaks, freely selected by the worker if at all possible, can be an important means to improve feeling well, job-related attitude, and hence performance in all kinds of work, mental and physical.

Fitting steps

Step 1: Make work easy to do; avoid fatiguing postures and highly repetitive tasks.

Step 2: Ask people for their opinions and judgments of the conditions of their work.

Step 3: Strive to improve.

Further reading

BAILEY, R.W. (1996). *Human Performance Engineering* (3rd Edn). Upper Saddle River, NJ: Prentice Hall.

CHENGULAR S.N., RODGERS, S.H., and BERNARD, T.E. (2003). *Kodak's Ergonomic Design for People at Work* (2nd Edn). New York: Wiley.

Notes

The text contains markers, °, to indicate specific references and comments, which follow.

11.1 Physiological and psychological principles:

Measuring workload effects: Karwowski 2001, Kumar 2004, Kumar et al. 1996, Stanton et al. 2005.

11.2 Tiredness, boredom, and alertness at work:

Alertness etc: see Bailey 1996, Proctor et al. 1994, Stanton et al. 2005, Wickens et al. 2004.

Frederick Taylor: 1856–1915. "Taylorism": Ruthlessly cutting cost and effort by planned management engineering. See Bjoerkman 1996.

Ego involvement, work performance: See Bailey 1996, Kroemer et al. 2001.

11.3 Suitable postures at work:

Tiring body postures: Chaffin et al. 2006, Delleman et al. 2004, Kroemer et al. 2003.

Ramazzini: see the translation by Wright 1993.

Repetitive tasks: see Kroemer 2001; Kroemer et al. 2003; Kumar 2004, 2008.

CHAPTER TWELVE

Workload and stress

What is stress?

The physician Hans Selye (1907–1982) introduced the term "stress" in the 1930s. He described it as the reactions of the human to good situations, *eustress*, and to bad situations, *distress*. Today, the emphasis is almost completely on the negative condition, when we feel "stressed out" by a stressor.

Stressor causes stress

Selye's choice of terms was unfortunate: he did not realize that engineers consider *stress* an impulse (such as from a heavy load on the back) that causes *strain* in the receiving structure (e.g., in bones and muscles of trunk and leg). These two opposite meanings of stress, either result (in psychology) or cause (in engineering), have led to much confusion. In this book, we follow the convention that a stressor causes stress.

12.1 Stress at work

Physiological reactions to stress

Selye° discovered that stress differs from other physical responses: either a positive or a negative impulse can cause similar reactions in the autonomic nervous and the endocrine systems. Typical responses are increased secretion of hormones in the adrenal gland, especially of adrenalin and noradrenalin, which set the whole organism in a state of heightened alertness, an ergotropic setting of the body, ready to fight or flee.

Stress causes emotions

Today, psychologists posit that the basic experience of stress is emotional°. Emotions obviously can affect certain body functions, such as heart rate, sweating, and sleep patterns; they even may precipitate or alter the course of major physical diseases. (However, it is not clear how stressors do this.) Stress can cause

235

anxiety, which then triggers the body's endocrine and autonomic nervous systems to speed up heart rate, and to increase blood pressure and rate of sweating. Stress can also cause muscle tension, which may lead to pain in the head, neck, back, or elsewhere.

Stress at work and leisure

The mind–body interaction works both ways: psychological factors can contribute to the onset or aggravation of a wide variety of physical disorders and, conversely, physical disorders can affect a person's thinking or mood. The intensity and effects of stress at work depend on the dynamic interactions between the individual and the specific conditions of the work environment; of course, problems outside work, which affect a person, make dealing with work stress more difficult. The emotional responses of individuals to the same stressors are different, and so are their bodily reactions.

Coping with stress

Distress, anger, anxiety, and even depression may arise when the person perceives that he or she cannot adequately cope with the demanding conditions. Because stress is a subjective phenomenon, stress management techniques must be tailored to each individual case. To deal with stress means to activate individual cognitive and behavioral strategies. They may aim directly at the work demands by such strategies as time management, work style adaptation, assertive communications, setting of limits, and other control tactics. Another approach focuses on individual emotions, such as by cognitive re-evaluation of the situation (reframing the problem), use of humor, relaxation exercises, or off-work activities and hobbies. Often, the person's physical health, nutrition, and habits such as smoking or drinking require attention. This often means a change in lifestyle. A radical solution is to quit the stressful job and look for work that is more suitable; however, this requires that another better job is at hand.

Eliminating stressors at work

Job stress is primarily a mismatch between demands and a person's ability to meet them. Instead of trying to adapt the person, it often makes more sense to eliminate stressors at work. The following list contains examples of stressors in the work environment.

1. Job content
2. Demand intensity
3. Workload complexity
4. Repetitive monotonous work

5. Excessive responsibility

6. Unreasonable unfriendly supervisor

7. Environmental circumstances such as noise, unpleasant climate (indoor or outdoors), crowded office space

8. Lack of recognition, missing rewards for achievement

9. Lack of control over one's job

10. Lack of job security

11. Lack of social support from supervisors and peers

12. Threatening, bullying boss or coworkers

Effects of stress Intensive or sustained stress often changes the ways in which people behave and feel. Irritability, general dissatisfaction, impaired attention, damaged interpersonal relations, anxiety, and depression are common symptoms. Healthwise, stress may bring about disturbed sleep, cardiovascular diseases, gastrointestinal complaints, musculoskeletal impairments, and weakened immune functions. Work-related stress may also lead to maladapted behaviors such as excessive smoking, or alcohol and drug abuse. Stress may also negatively affect the organization and its functioning through increased employee absence, poor work attitude, or reduced productivity and quality of work. Hence, stress can be costly to the organization° as well as to the individual.

Is stress always harmful? Obviously, stress is part and parcel of our life; it is a necessary condition for all living creatures who must react to new situations in an appropriate way. Life without stressors would not only be unnatural but boring as well. Paracelsus, a physician in the 16th century, said, *"Dosis sola facit venenum"*: it is only the dosage that determines whether something is toxic. The same is true for stress: the amount of stress determines whether it has adverse effects on well-being or whether it increases a person's ability to cope with life events. Where the boundary is between healthy and pathological stress varies from one individual and one situation to another. However, there is no question that surprises, joyful events, comical situations, unexpected good deeds, and friendly words make us feel alive and perky. Often, challenging tasks appear captivating and interesting, and we feel satisfied and rewarded when we succeed in meeting the challenge. Thus, good stress is part of a good life.

Measurement of stress A great variety of psychological techniques° is at hand to scale a person's perception of stress. They include well-being surveys, stress-arousal checklists, mood assessments, emotional intelligence questionnaires, and assessments of coping capabilities; more refined approaches appear continually in the psychosocial literature. Furthermore, a great number of health examinations exist in the medical field, as do various approaches in the industrial engineering and management disciplines. Clearly, such measurements require that well-trained experts take the lead in designing, supervising the tests, and in evaluating the results; this is not for laypersons.

Computer adaptation syndrome

The introduction of computers into offices provides examples of stress felt by office personnel. Many experienced office workers were concerned about their ability to handle the requirements of using computers, especially if they feared that lack of performance might lead to job insecurity. Furthermore, the perception was that computers required a high level of technical knowledge; in addition, there were concerns about radiation health risks associated with the cathode ray tubes in the monitors. The computer-use stress problem° ceased to exist within a rather short time, for a number of reasons: computers became much easier to use, mostly because of smarter software and well-designed training and instruction to help people become familiar with computer use. As many of the older office workers became proficient, they found that computers would make certain tasks easier, others more interesting and challenging. Younger office personnel grew up with computerized electronic devices as everyday gadgets, so they had no qualms about using computers at work. Thus, the computer adaptation syndrome of the 1980s and 1990s completely vanished.

12.2 Mental workload

Blue- and white-collar work

Well into the middle of the 20th century, a common distinction existed: so-called blue-collar workers performed physical work and white-collar workers primarily used their brains. Today, that simplistic division no longer exists because work tasks and job requirements have changed, as did clothing habits. A good example of those changes is the increased use of electronic devices, including computers, for information retrieval and storage even in work categories that one would associate with physical efforts, such as automobile repair. Still, there are occupations that mostly rely on the use of mental capabilities; pilots, medical personnel, process control operators, teachers, and journalists come to mind. It is reasonable to assume that

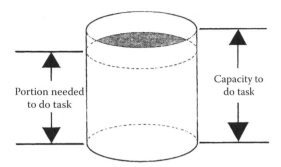

FIGURE 12.1 Resource model. (Adapted from Kroemer, K.H.E., Kroemer, H.B., and Kroemer-Elbert, K.E. (2003, amended reprint). *Ergonomics: How to Design for Ease and Efficiency* (2nd Edn). Upper Saddle River, NJ: Prentice-Hall/Pearson Education.)

such jobs pose special demands on mental capabilities; therefore, one should be able to measure both the mental workload and the related mental capabilities.

What is mental workload?

Surprisingly, a generally accepted definition of mental workload does not exist. In 2006, Megaw repeated the 1989 statement by Linton et al.: "The simple fact of the matter is that nobody seems to know what workload is. Numerous definitions have been proposed, and many of them seem complete and intuitively 'right'. Nevertheless, current definitions of workload all fail to stand the test of widespread acceptance or quantitative validation". Even ISO Standard 10075, which deals with mental workload, provides no definition.

Demand and resource

For practical purposes, we can rely on the concept of task demands and related operator capabilities already used in the discussion of stress. The assessment of workload, whether mental or physical, commonly relies on the resource construct: it assumes a given quantity of capability of which the job demands a certain percentage. This is shown in Figure 12.1. Accordingly, the workload is the portion of resource that is expended in performing the given task. If less is required than available, a reserve remains. If more is demanded than provided, an overload exists.

Overload versus underload

In the case of an overload, performance of the task is incomplete or even impossible and the operator is likely to suffer, physically or psychologically or both, from the overload stress. Measurement of the quality of performance provides clues for the quantity of overload. In an underload condition, the operator is capable of performing more, or better, according to the remaining reserve capability. Measurement of that residual capability provides an assessment of the actual workload. However, that simple model must be used with caution because there are often complex nonlinear relationships between performance and workload measures.

Task performance

Figure 12.2 shows, schematically, the relations between task demands, operator workload, and task performance. The task demands may be defined in terms of intensity, complexity, time, or other measures of difficulty. The operator workload is a function of the person's task-related capabilities that exist at the moment of demand. One result of these combined conditions is task performance. In an underload condition, satisfactory performance of the primary task does not require all of the operator's capabilities, and so an ability exists to perform a secondary task simultaneously. Measuring the performance on the secondary task is one way to assess the workload posed by the primary task; more about this below.

Of course, these simple concepts do not reflect the complexity of many real situations°. Operator workload is not only a direct result of task demands but also relies on how the operator perceives the task, which, in turn, depends on the operator's

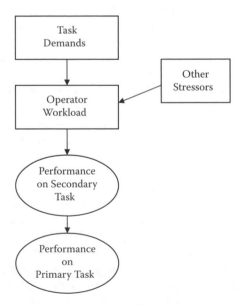

FIGURE 12.2 Workload and task performance

skill and experience. These circumstances may determine strategies that the operator adopts; this may in effect reduce the workload, particularly if efficient strategies are adopted. Furthermore, motivation, arousal, or fatigue can influence operator workload. While performing the task, operators monitor their own work and that feedback and judgment may make them perceive the task load level differently, after which they may alter their work strategies: all of this can have motivational consequences. Performance outcomes can modify the task themselves and, consequently, the resulting workload. A typical example is recognizing an outcome fault that calls for a diagnostic strategy, which often brings with it an increase in subsequent task demands.

12.3 Physical workload

If mental work uses brain power, then physical work employs muscle power. Heavy physical work regularly brings with it high energy consumption and severe demands on heart and lungs. Energy consumption and cardiac effort often set limits to the performance of hard work, therefore, these two body functions can serve as indicators of the severity of a physical task; see Chapter 10.

Mechanization has reduced the demands for exhausting work in many occupations; nevertheless, heavy work, which may overload the operator, still is common in mining, building, agriculture, forestry, and transport, including, for example, baggage handling by airline personnel. It remains a major ergonomics issue in developing countries.

12.4 Underload and overload

Too little, too much

Most people feel stressed because of an overload at work, which requires too much from them in terms of task intensity, complexity, diversity, time pressure, and such. However, some jobs demand too little effort, which leaves persons idle, bored, and their capabilities underused. This experience has led to the notion than one can describe the relations between task demands and the experienced workload as a function that looks like the letter U: in its center part, a balance exists between demand and load that is just right. However, either requiring too high a workload, or not demanding enough, leads to worker

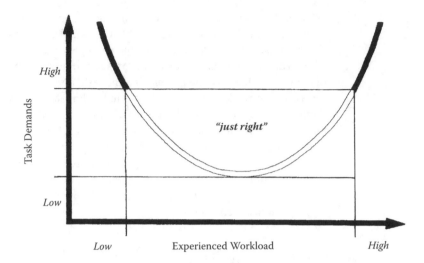

FIGURE 12.3 Assumed U-function between task demands and experienced workload

overload or underload, respectively. Task performance suffers under both overload and underload conditions. Figure 12.3 shows this postulated U-function. Of course, what is fitting depends on the situation and the individual operator.

Monotony and boredom

Monotony is the property of an environment that does not change over time, or where changes occur in repetitive and predictable fashion over which the operator has no or little control. A varied environment usually provokes interest and even excitement; in contrast, monotonous conditions produce boredom. *Boredom* at work is a state of human emotion produced by a dull, uninteresting job environment. A bored person often complains of feeling tired or fatigued, regularly accompanied by yawning, which one would expect from a person worn out by heavy physical work.

12.5 Psychophysical assessments of workloads

The understanding and definition of workload depends on two factors: one is the intensity of the task, the other the related capability of the person to perform the job. That relationship between demand and ability is further modified by the person's perceived stress and willingness to perform.

There are some simple cases where one is able to measure both the load and the operator's resulting physical strain:

Table 12.1 Typical methods of workload assessments

Methods	Techniques, examples
Archives and records	Description, statistics, trends
Indirect observations	Interview, discussion, survey
Direct observations	Chart, checklist, critical incidents
Expert analyses	Walk-through, judgment, scoring
Task analysis	Flow/time chart, time and motion study
Simulation	Mathematical/physical mock-up
Modeling	Task network, control processing
Performance measures	Work rate, output in amount and quality
Biomechanical measures	Posture, force/moment/work analysis
Physiological measures	Oxygen consumption, heart rate
Psychological measures	Perceived exertion, attitude, rating
Psychophysical measures	Subjective assessments—see below

examples are a baggage handler loading an airplane cargo hold, or a mason carrying bricks up a ladder at a building site. Here, an investigator may determine, in physics terms, what the job requires, and in physiological terms, how this loads the person's metabolism and circulation, such as discussed in Chapter 10.

Complex jobs, complex assessments

Such simple cases with only a few and easily assessable components are rare. Usually, the task demands are complex and environmental conditions often contribute intricately to the demands. Furthermore, it is often impossible to measure the responses of a person's body and mind to the demands by using just a few established physiological techniques, which assess only one of several strains on the body (via oxygen uptake, for example) and whose apparatus often hinders work performance. The literature reflects a long-ranging discussion° about the usefulness of so-called objective (analytical, quantitative) measures, compared to assessments that rely on subjective (qualitative, empirical) judgments, made by either an observer or the worker. Table 12.1 lists methods useful for workload assessments.

Listen to the worker

The last entry in Table 12.1 mentions psychophysical measures. These usually combine elements of biomechanical, physiological, and psychological assessments in the form of subjective judgments, made either by an observer or by the operator. The underlying premise is that an experienced person can perceive and integrate all existing loadings and then summarize the overall effects in the form of a judgment. That concept appears plausible. The reliability of judgments should increase when they are formulated within a standardized and controlled framework, for example, in preselected words or in numbers that reflect rankings (less than, more than) or mathematical relations, for example, double or triple as much as a reference.

Borg scales

Borg's well-known psychometric scales° incorporate such organized judgments. Table 12.2 contains the RPE and the CR 10 scales. The steps are "verbally anchored" and arranged to follow mathematical formulas. The ratings of perceived exertion (RPE) scale ratings have equidistant steps, which are numbered so that, multiplied by ten, they correspond to heart rates measured at these work levels. The CR 10 Scale has ratings in steps set to constant ratios.

Just as any other measurements, these Borg scales need rigidly controlled procedures to yield credible results. For example, his CR 10 Scale is put in front of the person who will rate the perceived intensity of work, with instructions such as,

> You do not have to specify your feelings, but do select the number that most correctly reflects your perception of the job demands. If you feel no loading, you should answer zero, nothing at all. If you start to feel something just noticeable, your answer is 0.5, extremely weak. If you have an extremely strong impression, your answer will be 10. So, the more you feel, the higher the number you are choosing. Keep in mind that there are no wrong numbers; be honest, do not overestimate or underestimate your ratings.

Borg claims that his scales have high reliability (meaning that a repeated test yields the same results as the initial test) and high validity because the test results correlate well with measures of heart rate.

Such psychophysical assessments (and there are many more, including some complicated ones°) have several practical advantages: they are relatively easy and inexpensive to carry out, do not usually interfere with work performance, and provide a comprehensive summary of the effects of all stressors at work. However, they need to be administered carefully, or their results may be very misleading.

Table 12.2 Borg scales

Number	Wording
Borg RPE scale	
6	No exertion at all
7	Extremely light
8	
9	Very light
10	
11	Light
12	
13	Somewhat hard
14	
15	Hard, heavy
16	
17	Very hard
18	
19	Extremely hard
20	Maximal exertion
Borg CR 10 scale	
0	Nothing at all
0.5	Extremely weak, just noticeable
1	Very weak
2	Weak, light
3	Moderate
4	Somewhat strong
5	Strong, heavy
6	
7	Very strong
8	
9	
10	Extremely strong, almost maximal
11 or higher	The individual's absolute maximum, highest possible

Source: Adapted from Borg, C. (2001) in Karwowski, W. (Ed) *International Encyclopedia of Ergonomics and Human Factors,* London: Taylor & Francis, pp. 538–541, and from Borg, G. (2005). Scaling experiences during work: Perceived exertion and difficulty in Stanton, N., Hedge, A., Brookhuis, K., Salas, E., and Hendrick, H. (Eds.) *Handbook of Human Factors and Ergonomics Methods* (pp. 11-1 to 11-7) Boca Raton, FL: CRC Press.

Summary

Work-related stress is an individual's emotional reaction to aspects of work demands, work environments, and work organization that the person feels are adverse and noxious. Although people react differently to job conditions, in general, stress results from a mismatch between demands and a person's ability to meet them. Instead of trying to adapt the person, it often makes more sense to change the demands.

Fitting steps

Step 1: When setting up a new task, carefully design out excessive task demands and adverse conditions.

Step 2: When workers complain of overloading, or of a boring task, rearrange to their satisfaction.

Step 3: Listen to the working people. They can tell you what's good and what needs improvement.

Further reading

STANTON, N., HEDGE, A., BROOKHUIS, K., SALAS, E., and HENDRICK, H. (Eds.) (2005). *Handbook of Human Factors and Ergonomics Methods.* Boca Raton, FL: CRC Press.

WILSON, J.R. and CORLETT, N. (EDS.) (2005). *Evaluation of Human Work* (3rd Edn). London: Taylor & Francis.

Notes

The text contains markers, °, to indicate specific references and comments, which follow.

12.1 Stress at work:

Hans Selye wrote *Stress without Distress*, 1974, and *The Stress of Life*, 1956.

Stress is an emotion: Carayon et al. 2006, Cox et al. 2005.

Stress is costly to the organization: Cox et al. 2005.

Measurement of stress: Megaw 2005.

Computer adaptation syndrome: Grandjean 1987, 1988.

12.2 Mental workload:

Real work situations: Megaw 2005.

12.5 Psychophysical assessments:

Discussion of measures: See *J. Ergonomics*, 45: 14, 2002; Wilson et al. 2005.

Borg Scales: Borg 2001, 2005.

Complicated scales: Wilson et al. 2005.

Organizing and managing work

This fourth section of *Fitting the Human* first addresses the issues that primarily determine how people get along with each other at work and, as one of the consequences, how much effort they put into their work.

How individuals feel about their company, and hence about their work, depends to a large extent on the organizational setup. The design of the work organization and the manner in which the organization is run are so-called macroergonomic issues, in contrast to the concerns about human factors details, for example, the workspace of the hands.

Finally, Section IV deals with another set of managerial responsibilities that concerns the organization of the work itself: how many hours of work per day, what is the best time for work during the day, and how to organize shift work suitably.

- Chapter 13 Working with others
- Chapter 14 The organization and you
- Chapter 15 Working hours
- Chapter 16 Night and shift work

CHAPTER THIRTEEN

Working with others

A large Chicago-based professional services company moved from an older office tower downtown into a newly built-out building. The company, a venerable institution with a rich sixty-year history, employed thousands of people. Employees often described themselves as part of a family, with extensive interaction among all levels of the corporate hierarchy and a strong prevailing team spirit.

The old offices had sentimental value for many employees; this is where many had begun their careers. However, the new office building was easier to access via public transportation, allowed for increased company growth, and was more reflective of the company's success, with luxurious interior appointments.

Once the relocation was complete, many company members soon regretted it. The move carried with it some unintended and unfortunate consequences. The configuration of the new space was different: in the former building, executives' offices and cubicles were interspersed among staff cubicles, whereas in the new space all executives' offices were located on two separate floors. Initially, the concept behind establishing executive floors centered on easier communications among the directors; additionally, the company wanted to reward executives' performance by providing them with especially posh suites. However, some of the effects of the new configuration on the corporate culture were utterly unforeseen and, in the long run, damaging. The new configuration sharply reduced the casual interaction between executives and employees that had existed before, when many of the managers followed the strategy descriptively known as "management by walking around." Employees and executives interfaced daily at the old building, meeting routinely at the coffee machines, in the hallways, and in the cubicles and offices; now, interaction was reduced to business-only discussions at formal meetings. Where before employees welcomed

executives' casual "visits" and drop-ins in their cubicles, invit-
ing the friendly and open banter, they now felt anxious and
vaguely frightened during formal business meetings.

After some time in the new space, employees felt disconnected from company leadership, even disenfranchised
from the company as a whole. There was now far less of a
team atmosphere and more of a "them versus us" philosophy. Interestingly, executives too felt out of the loop with the
employees, sensing a formidable new barrier between themselves and the staff. The team spirit that had once prevailed
was sharply and unexpectedly curtailed by the office move; an
important part of the company culture had been inadvertently
and irrevocably destroyed.

This example° shows how the layout of the workspace for individuals and work teams and the set-up of the whole organization can strongly affect people's work attitudes and output. The
topic of organizational design is treated in some detail in the
next chapter, 14.

13.1 Getting along with others

The example refers to many kinds of personal interactions:
"Employees and executives interfaced daily at the old building,
meeting routinely at the coffee machines, in the hallways, and
in the cubicles and offices [for] ... friendly and open banter ... ."
Indeed, the most basic and perhaps most important everyday
interactions take place between individuals, often informally.

Distances between persons

In different regions of the earth, in different populations with
differing traditions the expectations for personal interactions
do differ; yet, in the "old world" of Europe and North America,
social relationships between individuals generally take place at
certain distances, which the persons preserve between them.
How people maintain and use the personal space around them,
especially the distances they keep from other people, conveys
psychological and social messages.

Personal space

Most of us feel that there is volume around us that, when entered
by another person, can give rise to strong emotions. Selection
of the location within the personal space° where people deal
with each other reflects the nature of their relationship. Most
personal interactions take place within about an extended arm's
length whereas the whole so-called social distance extends out

to 3 or 4 m. The narrow "intimate distance" often involves body contact. Close friends and good acquaintances usually stay within half a meter and a meter and a half. Within a distance of about two meters, interactions between unacquainted persons and business transactions may occur. Beyond about two meters, there is usually no sense of friendship, and the interactions are formal. The "public distance" where voices may have to be raised for communication, is beyond about 4 m— yet, all these metered distances do vary with local customs.

Distances in personal relationships

When somebody who, for personal or social reasons, is not supposed to be close but invades a person's personal space nevertheless, the owner of that space usually experiences arousal and discomfort. However, if coworkers, good acquaintances, and friends stay too far away, disappointment and alienation are likely and formerly close relationships may deteriorate. At work, when group members are expected to cooperate as teams, performance should be best if they are in close proximity. Competition, however, is fiercer when persons remain outside the distances of personal relationships.

Owning an area

Some persons a display behavior pattern called territoriality, which is the inclination to occupy and control a defined physical space, such as property, a workplace, or just a section of a workbench. Preventive or reactive defenses may keep intruders at bay; an invader may be met with offensive behavior or the space may be abandoned. Often, the "owner" of an area personalizes it and demarcates it in some ways, for example, with flowers or photographs; these measures convey the feeling of control and security.

Crowding is the negative experience of too many people within a given space. As do the other feelings associated with personal space, the perception of crowding depends on individual characteristics, on the assessment of one's coping capabilities, and, of course, on the physical and social settings. Individual reactions to crowding may be aggression or withdrawal.

Teamwork

Working in groups is a common requirement in offices, in production and assembly, and in maintenance and repair tasks. It is a necessity in such threatening situations as rescue, firefighting, and military operations. Teams are two or more individuals who pursue a common goal by performing their tasks in interdependent, complementary, or distinct manners°. Team members must communicate with each other, and adapt and adjust their work to perform their tasks in a timely and integrated fashion. At the core of teamwork is a set of interrelated kinds of knowledge, skills, and attitudes. Behavioral and organizational aspects of teamwork have been mostly researched. In general, team members must be willing to communicate and to be flexible in adapting to the changing allocations of functions within the team. Not all individuals are able and eager to subordinate their individual work habits and specific skills to the rules of teamwork; furthermore, certain regional cultures emphasize working together, whereas other social environments further individual work.

13.2 Motivation and behavior

Happy employees

The field of study that helps us understand and deal with the interpersonal and managerial challenges in the workplace is called *organizational behavior*°. The "lean and mean" years of the last century are behind us; they made many employees in North America fear for their jobs and uncertain about their professional futures. Instead, many companies are trying again to keep their employees happy, satisfied and motivated. The reason is not sheer altruism but the recognition that improving employee satisfaction will improve profit.

Happy employees are productive; they treat customers better, work harder, and even take fewer sick days. Moreover, they tend to stay with the organization, which reduces one of the most significant costs: employee turnover related to dissatisfied employees. Leaving employees often generate huge expenses for recruiting and training new employees, and a new hire accomplishes substantially less initially than an experienced worker. The outright expense of replacing a valuable employee can range from half to several times a year's pay. Finally, there is the less tangible cost of unhappy employees who stay on the job: they are likely to show low productivity and poor customer service, but they also suffer personally from stress (see Chapter 12) and, potentially, from serious stress-related illnesses as they trudge through a dreary work routine.

Psychosocial work factors

How we feel about ourselves as members of a work organization depends on a variety of psychosocial work factors°, listed in Table 13.1. These relate to the demands of the job, what we have to do and how we do it, how we feel about our co-operation with peers and superiors in the organization, and what we expect in terms of our professional and employment future.

Table 13.1   Psychosocial job factors

Job demands	Physical workload
	Variations in workload
	Work pressure
	Cognitive demands
Job content	Challenges, interest
	Repetitiveness, monotony
	Development and use of personal skills
Job control	Control over work pace
	Control over physical environment
	Task/instrumental control
	Organizational control
Social interaction	Interactions with colleagues
	Interactions with supervisors
	Dealing with clients and customers
Job future and career issues	Ambiguity about job future
	Fear of losing job
Organizational and management issues	Employee participation
	Management style

Source: Adapted from Carayon, P. and Limm, S.Y. (2006). Psychosocial work factors. In Marras, W.S. and Karwowski, K. (Eds.) *The Occupational Ergonomics Handbook* (2nd Edn) *Interventions, Controls, and Applications in Occupational Ergonomics*, Chapter 5. Boca Raton, FL: CRC Press.

Division of labor

Even Taylor's[o] frequently (and often appropriately) maligned concept of division of labor, which from the early 1900s on led to work mechanization and specialization, promoted motivating the individual and sharing responsibilities between management and labor. (It also generated efficient work systems by careful design of work, provision of the right tools, and selection of able persons. In this course, a worker's task was often reduced into small components, which reduced skill requirements and facilitated performance evaluations. Thus, many jobs became simplified and standardized, repetitive, and monotonous.)

Quality of work life

Half a century later, with a more educated and liberated workforce, employees became more aware of their working conditions and environment and sought to improve their quality of work life. This raised the issues of recognition of personal value and improvement of working conditions and, consequently,

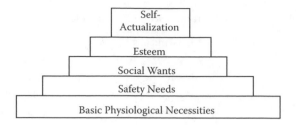

FIGURE 13.1   Maslow's needs hierarchy

theories were developed about the individual worker's motivation, productivity, and well-being. Individual needs and wants were emphasized and workers' behavior and their attitudes toward work considered important.

Motivation and performance

Motivation° incites, directs, and maintains behavior toward goals. Motivation and job performance are, of course, related; a motivated person who desires to do well at work is willing to expend effort to do so. Performance is the product of motivation and ability. Performance is moderated by situational constraints at work, which are factors that can stymie or enhance performance; examples are the climate (both psychological and physical) and up-to-date equipment.

Maslow's needs hierarchy

Maslow (1954, 1970) and his disciples developed the so-called Needs Hierarchy°, sketched in Figure 13.1. According to this model, motivation is a function of meeting personal needs, which step up from basic physiological necessities such as food and shelter to higher-order wants. Apparently, the basic needs must be fulfilled first. This achieved, we strive to meet our safety needs, which focus on economic and physical security. The third step up brings us to our social needs, above which are our wants for esteem, where we gain self-confidence through recognition, appreciation, and respect. The highest-order needs are of those for self-actualization, where we achieve our full potential and esteem.

Maslow says that behavior is motivated by the urge to satisfy needs of increasingly higher orders. This tenet has been modified in various ways to explain such observations as people moving forth and back among needs: what are steps according to Maslow may in fact be a continuum where needs and desires overlap and merge. This concept explains shifts among our concerns that relate to

> *Existence:* Food and essential supplies, compensation, working conditions

> *Relatedness:* Relationships and interactions with family, friends, colleagues
>
> *Personal Growth:* The desire for personal development, advancement, recognition

Job satisfaction Satisfaction with one's job, or dissatisfaction, can greatly influence motivation and task performance. As usual, there are large differences in individual perceptions of what makes a job satisfying or not. For example, somebody might value personal growth and appreciation of work over monetary rewards; this person would be satisfied with a rather low-paying job as long as there was an appreciation of performance and success in professional challenges. Another person, however, might be mostly interested in job security and good pay and be satisfied even if the position is seen as being of low rank.

Herzberg's two-factors theory In 1966, Herzberg explained his concept of job satisfaction with his so-called Two-Factors° theory. It assumes that positive *content* factors of a job mostly explain satisfaction whereas negative *context* factors lead to dissatisfaction. As Figure 13.2 illustrates, Herzberg and his disciples isolated five *content* factors that act as satisfiers or motivators:

1. Achievement
2. Recognition
3. The work itself
4. Responsibility
5. Advancement

If these factors are present, they can create job satisfaction and motivation to strive at work; their absence would make an

Content Factors: When positive, they promote satisfaction and motivation	Promotion chances
	Personal growth chances
	Recognition
	Responsibility
	Achievement

Context Factors: When positive, they prevent dissatisfaction	Supervision quality
	Pay
	Company policies
	Relations with others
	Working conditions
	Job security

FIGURE 13.2 Herzberg's Two-Factors theory: content satisfies and context dissatisfies

employee feel neutral or indifferent. On the opposite side are *context* factors (also called *hygiene* factors):

6. Company policies and administration

7. Compensation

8. Supervision and management

9. Interpersonal relationships

10. Physical conditions at work

11. Job security

If these context factors are negative, an employee becomes dissatisfied and lacks motivation. (The following Chapter 14 addresses some of these issues.)

In summary, Herzberg's theory suggests that a job should have positive context factors to prevent dissatisfaction and include plenty of positive content factors, which produce satisfaction and motivation.

13.3 Task demands, job rewards

Work conditions that motivate

Theorists have thought about many work conditions that motivate people to perform their tasks efficiently. The past has shown that, under austere conditions and for short periods of time, even such negative factors as fear and greed can drive people to outstanding task performance. However, as a society of valuable humans, we strive for a high quality of work life; everyone has a right to a fulfilling, rewarding, safe, and secure job. This goal matches well with the business intent to prosper because job satisfaction is associated with critical revenue-affecting variables such as turnover, absenteeism, and job performance.

Job enlargement and enrichment

In contrast to earlier work specialization that evolved from Tayloristic principles, the ideas of job enlargement and job enrichment form the basis for many current job design theories. Job enlargement means a larger variety of tasks and activities, usually at about the same professional level; job enrichment means expanding work skills and increasing work responsibility, often taking in tasks at higher professional levels. Job design, in general and in details, can strongly affect our attitudes and hence our performance.

The Hawthorne effect

Sociologists and industrial psychologists have performed innumerable experiments in which they changed working conditions and observed their effects, if any, on work attitude and output. Among the best-known experiments are those done in the 1920s in the Hawthorne° Works near Chicago. There, the intervention consisted of improving the lighting conditions in a manufacturing/assembly task. Each rise in the lighting level brought about improved output; but when the illumination level was finally lowered, performance still improved. In somewhat simplified terms, one may conclude that paying attention to the workers, taking their comments and activities seriously into account, listening to them—in short, treating them as important—led to improved output regardless of the magnitude, even the direction, of the overt work intervention taken.

The term "Hawthorne effect" has become proverbial: it describes a work situation where an introduced change triggers an increase in productivity but not because of the change itself but because the participating workers find themselves in the spotlight of attention, which encourages strong motivation and extra effort. These effects wear off and are no longer present later on. Even though short-lived, they point to the effectiveness of positive engineering and managerial measures.

Goal setting and rewards
Obviously, the rewards inherent in a job must be meaningful and valuable to the employee, and with individuals being so different, no universal reward system exists. Goal-setting theories posit that people set targets and then purposefully pursue them; their premise is that conscious ideas underlie our actions, and the goals that we have set motivate us, direct our behavior, and help us decide how much effort to put into our work. For some persons, the more difficult goals foster higher levels of commitment; and the more specific the goal, the more focused the efforts of the individual to attain it.

Motivation and work behavior
In spite of how compelling these and other existing theories are, there is no clear-cut correct answer to the question of what motivates people. Instead, bits of theories apply to different people, under different circumstances. Motivation is both intrinsic and extrinsic: factors within us and external to us drive our behavior. What most likely occurs is that we consider our own needs and wants and either consciously or subconsciously determine our goals; then we act to increase the chances of obtaining what we want. We do know from Maslow's, Herzberg's, and others' theories that we will strive first to fulfill basic survival needs such as securing food and shelter; beyond this, needs are still real but vary widely among individuals in terms of priority and strength. Other people influence us and shape our motivation

and behavior because we are all, to some degree, social creatures. How hard we work depends on what our work will bring us: if we feel rewarded in ways that are meaningful and valuable to us, and if we perceive a definite link between the strength of our efforts and performance, we will work strenuously to perform well and gain those desired rewards.

What motivates people can shift. In Europe and North America, for example, pay and stability used to be among the most important motivators in a job; now, with an increasingly diverse, lifestyle-conscious workforce, there are many other nonfinancial ways to reward employees and keep them motivated and happy. On-site child care, flexible work hours, and vacation time are a few examples of the rewards that employees might seek; furthermore, they probably put high value on having friendly coworkers and a nurturing corporate atmosphere: my company is "a good place to work" and "my buddies on the job are like family".

Summary

Today's educated and liberated persons seek pleasant conditions at work in which they can use and expand their skills; they consider work as an important part of their total quality of life. Careful design of work, provision of the right tools, and a suitable work environment are basic ergonomic measures, which must be augmented by an organizational "culture" (see Chapter 4.2) that encourages personal development.

An individual's motivation, productivity, and well-being are linked: performance is the product of motivation and ability. How hard a person works depends on what that effort yields. If there is a definite link between the strength of exertion and performance, and if one feels rewarded in ways that are meaningful and valuable, individuals will work strenuously to perform well and gain those desired rewards.

Fitting steps

Step 1: Carefully design work, provide the right tools and a suitable work environment.

Step 2: Provide opportunities for the individual to develop skills and fulfill aspirations.

Step 3: Be innovative, flexible, and adaptive.

Further reading

LANDY, F.J. and CONTE, J.M. (2006). *Work in the 21st Century: An Introduction to Industrial and Organizational Psychology* (2nd Edn). Malden, UK: Blackwell.

MUCHINSKY, P.M. (2007) *Psychology Applied to Work* (8th Edn). Belmont, CA: Wadsworth.

Notes

The text contains markers, °, to indicate specific references and comments, which follow.

13.1 Working with others:

Example: Adapted from Kroemer et al. 2001.

Personal space: Proctor et al. 1994.

Teamwork: Muchinski 2007, Stanton et al. 2005, Wilson et al. 2005.

13.2 Motivation and behavior:

Organizational behavior: This section adapted from Chapter 2 in Kroemer et al. 2001, 2005.

Psychosocial work factors: Smith et al. 1989.

Taylor's Two-Factors Concept: Taylor 1911.

Motivation: See Muchinsky 2007 and many other related books.

Maslow's Needs Hierarchy: Maslow 1954, 1970.

Herzberg's Two-Factors Theory: Herzberg 1966.

13.3 Task demands, job rewards:

Hawthorne Effect: Roethlisberger et al. 1939, Parsons 1974.

CHAPTER FOURTEEN

The organization and you

For smooth and successful operation of any business, good fit is necessary not only between work and worker, but between workers and their employer's organization as well. Organizations are networks of related parts, and all the elements must work together to enable the organization to function as a whole. This chapter treats various kinds of organizations. Of course, to run a small farm by family requires much less of an organization than setting up a manufacturing business, or managing a fire-fighting or military team.

14.1 Structure, policies, procedures, culture

Elements of an organization

There are many ways to chart organizations. The model° shown in Figure 14.1 depicts basic components of the organization that determine the roles of its human players. In this case, five main elements interact to define a company: its *strategy* (the company's plan for success), its *structure* (corporate hierarchies), its *conduits* (channels for allocating, controlling, and tracking corporate resources), its *rules and regulations, policies, and procedures*, and its *social climate* (employees' feelings about the company) and *culture* (behaviors and feelings within the company). All affect one another, and as one changes, the others too must adapt.

Strategy

Figure 14.1 shows "strategy" as one of the main components of an organization. Strategy is the plan—stated or implicit—that the company has to succeed in achieving its goals. It guides the company's operations and determines specific tactics that the company will use. Examples include the repair service that

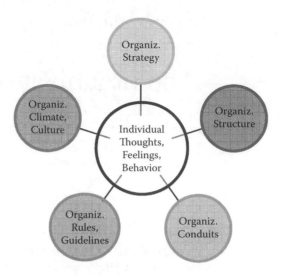

FIGURE 14.1 A basic human-centered organizational model (Adapted from Kroemer, K.H.E. and Kroemer, A.D. (2005). *Office Ergonomics*, authorized translation into Korean. Seoul: Kukje)

pledges to dispatch a plumber within two hours of a request, the grocery store that wants to offer a wider selection of fresh produce and gourmet items than competitors, and the manufacturer intending to produce the best airplanes.

Structure

Structures outline the hierarchies within an organization. In larger companies, they are usually depicted as detailed diagrams linking the elements in the organizational chart. An organization's structure determines accountability and authority within its ranks; essentially, it defines the official relationships that exist between employees. Each level in a structured organization has its own degree of authority and responsibility; moving up in the hierarchy increases one's authority and responsibility.

Obviously, the structure of an organization determines each employee's position, the ways in which instructions come from the top down, reports travel upward, and how persons at the same level communicate with each other. In a "Mom and Pop shop", formal organizational structures are usually unnecessary; yet, some division of labor and responsibilities develops naturally according to personal preferences and skills. Those down/up/sideways relations are of utmost importance in any more complex organizations, such as manufacturing plants, airlines, big-city firefighter companies, or a military department.

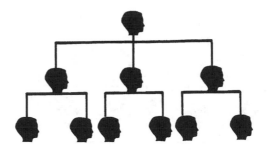

FIGURE 14.2   A basic layered organizational chart

Fixed structure

Volumes have been written about organizational designs°. Under normal circumstances, each employee should only be accountable to one boss; this is known as the "unity of command" principle. Figure 14.2 shows a basic example of a conventional layered organizational chart. Among its chief advantages is the clarity of communication down and up.

Flexible organization

However, employees and work projects may not succeed in this simple hierarchical structure, but do better in a team network. Figure 14.2 depicts a conventional organizational structure with several departments serving under one boss. Each department has several employees at the same level who perform specific tasks. However, job enlargement and enrichment (see Chapter 13), often associated with teamwork, generally do not flourish under this conventional setup. For example, originally an electrician was responsible exclusively for the maintenance of all electrical equipment; but in a new and flexible, team-oriented organization, this person now has multiple roles, which also include testing and designing new equipment, developing software, task planning, and even some financial accounting. Similarly, other employees also have multiple roles that overlap the formerly separate areas of production, equipment, maintenance, delivery, and funds allocation. This can lead to a complicated amoeba-type organizational structure, sketched in Figure 14.3, which is suitable for autonomous, problem-solving teams (see Chapter 13) able to self-organize for providing, collectively, the abilities to solve varying problems and meet changing tasks.

Conduits

Conduits are the supply systems by which information travels within the organization; in theory, these pipelines (occasionally "grapevines") follow the official structures that provide the channels by which people (human resources, employee

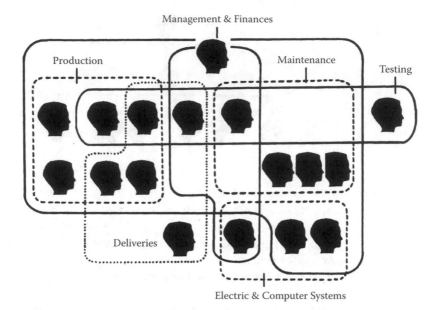

Management & Finances

Production

Maintenance

Testing

Deliveries

Electric & Computer Systems

FIGURE 14.3 A flexible work organization with overlapping worker teams. (Adapted from Nagamachi, M. (2001). Relationships among job design, macroergonomics, and productivity. In Hendrick, H.W. and Kleiner, B.M. (Eds.) *Macroergonomics. Theory, Methods, and Applications,* Chapter 6. Mahwah, NJ: Erlbaum.)

appraisals), money, and things (machines, equipment, supplies) are properly allocated, controlled, and tracked.

Rules and guidelines

Policies and procedures govern an organization's conduct. In larger companies, they appear written up in detail in an employee handbook. Examples of policies include the amount of medical benefits a company provides, retirement plans available, or the number of paid vacation days an employee receives. Procedures and rules prescribe the ways by which employees and supervisors interact. Guidelines are often not documented, but are nevertheless widely understood, for example, how often e-mails are checked or whether coffee is provided free of charge to the employees.

Company climate and culture

Organizational *climate and culture* appear in the norms, beliefs, unwritten rules, rites and rituals, behavior, and practices among the members of an organization. "Culture" is created by underlying values that are unseen whereas, on the surface, the behaviors of the system members, its actors as heroes and villains, are visible. Social climate and culture have become very important, especially in bureaucracies, as major determinants of employees' overall happiness within an organization. (That was well known, but then apparently forgotten in the "lean and mean management" decade of the late 20th century.) At this time of casual work environments, telecommuting, job-sharing, and emphases both on individuality and participatory work, companies' cultures vary widely—and apparently are more important to people than ever before.

Human–system integration

Macroergonomics° is a sociotechnical systems approach based on the recognition of a fundamental shift in the value systems of workforces. In North America and Western Europe, for example, people now expect to have control over the planning and pacing of their work, to exercise decision-making responsibilities, and to enjoy broadly defined jobs that provide the feelings of responsibility and accomplishment. This approach is often called human–system integration, HSI.

Macroergonomics encompasses a theory of new work organizations that are (a) less formalized, that is, less controlled by standardized procedures, rules, and detailed job descriptions; and (b) more decentralized so that tactical decisions are delegated to the lower-level supervisors and workers. These are profound changes from traditional bureaucratic work systems.

A central theme of macroergonomics is that the work systems should be participatory in nature. For this, people must have sufficient knowledge and power to influence both processes and outcomes in order to achieve desirable goals. Such participatory structure can take a variety of forms; in fact, different social cultures are likely to generate different forms of macroergonomic organizations.

The human is in the center

Good fit among workers, work, and organization make for smooth and efficient performance. The human must be at the center of the organization's operations, because ultimately, companies cannot exist by their strategy, structure, and machinery, but rather rely on living, working, and interacting people. Without humans, the organization is dead. People affect the organization and the way it functions and, in turn, the organization affects these individuals. Figure 14.1 shows these relations schematically. Why people act the way they do—and what we can do to keep them satisfied and successfully interacting with each other at work—is a fascinating puzzle that has kept psychologists and behaviorists° busy for many decades. Some explanations came into view in the foregoing Chapter 13; others follow here in this chapter.

Social contracts   Throughout history, the relations between individual workers and the organizations employing them have vacillated between extremes: from "I'll work hard for them until I die", to "Good enough for government work", on one side, and from heartless exploitation to near-parental care on the other side. These social contracts, usually unwritten but well understood nevertheless, included appropriate pay, a secure workplace as well as health and retirement care in exchange for reliable, thorough, and hard work, once proverbial among employees as in "A Krupp man forever" or, "Once IBM, always IBM". To the employees, these conditions provided the feel of security and belonging, of recognition and appreciation, which, in turn, generated interest in doing their work well, toiling carefully and thoughtfully in the interest of The Company. On the other hand, care for their workers, especially in the case of illness and after their working years, was costly to the organization but rewarded loyalty: in some cases, employees took advantage of the goodwill.

14.2   Design for motivation and performance

A company's social climate and culture strongly depends on its structural design. Both affect the emotional state of the people in an organization, how they individually feel about the company, their coworkers, and their jobs. (Chapter 8 in this book treats the climate in its physical sense, such as temperature and humidity. Of course, use of the term culture here has little to do with its historical understanding.) An organization's social climate and culture is a group phenomenon; it relates to the behaviors, beliefs, values, customs, and ideas that the employees as a group hold.

> *Consider two advertising agencies°; one a large and highly traditional company headquartered in a metropolis, the other a small "boutique" agency in a smaller town. Here are some of the characteristics of each: the first agency is 60 years old, has many long-standing blue-chip clients, and is highly structured, with formal dress codes, office procedure manuals, and written policies for all conceivable situations. Internal memos follow a set format and are heavily scrutinized by several administrative levels before they are issued; annual retreats and parties are scheduled and planned many months in advance. The second agency is 10 years old, with only a handful of young partners who employ highly creative individuals in an entirely*

unstructured office. Shorts and baseball caps are standard attire, no handbooks exist, and decisions are made quickly and with virtually no bureaucracy. Clearly, each of these companies—although both operate in the same industry—have very different cultures, with widely diverse values and behaviors and hence with different climates. Employees in a company acquire its culture when they first begin working there; they become socialized into the broader group by interacting with the different members.

One might be inclined to believe that under harsh conditions such as working in a foundry or when fighting a fire, social climate and culture would be of no importance; however, although the term seems improper, there are still social rules that govern how to behave, how to work with the others, and how to perform individual and team tasks.

The individual in the organization

The individual is the innermost and most important asset of an organization. When a company's stock market value is determined, hard assets such as property, plant, and equipment generally make up one-third to one-half of its value. The remainder consists of soft human attributes such as customer base and employee involvement. Human resources are a company's most valuable assets. Therefore, it is essential to consider the inner circle of the diagram in Figure 14.1 to find out what influences the employees' state of mind.

Individuals are unique. Their societal upbringing, social environment, life experiences, and personalities all make everyone special. It is a complex proposition to assemble individuals into groups in a work environment and to expect that all work together, according to company directives, given the tensions inherent in a job: hard or monotonous duties, long hours, and imposed relationships with other people. The organization has limited possibilities to select persons and to train them so that, as employees, they will fit in. However, instead of relying solely on those approaches (common tasks of industrial psychology), greater success should result from laying out the organization, and designing work tasks and conditions therein, in such ways that they accommodate and utilize individual capabilities and aspirations.

APCFB model

Several well-accepted concepts of behavior, motivation, and job satisfaction provide insights into why people act the way they do on the job. One may try to explain° personal behavior by considering the individual's assumptions, perceptions, conclusions, feelings, and behaviors. The APCFB model posits that a person's closely held assumptions color the perceptions of a given event, leading to highly individual conclusions; these in turn cause feelings that result in behaviors. Inasmuch as everyone has a different set of assumptions, individual behaviors or reactions even to one common event can be widely divergent.

Motivation

The attempt to understand the motivation of employees underlies behavior theories: motivation is the attitude toward attaining a goal. The most widely cited original theories to explain behavior and motivation include Maslow's Hierarchy of Needs, already mentioned in Chapter 13, and Alderfer's ERG theory. The basic premise of both theories is that individuals continuously strive to satisfy certain needs, and that this quest in turn drives behavior. Expectancy theory is a conglomerate of several researchers' ideas about motivation; it assumes that an individual's motivation (and resulting satisfaction) depends on the difference between what the work environment offers versus what the worker expects.

Motivation incites, directs, and maintains behavior toward goals. Motivation and job performance are, of course, related; a motivated person who desires to do well at work is willing to expend effort to do so. Performance is the product of motivation and ability. Performance is moderated by situational constraints at work, which are factors that can stymie or enhance performance; examples are an organization's climate (both psychological and physical) and up-to-date equipment.

Satisfaction

Job satisfaction closely correlates with motivation and, naturally, several theories exist to explain the degree of pleasure an employee derives from his or her job. Some approaches postulate that job satisfaction is determined within the individual, achieved when physical and physiological needs are met; other concepts center on the external factors of social comparison. Work conditions are likely to influence job satisfaction strongly; Herzberg's Two-Factor theory (see Chapter 13) is well known. He postulated that certain job conditions (he called them "hygiene" factors) such as pay and the work environment could generate dissatisfaction if perceived as negative. However, in Hertzberg's opinion, simply making these factors positive, (good pay, for example) would not result in satisfaction; only positive work content factors (such as achievement and subsequent recognition) produce job satisfaction.

Stress

In Selye's original theory°, stress had a positive side (see Chapter 12): on the job, it was a condition in which a person felt compelled to better a situation, improve one's performance, and achieve a higher degree of satisfaction. In today's parlance, we refer only to what Selye called "dis-stress": a worrisome condition with demands at or beyond our means to cope, nerve-racking and exhausting. Stress, as we understand the term today, can occur at all organizational levels and in all jobs. Stress can stem from situations outside work and interact with job-related stress.

Stress is the individual's psychological and physiological response to excessive demands; it can carry severe consequences in terms of both physical and mental health. Avoiding such stress is critical for the well-being of employees and accordingly for the continued operational soundness of an organization.

Because individuals react in different ways to environmental demands, stressors vary in terms of the severity of their impact on people. In general, however, the biggest stressors are the following.

- Demanding jobs with high work loads, pressure to perform, a too-fast work pace
- Lack of control over the work process, need of autonomy
- Task perceived as overly complex, or with conflicting demands
- Overwhelming responsibilities for others
- Lack of social support, isolation
- Monotony and underload on the job, with overly routine, repetitive, and boring duties and little content variety
- Poor supervisory practices, including nonsupportive superiors or incompetent management
- Technical problems, such as frequent computer malfunctions or equipment breakdowns

Proper organizational policies and prudent actions by those carrying responsibilities in the organization can eliminate all these stress producers.

14.3 A good place to work

According to need- or value-based theories°, job satisfaction is an attitude that is specific to each individual. Every person has physical and physiological needs that one strives to fill in order to obtain satisfaction. A satisfying job meets physical needs (such as food and shelter) through appropriate income, and meets psychological needs (such as self-esteem and intellectual stimulation) through growth opportunities and professional recognition.

A second group of theories focuses on social comparison. Such theories posit that people assess their own feelings of job satisfaction in relation to other persons; by inferring their feelings about their jobs, they compare themselves to the other people who work in similar capacities.

Work conditions are the focus of the third group of theories that attempt to explain job satisfaction; Herzberg's Two-Factor theory is the best known, although still being discussed—see Chapter 13.

Ultimately, the various theories contribute to an overall understanding of what kind of job it takes to make employees satisfied.

A positive place to work

As a summary statement of the various factors underlying motivation and job satisfaction, we might do best to take a comment from the trenches. One of the U.S. companies that continually made the "100 Best Companies to Work for" list offered these reasons for its success°: "This is a positive place. People are friendly . . . they feel challenged. They feel respected and valued. And they respond with loyalty." The company's philosophy included: "Treat others the way they want to be treated. Strive for mutual respect and for an atmosphere that makes people proud to work here. Provide career opportunities. Say thank you for a job well done."

Quality of life at and off work

Striking a balance between life at work and life outside the job is, of course, of primary importance to the individual; but this has become an issue of the employer as well. A leanly staffed company in a tight labor market is likely to set up a situation in which the employees might well be overloaded with work. Technological developments, such as pagers and cell phones and handheld computers, which connect employees to work around the clock, also influence the burden many employees feel. Employers realize that overtaxing employees may lead to reduced productivity and employee exodus; consequently, organizations may well take measures to keep their employees from feeling overwhelmed by work. These measures include limited hours of work, off-time free of work tasks and phone calls, lengthy and uninterrupted vacations, limited meetings, flexible schedules, and employee sabbaticals: "A rested employee is essential to a company's business".°

Summary

An organization cannot function by its strategy, structure, and machinery alone; it is powered by its working, thinking, aspiring, and interacting people. People affect the organization; they determine the way it functions; yet, in turn, the organization affects these individuals. Good fit among individuals, work, and organization make the system run smoothly.

Motivating employees is a complex yet, of course, highly desirable goal. Many theories exist that try to explain motivation; they can be roughly divided into two categories. The first group focuses on the individual and the inherent traits; the second group particularly considers the immediate environment and the overall work organization.

Taylorism and mechanization, then automation changed jobs into monotonous and unskilled drudgery. Jobs became simple, repetitive, and boring; organizations became sectional, hierarchical, and stuck in bureaucracy. These trends deprive humans of outcomes such as expertise, autonomy, and self-fulfillment.

Today's educated and informed employees expect to be recognized at work as skillful and resourceful individuals. They expect respect and recognition for their efforts. They expect better-designed jobs, greater participation in decision making, and the freedom to work in a less formal organization. They expect a pleasing environment; they expect to work with pleasant people as colleagues beside, below, and above in the system. These expectations translate into a human–system integration that is less hierarchical than in the past, less formalized and less centralized in its organizational structures.

Human-centered layout of the whole organization and ergonomic design of the work environment can lead away from the aloof "I don't know what I'm doing; I just work here," to the interested participation expressed by "If I don't show up, this place will come to a stand-still."

Here is a voice from Japan°:

- Body movement and thinking on the job stimulate alertness.
- A complex job enriches skill and motivation and generates satisfaction when completed.

- Flexibility and freedom of choice in the execution of a task result in efficiency and productivity.
- Opportunities for decision-making provide a feeling of responsibility that leads to performance-directed motivation and satisfaction.
- The organization that promotes participation enhances its members' self-development and self-realization and furthers its own success.

Fitting steps

Step 1: Lay out the organization to utilize people's skills and aspirations.

Step 2: Encourage the members of the organization to work the way they "like to do" as this furthers the strategic goals.

Step 3: Be innovative, flexible, and adaptive.

Further reading

GREENBERG, J. and BARON, R.A. (2003). *Behavior in Organizations: Understanding and Managing the Human Side of Work* (8th Edn). Upper Saddle River, NJ: Prentice-Hall/ Pearson Education.

PEW, P.W. and MAVOR, A.S. (EDS.) (2007) *Human-System Integration in the System Development Process*. Washington, DC: The National Academies Press.

Notes

The text contains markers, °, to indicate specific references and comments, which follow.

14.1 Structure, policies, procedures, culture:

Model: This text uses material from Chapter 2 in Kroemer et al. 2001.

Organizational designs: Booher 2003, Pew et al. 2007.

Macroergonomics: Hendrick et al. 2001a,b.

Psychologists and behaviorists: Greenberg et al. 2003, Stanton et al. 2005, Wilson et al. 2005.

14.2 Design for motivation and performance:

> *Advertising agencies*: Example from Kroemer et al. 2001.

> *Explain personal behavior*: Alderfer 1972, Greenberg et al. 2003, Muchinsky 2007, Landy et al. 2006.

> *Selye's original stress theory*: Read Selye 1978.

14.3 A good place to work:

> *Need- or value-based theories:* Alderfer 1972, Greenberg et al. 2003, Muchinsky 2007, Landy et al. 2006.

> *100 Best Companies to Work:* Candace Goforth, Knight Ridder/Tribune, 6 February 2000.

> *A rested employee is essential to a company's business:* Joann Lublin, Wall Street Journal p. B4, July 6, 2000.

Summary:

> *Voice from Japan:* Paraphrased from Nagamachi 2001.

CHAPTER FIFTEEN

Working hours

Mothers of young children are busy all day long, often even during the night. Farm work has always been associated with long working hours, often starting at sunrise and ending at sunset. During the so-called Industrial Revolution, around 1800 in Europe, home-based handwork as well as factory work often lasted 10 to 12 hours a day, 6 or even 7 days a week. Today, in cities, firefighters and ambulance personnel can be on call 12 hours a day, but much of that time is—they hope—just waiting time; yet, they are ready for any emergency.

The human body changes its physiological functions throughout the 24-hour day. In the waking hours, the body is naturally ready for physical and mental work; during the night, it relaxes and sleeps. Attitudes and behavior also change regularly during the day. New time signals and unusual periods of activity and rest (such as starting shift work; see the following Chapter 16) can upset the daily rhythm and have performance and health consequences.

15.1 Circadian body rhythms

Body rhythms

Daily rhythms are natural temporal programs controlling the lives of animals and plants. In the human, body temperature, heart rate, blood pressure, and hormone excretion, for example, follow a set of daily recurring fluctuations, called circadian° (or diurnal) rhythms, as sketched in Figure 15.1. Several programs intertwine with each other, such as core temperature, blood pressure, and sleepiness.

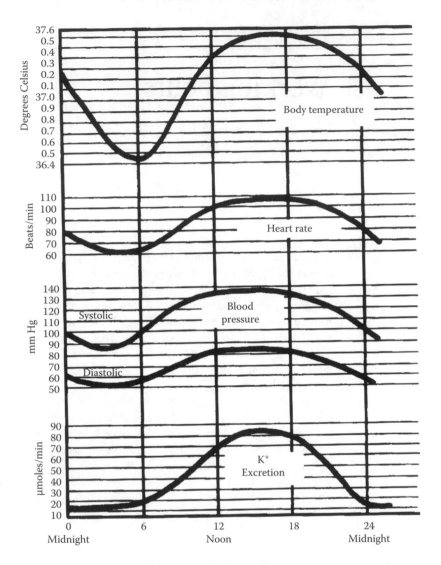

FIGURE 15.1 Typical variations in body functions over the day. (Adapted from Kroemer, K.H.E., Kroemer, H.J., and Kroemer-Elbert, K.E. (1997). *Engineering Physiology. Bases of Human Factors/Ergonomics* (3rd edn). New York: VNR–Wiley.)

Daily rhythms   During daylight, we are awake, active, and we eat and drink, while at night, we fast and sleep. Physiological events do not exactly follow that general pattern. For example, body core temperature falls even after the body has been sleeping for several hours; it is usually lowest around 6 o'clock in the morning. Core temperature then rises quickly when one gets up. It continues to increase, with some variations, until late in the evening. Thus, body temperature is not just a passive response to our regular daily behavior, such as getting up, eating meals, performing work, and doing social activities. Another example is skin temperature, which increases with the onset of sleep, regardless of when this occurs. The circadian physiological rhythms of the body are solid and self-regulated.

24-Hour cycles

When a person is completely isolated from external time markers and events such as night darkness and daylight, and no regular activities are set, the internal body rhythms are running free, under internal control° only. Many experiments have consistently shown that circadian rhythms continue when running free, but their time periods are slightly different from the regular 24-hour duration: most rhythms run freely at about 25 hours, some take longer. This experience indicates that body rhythms can persist independently of external stimuli and follow their own built-in clocks. However, when subjected to daily 24-hour activities and associated time markers, the internal rhythms synchronize at 24-hour cycles.

Individual differences

Some people have consistently shorter (or longer) free-running circadian rhythm periods than others. Those with short periods are likely to be "morning types" who get up early from sleep whereas those with longer internal rhythms are probably "evening types" who stay up into the late hours of the night. Rhythm amplitudes, for instance of body temperature, usually reduce with increasing age, which could explain a shift toward early morning activity among the elderly. Also, oscillatory controls seem to lose some of their power as one ages: this would indicate a greater susceptibility to rhythm disturbances with increasing age.

Synchronized biological clocks

Of course, under regular circumstances there is a well-established phase coincidence between external activities and the internal circadian events. For example, during the night, the low values of physiological functions, such as core temperature and heart rate, are primarily due to the diurnal rhythm of the body; however, they are further helped by the body's nighttime inactivity and fasting. During the day, peak activity usually coincides with high values of the internal functions. Thus, normally, the observed diurnal rhythm is the result of internal (endogenous) and external (exogenous) events, that concur. If that balance of concurrent events is severely disturbed, wellbeing, health, and performance decline.

The regulation of alertness, wakefulness, sleepiness, sleep, and of many physiological functions may be under the control of two internal clocks of the body°: one controls sleep and wakefulness, the other physiological functions, such as body temperature. Under normal conditions, the internal clocks are linked so that body temperature and other physiological activities increase during wakefulness and decline during sleep. However, this congruence of the two rhythms may be disturbed by night shift work, for instance, where one must be active during nighttime and sleep during the day. (More on shiftwork in the following Chapter 16.) As such patterns continue, the physiological clocks adjust to the external requirements of the new sleep/wake regimen. This means that the formerly well-established physiological rhythm flattens out and, within a period of up to two weeks, re-establishes itself according to the new sleep/wake schedule.

Resetting biological clocks

Strong and persistent external events can put the regular circadian rhythms into a new order. This happens commonly when we travel fast and far to the east or west. Upon arrival at the new location, we experience "jet lag" at first: our body functions at its normal timing, which differs from the local time settings of daylight and night darkness, of waking up and seeking sleep, of working hours and meal times. These events and activities act as time markers. Their regular repetition resets our internal chronobiological clocks, which regulate our diurnal rhythms.

Airline crews and other short-time visitors to a different time zone may decide to simply stay on their home time setting until they return to their regular time zone. Of course, this is not practical for long-term moves to a new time zone, where it is better to to realign. That rearrangement takes usually at least 3 days: the larger the jumps in time, the longer the adjustment time to overcome the rhythm lag.

Sleep deprivation

Traveling through several time zones, then being ready to get some sleep but forced to stay awake, can put us into a difficult situation: we need sleep. Why sleep is necessary°, and precisely what it does to our body and mind, is—surprisingly—still a topic of much discussion among researchers. Whatever the reason, we all know that performing everyday tasks can be very difficult while being deprived of sleep.

There are occasions that demand continuing work for long periods of time, such as a full day or longer. This not only means being busy for long periods of time but also encompasses loss of sleep. Hence, negative effects on operator well-being and performance associated with such extended working spells partly result from the long duration of work, but they are due to the lack of sleep also.

Performing tasks while sleep-deprived

Sleepiness and wakefulness appear in unpredictable waves; so, work performance may be fine for a while and then become difficult and faulty. Sleepiness has little effect on performance of demanding work tasks that are interesting and appealing, especially if they last less than half an hour, even if they include complex decision making. Accuracy in performing a job may be still quite good even after losing sleep, but it takes longer to perform the job. Self-paced work deteriorates less with sleepiness than externally paced tasks.

However, sleep loss negatively affects performance of work to be done over longer periods of time, particularly so if the task is repetitive, monotonous, or complex. Decision making takes longer if the task is disliked and unappealing. Long-term and short-term memories degrade when people are required to stay awake for long periods of time.

15.2 Rest pauses, time off work

Given the natural circadian rhythms in physiological functions, one expects corresponding changes in attitudes and activities during the day. Of course, the daily organization of getting up, working, eating, relaxing, and sleeping also affects, often strongly so, attitudes and work habits and work performance. In experiments, it is possible to separate the effects of internal and external daily rhythms. For practical purposes, it is reasonable to look at the results of the combined internal and external factors as they affect workload and work output.

Performance changes during daylong work

For many people, the morning hours seem to be best for strenuous activities, mental and physical, because fatigue arising from work already performed may reduce performance over the course of the day. Often, performance shows a pronounced reduction early in the afternoon: this is the post-lunch dip. However, there seem to be no similar changes in physiological functions; for example, body temperature remains unchanged at that time of the day. Hence, it appears reasonable to speculate that the interruption of the activities by a noon meal and the following digestive activities of the body, together with reduced psychological arousal and expectation and habits would bring about this reduction in performance. However, in other activities, primarily those with medium to heavy physical work, no such lunch dip seems to occur except when the food and beverage ingestion was very heavy and if true physiological fatigue had built up during the pre-meal activities.

Stress of long working hours

Some individuals are capable of continued heavy physical labor and, in fact, enjoy doing it whereas others prefer lighter physical tasks or mental activities; the types and intensities of activities pose varying demands on the worker. Section III of this book contains discussions of how to assess the actual workloads and, consequently, how long persons can perform such work. If metabolic, circulatory, or respiratory labor demands exceed sustainable limits even with interspaced rest periods, the work must be terminated; if these systems continue to run smoothly, work interspaced by pauses can go on.

The demands posed by mental activities such as visual searches or information processing, especially by tasks requiring concentrated attention, vary strongly by their nature and intensity; in combination with individual traits or motivation they create varying levels of stress; see Chapters 11 and 12. Persons expected to perform monotonous mental tasks which are low in novelty, interest, and incentive, or that are high in complexity, need breaks every half hour or so; otherwise, performance falls off severely. Accordingly, tasks that are both monotonous and complex should not be required over long periods of time. Varying or new tasks are more resistant to deterioration if they require decision making, problem solving, or if they deal with information that is highly interesting or rewarding; on the other hand, highly complex tasks can be overly demanding and exhausting.

Rest pauses

The same logic applies to physical work and to mental activities: suitable breaks permit a person to, literally, "catch my breath" or "take my mind off the task"; pauses and off-work time allow relaxing, recovering, and reloading.

The working human body needs to maintain a balance between energy consumption and energy replacement: put simply, between work and rest. This supply–use–supply process is an integral part of the operation of muscles, of the heart, and of the organism as a whole. Rest pauses are indispensable physiological and psychological requirements if performance must be maintained.

There are four types of work pauses:

1. Spontaneous pauses
2. Disguised pauses

3. Work-caused pauses

4. Pauses prescribed by management

Spontaneous pauses are taken by workers on their own initiative to interrupt the flow of work in order to "take a break and recharge". These pauses are in response to the effort; they are needed to recuperate. As discussed in Chapter 10, as a rule many short pauses have greater recuperative value than a few long ones.

Disguised pauses are those where the worker takes up an activity other than the main job; examples are replacing work tools or tidying the workplace. Many of these disguises serve to cover a necessary spontaneous pause. It would be better to take these pauses openly, so that the necessary recovery from the work to can be fully realized.

Work-caused pauses are interruptions that arise from the work itself, such as waiting for a machine to complete an operation or for a tool to cool down, or they may be caused by a breakdown in the flow of work. These breaks are beyond the worker's control, yet they provide time to recuperate.

Management-prescribed pauses are wisely provided so that workers can get away from the job and rest openly, to have time for eating and for other personal needs. As a rule, during shifts lasting up to 8 hours, one long pause of 30 minutes or longer should be provided during the middle of the work, and two shorter pauses, say of 15 minutes each, should break up the two halves of the working time.

Pauses are physically or psychologically helpful, often needed to maintain performance during the full work shift. Taking spontaneous pauses should be encouraged by management; disguising a pause with another activity reduces the recovery value of the break.

15.3 Daily and weekly working time

One prerequisite for our well-being is the maintenance of our circadian rhythms in spite of external disturbances. This state of balanced control, called homeostasis, regulates our body functions, makes us sleepy or awake, enables us to work, establishes our inclinations to perform certain activities, and determines our social behavior.

Days on/off work

Most persons are used to working during daylight. Physiologically, we should be able to do so every single day, provided that we have suitable conditions of work, including rest breaks. However, tradition and societal customs have made us expect to have days free of work regularly°: in many regions, this is every seventh day. In most of Europe, for example, Sundays have been off-work since at least the mid-1800s; free Saturdays followed in the mid-1900s; in the 1990s, a 4-day work week was widely proposed. The conventions of certain sequences of days with work, followed by days off work, have led to many theories and experiments° to determine suitable numbers of working hours during the day in relation to working days during the week (discussed below) and of shift work (discussed in Chapter 16).

History of working time

Weekly and daily working times are closely related. In Switzerland, the Factory Act of 1877 stipulated a 65-hour workweek, with 11 hours of work each weekday and 10 hours on Saturday. In 1914, a legal amendment shortened the weekly work to 48 hours. Figure 15.2 shows the history of working hours in the United States from 1850 to 1963. In 1850, outside agriculture the average working time was about 70 hours per week, lowered to around 50 hours in 1940. In 1963, the working time per week was around 40 hours in industry, where it has stayed, with little change, until now. About 40 hours of working time per week are common in most industrialized regions on earth.

The reduction of the weekly working time to 40 hours, to even less in some countries, was often associated with a

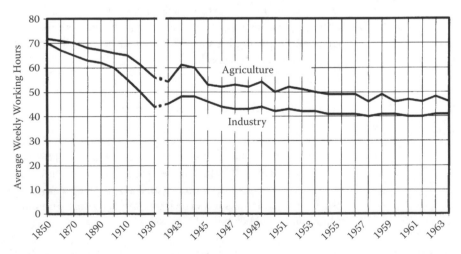

FIGURE 15.2   Weekly working hours in the United States

changeover from a 5-day workweek to 4 days of work each week. Employees and employers generally both welcomed that reduction in workdays; it often came with a slight increase in productivity and reduction in absenteeism°. It is of socioeconomic interest that this reduction in work hours and workdays allowed employers, in some countries, to hire more workers and thus reduce widespread unemployment.

Eight-hour workday

Apparently, the division of 24 hours into 3 about equal sections fits the current lifestyle in many societies. That setup is commonly acceptable because it offers about 8 hours for family and social interactions, reserves 8 hours for sleep, and still provides 8 hours for work. As customary, the work shift of 8 hours should be interrupted about halfway with a long break of at least 30 minutes, commonly used for taking in a meal. A truly long break could be used for an extended siesta around noon, as traditional in hot climates. Each of the 2 blocks of working time should be split by another scheduled break, lasting perhaps 15 minutes, and workers should be encouraged to take additional pauses as desired and needed.

The work shift of 8 hours has become widespread: it suits organizational purposes in trade and industry, especially because the 8-hour work period allows establishing a three-shift schedule that covers the 24-hour day; yet, so could do a two-shift arrangement with a 12-hour work period—more about shift arrangements in the following Chapter 16.

Brief work shifts

Work that is very demanding, mentally or physically, may be impossible to maintain for the customary 8 hours of working time a day, even with suitable breaks for recovery: instead, 6 hours may be more acceptable as shift duration. Even in such shortened work arrangements, a substantial break at about midtime is necessary, and each of the two halves of working time should be interrupted by at least one scheduled break. Workers should take additional pauses as needed.

Long work shifts

Long work shifts may be scheduled for a variety of reasons: for example, big-city firefighters spend much of their 10- or 12-hour work shifts waiting for the call of duty. In automated plants, supervisory presence makes sure that the process runs smoothly with occasional human interference as needed. Twelve-hour shifts may be desired to fill the 24-hour day—more about shift arrangements in the following Chapter 16. However, working periods of 10 or 12 hours are not suitable if they require continual effort from the worker. Humans cannot perform physical or mental work at a continuing high rate over long periods of time without physical or mental fatigue with its accompanying resulting negative effects on well-being and work performance: lapses in attention are unavoidable, mistakes are likely, and accidents possible.

Flextime

Flexible working hours are a recent form of work organization. Its specific feature is that a person can distribute the actual work time, say 40 hours per week, over the official work days as long as certain requirements are met, such as full presence on certain days, or during a core time of the day, for example, from 10 to 12 in the morning.

Sliding time (or gliding time) is a subgroup of flexible work hours where, instead of weekly working hours, the daily working hours can be slid (floated) over the day. As with weekly flextime, usually certain requirements must be met, such as being present at set times.

Both arrangements have been generally popular with employees and employers, although not all kinds of work are suitable for flexible arrangements. Furthermore, in nearly every organization, there are groups of employees who must be present during all the conventional hours, such as receptionists or cafeteria workers, or assembly workers organized in groups.

Working at home, especially in the home office usually associated with computer use, has the most flexible arrangements. Control of the number of working hours, and their scheduling, is on the individual; the overriding requirement is to get work done on time.

All these flextime arrangements abolish much of the traditional control over working time by the employer and put more responsibility on the individual worker. The emphasis is on performance, not on hours spent on the job. Yet, overly ambitious persons might spend more time working on their own than they would do if they were on regular schedules.

Compressed work week

Flextime is often, although not necessarily, associated with compressed work weeks that have more working hours per day, but fewer workdays per week. For example: the customary numbers of weekly work hours is 40, divided into 8 hours of daily work. This can be consolidated into 4 days per week, where each day has 10 hours of work. This allows the worker to have 3 instead of 2 free days each week. Apparently, this is attractive to many employees and employers for reasons such as these: there are more work-free days, best if together in one block; the number of trips to and from work is reduced, and there are

Table 15.1   Comments on flextime

Positive

Appeals generally to employees and employers

Makes available work-free time at the employee's choosing

Does not result in reduced employee pay

Reduces commuting traffic problems

Less fatiguing for workers

Increases job satisfaction

Recognizes and utilizes employees' individual differences

Reduces tardiness

Reduces absenteeism

Reduces employee turnover

Increases performance

Negative

Makes it difficult to cover jobs at all required times

Makes it difficult to schedule meetings or training sessions

Reduces communication within the organization

Increases energy and maintenance cost

Requires more sophisticated planning, organization, and control

Requires special recording of work time

Requires additional supervisory personnel

Source: Adapted from Knauth, P. (2007) Extended work periods. *Industrial Health,* 45, 125–136 and Kroemer, K.H.E., Kroemer, H.B., and Kroemer-Elbert, K.E. (2003, amended reprint). *Ergonomics: How to Design for Ease and Efficiency* (2nd Edn). Upper Saddle River, NJ: Prentice-Hall/Pearson Education.

fewer setups and closedowns at work. However, there are also concerns about increased fatigue due to long workdays and reduced performance and safety—see Tables 15.1 and 15.2.

The type of work to be performed mostly determines whether compressed work weeks can and should be used. Thus, long shifts have been mostly employed in cases where one is "on standby" much of the time. Also, activities that require only few or small physical efforts, especially if they are diverse and interesting yet fall into routines, have been done in long shifts. Examples are nursing, clerical, and administrative work, technical maintenance, computer supply operations, and supervision of automated processes. Little experience exists from long shifts that include manufacturing, assembly, and machine operations. Furthermore, information stems mostly from subjective

Table 15.2 Comments on extended work days/compressed work weeks

Positive
Appeals generally to employees and employers
Makes available more work-free days for family and social life
Provides more time per day for scheduling meetings or training sessions
Generates fewer startup and warmup expenses
Does not result in reduced employee pay
Reduces commuting traffic problems
Negative
Decreases job performance
Requires overtime pay
Tends to fatigue workers
Increases tardiness and early departure from work
Increases absenteeism
Increases on-the-job and off-the-job accidents
Increases energy and maintenance cost

Source: Adapted from Knauth, P. (2007) Extended work periods, *Industrial Health* 45, 125–136 and Kroemer, K.H.E., Kroemer, H.B., and Kroemer-Elbert, K.E. (2003 amended reprint) *Ergonomics: How to Design for Ease and Efficiency* (2nd Edn) Upper Saddle River, NJ: Prentice-Hall/Pearson Education.

statements of employees, results of psychological tests, and scrutinizing performance and safety records in industry.

Working very long work shifts, such as 12 hours, is likely to lead to fatigue, cause drowsiness and lead to reductions in cognitive abilities and motor skills, and to generally reduce performance during the course of each work shift as the work-week progresses. There is the potential that a fatigued worker might take careless shortcuts to complete a job, and that work practices may be less safe in tasks that are tedious because of high cognitive or information-processing demands, or in highly repetitive work.

Summary

Human physical and psychological functions as well as behaviors follow a set of regular daily fluctuations, the circadian (diurnal) rhythm. Under regular living conditions, this temporal program is well established and persistent. Upsetting the rhythm by strong external time markers and events, such as work in a different time zone, can lead to reduced performance and well-being. Regular work shifts should be done during daylight.

The suitable length of the working time per day primarily depends on

- the type and intensity of work and how these relate to worker capacities
- the ways work is organized within societal and social customs

A working time of eight hours a day is generally acceptable for psychological, physiological, social, and organizational reasons; yet, scheduled and freely taken rest periods are necessary. Under certain circumstances, shorter or longer daily work times are reasonable.

Apparently, the common division of the 24-hour day into three about equal sections fits the current lifestyle in many societies because it reserves about 8 hours for family and social interactions, 8 hours for sleep, and still leaves 8 hours for work.

Fitting steps

Step 1: Determine the suitable length of the workday with rest pauses.

Step 2: Establish whether fixed or flexible arrangements are desirable.

Step 3: Rearrange as experiences suggest.

Further reading

KNAUTH, P. (2007). Extended work periods. *Industrial Health,* **45**, 125–136.

SIEGEL, J.M. (2003). Why we sleep. *Scientific American*, Nov., 92–97.

Notes

The text contains markers, °, to indicate specific references and comments, which follow.

15.1 Circadian Body Rhythms:

Daily rhythms: Called *circadian* from the Latin *circa*, about, and *dies*, the day; *diurnal* from *diurnus*, of the day.

Rhythms "running free" under internal control: Folkard et al. 1985.

Model of two internal clocks: Horne 1988.

Why sleep is necessary: Horne 1988; Siegel 2003.

15.3 Daily and Weekly Working Time:

Have days free of work regularly: from a formal/decimal point of view, there is little sense in such numerical arrangements as 12 months a year, 7 days a week, 24 hours a day, 60 minutes per hour.

Theories and experiments to determine suitable working hours: Read the discussions in Folkard et al. 1985, 2003.

Increase in productivity and reduction in absenteeism: Summarized in Kroemer et al. 2003.

CHAPTER SIXTEEN

Night and shift work

Physiological functions of the human body wax and wane rhythmically throughout the 24-hour day. During daylight, the body is prepared for work; sleep is normal at night. Our attitudes and behavior fluctuate accordingly and so does our social life. A new set of time signals, of activity and rest, such as those associated with shift work schedules, can upset the normal daily rhythms. Shift work should be arranged to least disturb physiological, psychological, and behavioral rhythms to avoid negative health and social effects and to prevent reductions in work performance.

16.1 Need for sleep to be awake

Physiological and psychological functions of the human follow daily fluctuations called circadian rhythms. They are easy to observe in body temperature, heart rate, blood pressure, and hormone excretion, as sketched in Figure 15.1 in the preceding chapter. These natural temporal programs keep our body and mind healthy°; they make us able to work and to function in our social environment.

The circadian programs intertwine: for example, core temperature and blood pressure are coupled with sleepiness and wakefulness. Self-sustained "internal clocks" control the rhythms, which run on a 24-hour cycle synchronized by our daily time markers: real clocks, getting up, eating, working, and going to sleep.

The circadian rhythms show a high persistence under varying external conditions. However, strong and repeated external events and time markers can upset the regular circadian rhythms and disturb the daily balance of wakefulness and sleep, activities and rest, alertness and drowsiness. Such balance is a prerequisite for human health and performance.

Getting enough sleep is part of that requirement. Adult individuals differ in the number of hours needed for sufficient sleep: 8 hours is commonly enough, but many persons, especially the elderly, get by with less. Although it is obvious that humans need sleep, exactly why they need sleep is not clear°. There is the general opinion that sleep has recuperative benefits, allowing some sort of restitution or repair of tissue or brain following the wear and tear of wakeful activities. However, what is meant by restitution or repair is usually not clearly expressed, nor fully understood. Certainly, sleep is accompanied by rest and, consequently, by energy conservation. But, when not forced to be active, a human can attain similar relaxation during wakefulness. Regarding restitution of the body, it is an everyday experience that muscle ache resulting from a day's effort is gone the next morning; yet, it might go away just as well during a long rest while awake.

Limited sleep deprivation does not impair the *physiological* ability to perform physical work. Apparent reductions in physical capability, blamed on sleep deprivation, may be mostly due to reduced psychological motivation. The lack of experimental findings regarding the physical benefits of sleep is somewhat surprising because this seems to contradict common experience.

In contrast, the restorative benefits of sleep to the brain are well established. Two or more nights of sleep deprivation bring about psychological performance detriments, particularly reduced motivation to perform (but apparently not a reduction of the inherent cognitive capacity), behavioral irritability, suspiciousness, slurred speech, lack of attention, drowsiness, memory problems, and other performance reductions. As mentioned in the previous Chapter 15, mental performance is less diminished on stimulating and motivating tasks than on boring and repetitive and rewardless tasks. More than two nights of sleep deprivation reduce all task performance. These performance degradations indicate function impairments of the central nervous system owing to sleep deprivation, hence a need for the brain to sleep.

Sleep phases

The brain and muscles are the human organs that show the largest changes from sleep to wakefulness: their electrical activities provide well-established techniques of observation. Electrodes attached to the surface of the scalp can pick up electrical activities of the cortex, called the encephalon because it wraps around the inner brain. Thus, the name of this measuring technique is electroencephalography, EEG. The EEG signals provide information about the activities of the brain. The other common technique records the electrical activities associated with the muscles that move the eyes and those in the chin and neck regions. The electrical recording of muscle (from the Greek, *myo*) activities is called electromyography, EMG. Both techniques are often applied together to record and describe events during sleep.

The amplitude of EEG signals is measured in microvolts; the amplitude rises as consciousness falls from alert wakefulness through drowsiness to deep sleep. EEG frequency is measured in hertz; the frequencies observed in human EEGs range from 0.5 Hz to 25 Hz. Fast waves are frequencies above 15 Hz; frequencies under 3.5 Hz are slow waves. Frequency falls as sleep deepens; slow wave sleep is of particular interest to researchers.

Certain frequency bands carry Greek letters.

- Beta, above 15 Hz. These fast waves of low amplitude (under 10 microvolt) occur when the cerebrum is alert or even anxious.

- Alpha, between 8 and 11 Hz. These frequencies occur during relaxed wakefulness when there is little information input to the eyes, particularly when they are closed.

- Theta, between 3.5 and 7.5 Hz. These frequencies are associated with drowsiness and light sleep.

- Delta, slow waves under 3.5 Hz. These are waves of large amplitude, often over 100 mv, and occur more often as sleep becomes deeper.

Also, certain occurrences in the EEG waves have descriptive names, such as vertices, spindles, and complexes, which appear regularly associated with certain sleep characteristics.

REM and non-REM sleep phases

EMG outputs of the eye muscles often serve as the main descriptors: sleep is divided into periods associated with rapid eye movements, REM, and those without, non-REM. Non-REM conditions are further subdivided into four stages according to their associated EEG characteristics. Table 16.1 lists these.

The REM sleep phase shows irregular breathing and heart rate, as well as low voltage, fast brain activities visible in the EEG. In the other main sleep phase, non-REM, regular and slow breathing and heart rates occur, and the EEG activity is

Table 16.1   Sleep stages

Condition	Sleep stage	Muscle EMG	Brain EEG	Sleep percentage, on average
Awake	—	Active	Active , alpha and beta	—
Drowsy, transitional light sleep	1 Non-REM	Eyelids open and close, eyes roll	Theta, loss of alpha, sharp vertex waves	5
True sleep	2 Non-REM		Theta, few delta, sleep spindles, k-complexes	45
Transitional true sleep	3 Non-REM		More delta slow wave sleep	7
Deep true sleep	4 Non-REM		Predominant delta slow wave sleep	13
Sleeping	REM	Rapid eye movements, other muscles relaxed	Alert, much dreaming, alpha and delta	30

Source: Adapted from Kroemer, K.H.E., Kroemer, H.J., and Kroemer-Elbert, K.E. (1997). *Engineering Physiology. Bases of Human Factors/Ergonomics* (3rd Edn). New York: VNR–Wiley.

slow but shows high voltage. These phases change cyclically during sleep. The REM/non-REM cycles occur in roughly 1.5-hour timings; however, this duration has large within- and between-subjects variability and appears to shorten in the course of a night's sleep, accompanied by a lengthening of the REM portion. Figure 16.1 depicts a diagram of typical sleep phases during the night.

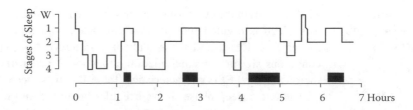

FIGURE 16.1   Typical phases during undisturbed night sleep. W indicates wakefulness; the black bars represent REM sleep.

16.2 Performance and health considerations

Performance levels are generally lower during nighttime activities, when body and brain usually rest, than during daylight time. This has long been known: Figure 16.2 shows a summary graph of reading errors recorded between 1912 and 1931. During that time, about 175,000 readings of gas meters were taken around the clock. The number of errors was by far highest around two o'clock in the morning. Another peak in errors, although noticeably lower, happened after lunch. Similar relations between mishaps and time of the day have been reported ever since. Figure 16.3, with data from the mid-1950s, illustrates

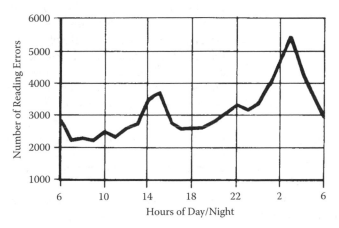

FIGURE 16.2 Frequencies of errors in gas meter readings from 1912 to 1931

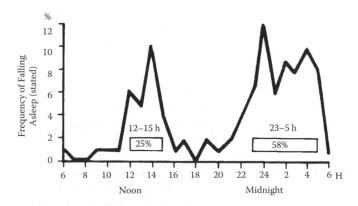

FIGURE 16.3 Frequencies at which truck drivers fell asleep at the wheel in the 1950s

the frequency at which truck drivers fell asleep at the wheel. That occurred most often after midnight and, again, around lunchtime. Sleep loss and its resulting sleepiness are likely to aggravate such naturally occurring fluctuations in performance.

Sleep loss and tiredness

Apparently, only the brain assumes a physiological state during sleep that is unique to sleep and cannot be attained during wakefulness. Muscles, for example, can rest during relaxed wakefulness, but the cerebrum remains in a condition of quiet readiness prepared to act on sensory input, without diminution in responsiveness. Only during sleep do cerebral functions show marked increases in thresholds of responsiveness to sensory input. In the deep sleep stages associated with slow–wave non-REM sleep, the cerebrum seems to be functionally disconnected from subcortical mechanisms. The brain needs sleep for restitution, a process that cannot take place sufficiently during waking relaxation. It appears that not all the regular full sleep time is essential for brain restitution. Following sleep deprivation, usually not all lost sleep is reclaimed; a sleep period that is up to 2 hours shorter than usual can be endured for many months without negative consequences. Evidently the first 5 to 6 hours of regular sleep (which happen to contain most of the slow-wave non-REM sleep and at least half the REM sleep) are obligatory to retain psychological performance at normal level[o].

Normal sleep requirements

Certain age groups in the western world show rather regular sleeping hours: young adults sleep, on average, about 7.5 hours; yet, some people are well-rested after about 6 hours of sleep, whereas others habitually take 9 hours and more. If people can sleep for just a few hours per day, many are able to keep up their performance levels even if the attained total sleep time is shorter than normal. The limit seems to lie around to 5 hours of sleep per day, with even shorter periods still being somewhat useful. The amount of slow-wave sleep in both short and long sleepers is about the same, but the amounts of REM and non-REM sleep periods differ considerably. Individuals naturally sleeping less than 3.5 hours are very rare among middle-aged people; no true nonsleepers have ever been found among healthy persons.

Overcoming sleepiness

A sleepy person may experience so-called microsleeps, which occur more frequently with increasing time at work. One falls asleep for a few seconds, but these short periods (even if frequent) do not have much recuperative value because one still feels tired and performance still degrades. Another commonly observed event during long working times coupled with lack of sleep are periods of no performance, also known as lapses or gaps. These are short periods of reduced arousal or even of light sleep.

Naps

If a person does not get the usual amount of sleep, the apparent result is tiredness, and the obvious cure is to get more sleep. Shift workers in particular experience this problem, when they have to make up in daytime for missed sleep due to working during regular sleep time. Napping° for a while, up to an hour or two, between periods of work helps to overcome sleepiness and related performance decrements already discussed in Chapter 15. If a person is awakened during a deep-sleep phase, "sleep inertia" can appear, but that is not of great concern in light of the benefits derived from the nap. Naps taken at any time during the circadian rhythm have positive effects. Many persons say that a brief nap after lunch helps them to get going again whereas others do not feel a need for it.

If working long periods of mental activities are unavoidable, one may try to perform physical exercises between sections of work. However, this is not a sure way to prevent performance deterioration. Also, intermittent white noise may improve performance slightly and stirring music may help. Drugs, particularly amphetamines, can restore performance to a normal level when given after even three nights without sleep; however, using drugs instead of employing suitable work shifts is not a recommended remedy.

Recovery from sleep deprivation is quite fast. A full night's sleep, undisturbed, probably lasting longer than usual, restores performance efficiency almost fully.

16.3 Organizing shift work

Shift work

One speaks of shift work if two or more persons, or teams of persons, work in sequence at the same workplace. Often, each worker's shift repeats, in the same pattern, over a number of days. For the individual, shift work means attending the same workplace regularly at the same time (continual shift work) or at regularly recurring times (discontinuous, rotating shift work).

Shift work is not new. In ancient Rome, a decree required that deliveries were to be done at night to relieve street congestion. Bakers have traditionally worked through the late night and early morning hours. Guards and firefighters always did night shifts. With industrialization, long working days became common with teams of workers relaying each other to maintain blast furnaces, rolling mills, glass works, and other workplaces where continuous operation was desired. Covering the 24-hour period with either two 12-hour work shifts, or with three 8-hour work shifts, became a common practice. In the first part of the 20th century, the then-common 6-day work weeks with 10-hour shifts were shortened, as shown in the preceding chapter in Figure 15.2. Since the middle of the 20th century, many work systems use a shift arrangement of 8 hours per day on five workdays per week.

Circadian rhythm and shift work

With the increase in shiftwork, circadian rhythm adjustments, especially the lack thereof, and the fear of consequent negative health outcomes became important topics in industrial medicine. The first large surveys came from Norway, around 1960. They showed that shift workers had significantly more digestive ailments and nervous disorders than their colleagues on regular day shifts. Many shift workers disliked that arrangement and tried to return to regular day work.

Health concerns of shift work

Half a century later, these early findings of negative health effects of shift work are still valid. Working at unusual times, especially during the night; working long shifts, such as 10 or 12 hours; and changing timings of work, especially when they lead to sleep loss, all cause health concerns especially if these shift work conditions combine to cause extraordinarily long working times and deficits in rest and sleep. Prominent on the list of concerns for human well-being are:

- Interruption of natural physiological and psychological rhythms
- Fatigue, drowsiness, sleepiness
- Increased accident risks, both on and off the job
- Sleep disorders
- Gastrointestinal disorders
- Cardiovascular diseases
- Nervous disorders
- Disturbances of family and social life

Obviously, the main reasons underlying these concerns are physiological and psychosocial: *physiological* in that such work needs to be performed at times when the body is set to sleep and rest, and *psychosocial* because work must be done at times that are commonly employed for family and social interaction, and for leisure activities.

Three basic solutions

There are three basic solutions that help to avoid serious complications related to shiftwork done during the evening or night, or at other unusual timings. The first solution is to insert just one odd work shift and then return to the normal schedule. The second solution is to adopt a permanent work arrangement at the unusual time. The third solution, which however retains some of the complications, employs a limited number of work shifts at the odd time, followed by an interruption (usually on the weekend) and then continues again with a limited number of work shifts at another timing.

Just one odd shift

The fundamental solution to avoid serious complications related to shift work at odd timing is not to upset the circadian rhythm. This is achieved by doing just one unusual work shift (or two shifts), to muddle through and take a good rest thereafter, and then return to the regular work schedule. However, as discussed in the foregoing Chapter 15, getting through the unusual work period is fatiguing and strenuous for the worker and may seriously reduce work performance.

Permanent shiftwork

There are persons who volunteer to work continually evening or night shifts; they are willing to readjust their circadian rhythms and their habits of private and social interactions permanently. Such volunteers are, for example, bakers and night guards of olden times; today they are firefighters, nurses, and physicians; drivers of trains, buses, trucks, and taxis; workers on offshore oil rigs; and many other persons who agreed to be on evening or night shifts for weeks, months, even years on end. These persons readjust their physiological and social rhythms fully and permanently; therefore, they are not likely to suffer from disorders related to rhythm disturbances.

Societal expectations of free time

Permanent evening or night shift workers set up their own personal schedules for sleep, leisure times, and social networking, possibly at odds to the habits of a society that values free time in the evening and on weekends more highly than free time at other times of the day or week. In a theoretical view, the societal expectation of free time especially on weekends generates a major hindrance for permanent readjustment of daily rhythms. Many shift workers return to their ingrained temporal societal customs and expectations over the weekend and thus fall out of sync with their work shift rhythm. That phenomenon seems to underlie the widespread organizational preference for starting new shift schedules after the weekend.

Shift patterns

For practical reasons, shiftwork seems to be unavoidable in many organizations. Shiftwork appears in several basic patterns, but any given shift system may comprise aspects of several patterns. Four particularly important features° are:

1. Do shifts extend into hours that would normally be spent asleep?
2. Do shifts cover the entire seven-day week and include days of rest, such as a free weekend?
3. How long is each shift?
4. Do shift crews either rotate shifts or do they work the same shifts permanently?

All of these aspects are of concern regarding the welfare of the shift workers, their work performance, and the organizational scheduling.

Shift rotations

In terms of organizing the schedule, it is easiest to establish a permanent set-up or a weekly rotation schedule. The literature discusses many such solutions, most of which rely on traditional societal expectations, especially regarding weekends, that originated in Europe and North America. In most systems used today, the same 8-hour shift is being worked for 5 days, usually followed by 2 free days during the weekend. This regimen, however, does not cover all 21 shift periods of the week, thus additional crews are needed to work on weekends. If one uses three shifts a day (day D, evening E, night N,) with four teams, the shift system (for each team) is D-D-E-E-N-N-free-free with a 6-work/2-free day ratio and a cycle length of eight days; this is called the metropolitan rotation. The continental rotation, which also assumes three shifts per day and four crews, has the sequence D-D-E-E-N-N-N, free-free-D-D-E-E-E, N-N-free-free-D-D-D, E-E-N-N-free-free-free; its work/free day ratio is 21/7, its cycle length 4 weeks and free days follow each set of night shifts.

Table 16.2 shows a schedule for rotating shifts that seems to be particularly suitable for Europeans. Each person works the same shift twice, then rotates forward to the next later shift for two periods, then rotates forward again to the next set of two shifts. Then follow four free days, which, however, do not always include weekend days. The system has a work/free day ratio of 6/4, its cycle length is exactly 10 weeks, on average, each week has 33.6 work hours.

Table 16.2 Proposed 6:4 rotation shift schedule

Week	Mo	Tu	We	Th	Fr	Sa	Su	Mo	Tu	We	Th	Fr	Sa	Su
1, 2	D[a]	D	E[b]	E	N[c]	N	free	free	free	free	D	D	E	E
3, 4	N	N	free	free	free	free	D	D	E	E	N	N	free	free
5, 6	free	free	D	D	E	E	N	N	free	free	free	free	D	D
7, 8	E	E	N	N	free	free	E	E	D	D	E	E	N	N
9, 10	free	free	free	free	D	D	E	E	N	N	free	free	free	free

Source: Adapted from Knauth, P. (2007b). Schicht- und Nachtarbeit, shift work and night work. In Landau, K. (Ed) *Lexikon Arbeitsgestaltung.* Stuttgart: Gentner.

[a] D: Day shift.
[b] E: Evening shift.
[c] N: Night shift.

Suitable shift systems?

Two sets of criteria exist by which to judge the suitability of shift plans: they concern the worker and the organization. The human is by nature used to daylight activity, with the night reserved for rest. Working evenings or during the night, and possibly working on weekends and holidays, generates physical and psychosocial stress, which may be light or severe, depending on the circumstances and the person. Organizational criteria include the number of shifts per day, the length of every shift, the coverage of the day and week by shifts, work on holidays, and the performance of work. Obviously, any shift work schedule is a compromise of various, occasionally contradicting, expectations.

Consideration of the following points is part of the decision to select one of the many possible shift plans.

- Daily work duration normally should not be more than 8 hours.
- The number of consecutive evening or night shifts should be as small as possible; it is best to intersperse only one single late shift between day shifts. (The alternative solution is to stay permanently on the same late shift.)
- A full day of free time should follow every single night shift.
- Each shift plan should contain consecutive work-free days, preferably including the weekend.

During evening or night shifts, management should provide high illumination levels at the workplace to suppress production of the hormone melatonin which causes drowsiness. Furthermore, environmental stimuli can help to keep the worker alert and awake, such as occasional stirring music and provision of snacks and (possibly caffeinated) beverages. Interaction with co-workers, even taking of short naps should be encouraged as the type of work allows. The work task should be interesting and rewarding, because boring and routine tasks are difficult to perform efficiently and safely during the night hours.

The shift worker can use coping strategies for setting the biological clock, obtaining restful sleep, and maintaining satisfying social and domestic interactions. For example, sleep should be taken directly after a night shift, not in the afternoon. Sleep time should be regular and kept free from interruptions. Shift workers should seek to gain their family's and friends' understanding of their rest needs. Certain times of the day should be set aside specifically and regularly to be spent with family and friends.

Summary

Human body functions and human social behavior follow circadian rhythms. Shift work required at unusual times can put these synchronized rhythms and the associated behavior of sleep (usually during the night) and of activities (usually during the day) out of order. Resulting sleep loss and tiredness usually have negative health effects on the shift worker, diminish family and social life, and reduce work performance.

Shift work is often desired for organizational or economic reasons. Rules for acceptable regimes of shift work include:

- Job activities should follow entrained body rhythms.
- It is best to work during the daylight hours.
- Evening shifts are preferred to night shifts.
- If evening or night shifts are necessary, two opposing recommendations apply:
  - Either work only one evening or night shift then return to day work; or
  - Stay permanently on the same shift, whichever that is.

Task performance is influenced by four major interrelated factors: the internal diurnal rhythm of the body; the external daily organization of work activities, subjective motivation and interest in the work, and type and conditions of work. Physical and mental performance are particularly weak during "low" periods of the circadian cycle such as in the early morning hours. Performance deteriorates during long-continued work, especially when accompanied by sleep loss. Deterioration can be counteracted by interesting and rewarding work, interaction with co-workers, breaks (naps), snacks and beverages, and by environmental stimuli such as bright illumination.

One long night's sleep usually restores performance to a normal level, even after extensive sleep deprivation.

Fitting steps

Step 1: Attempt to work only with day shifts.

Step 2: If two or three shifts a day are necessary, attempt to establish permanent assignment by requesting volunteers.

Step 3: If shift rotations are necessary, carefully plan schedules to suit workers' preferences.

Further reading

HORNBERGER, S., KNAUTH, P., COSTA, G., and FOLKARD, S. (EDS.) (2000). *Shiftwork in the 21st Century*. Frankfurt: Lang.

MONK, T.H. (2006). Shiftwork, in Marras, W.S. and Karwowski, K. (Eds.) *The Occupational Ergonomics Handbook* (2nd Edn) *Fundamentals and Assessment Tools for Occupational Ergonomics,* Chapter 32. Boca Raton, FL: CRC Press.

Notes

The text contains markers, °, to indicate specific references and comments, which follow.

16.1 Need for Sleep to Be Awake:

These natural temporal programs keep our body and mind healthy: Folkard et al. 1985.

Why we need sleep is not clear: Horne 1988; Siegel 2003.

16.2 Performance and Health Considerations:

Importance of REM and non-REM sleep: Horne 1988.

Napping: Driskell et al. 2005.

16.3 Organizing Shift Work:

Important shift features: Other identifiers of shifts and shift patterns are the starting and ending time of a shift; the number of workdays in each week; the hours of work in each week; the number of holidays per week or per rotation cycle; the number of consecutive days on the same shift (cycle length), which may be a fixed or variable number; the schedule by which an individual worker either works or has a free day or free days; the number of free days per week and year; the number of shift teams. For more details, read Costa 1996; Folkard et al. 2003; Knauth 2006; Tepas 1999.

Human engineering

The previous sections of this book dealt with specific categories of knowledge about the human body and mind, abilities and limitations. In many cases, the discussions also concerned effects of the environment on the human ability to perform.

This last section of the book assembles detailed knowledge, taken from the foregoing chapters in this book, and applies it to several examples of intricate multipart designs, such as of homes and workplaces. These are typical cases of human engineering (occasionally called human factors engineering), which require gathering and using information from many special knowledge areas:

- Chapter 17 Designing the home
- Chapter 18 Office design
- Chapter 19 Computer design
- Chapter 20 Workplace design
- Chapter 21 Making work pleasant and efficient

CHAPTER SEVENTEEN

Designing the home

A home suitable for living in Norway would not serve its purpose well in Egypt; a community house from Fiji would look strange in Scotland. Different house designs provide shelter from different climates. The inhabitants may have rather divergent opinions about comfort and privacy. Accordingly, there is not just one type of dwelling that is right from the ergonomic point of view; instead, there are many good solutions, depending on climate, availability of materials, and societal expectations.

The following text discusses human engineering solutions for homes suitable for moderate climates. Some of the ideas to make living easy and safe apply universally but others are restricted to the western architectural traditions originally developed in Europe.

The main purpose of having a home is to be sheltered from unpleasant features of the environment. This achieved, other attributes come to mind, including privacy, safety, comfort, convenience, pleasure, and aesthetics. Different inhabitants are likely to have dissimilar opinions about the importance of these attributes, but from an ergonomic point of view, usability of the design of the home is very important. For young and healthy persons, rather unusual designs such as spiral staircases and lofts might be quite attractive, but such features are laborious to maintain and difficult to use, even dangerous for pregnant women and children, and for elderly and impaired persons.

17.1 Designing for mother and child

Child-proofing

Most homes serve, usually for many years, as a safe haven for children and their mothers. This requires the design of the interior to be child-friendly, mostly in terms of safety for young children. Examples of "child-proofing" are not having stairs, or at least blocking them off; not having protruding hard and sharp objects; avoiding hot items that can burn skin, such as a stove surface in the kitchen, or scalding-hot water in the bath, all measures that make the home safe for everybody else.

Design for pregnancy
During pregnancy, many everyday tasks become more difficult for the mother-to-be. Most of the difficulties derive from reduced reach and mobility, often related to back pain. Owing to the increasing bulk of the trunk with pregnancy, objects on the ground near the feet are hard to see, and stumbles and falls are feared dangers, especially if the sense of balance is affected. Frequent urination is a common symptom of pregnancy, which requires convenient access to a toilet and washroom.

With advancing pregnancy, in general the ability to perform work that requires great exertion extending over long periods and involving a great deal of mobility, such as low bending and far reaches, decreases. Typical are the following responses to a survey in which pregnant women listed their most difficult activities.

- Picking objects up from the floor
- Walking upstairs
- Driving a car
- Getting in and out of a car
- Using seat belts in a car
- Ironing
- Reaching high shelves
- Getting in and out of bed
- Using public toilets

Overall, the answers (not counting those referring to car use) indicate that proper overall layout of the home, and the specific design measures already mentioned, contribute to achieving a design that helps pregnant women and suits all inhabitants°.

17.2 Designing for impaired and elderly persons

During a resident's illness or recuperation from an injury, the home serves as a temporary care facility, which may have to accommodate a wheelchair, possibly even a gurney. Such use makes thoughtful layout and detail design highly important, especially of the bedroom, toilet and bath, and hallways. The same carefully incorporated design features also accommodate aging occupants° who are losing some of the physical and mental abilities that they had in younger years. Continuing to live in one's own home has the major advantage of being in a familiar setting with all its physical and emotional implications. These include feeling at home, being comfortable, enjoying privacy, and having the satisfaction of self-sufficiency and independence.

Remodeling

Unless by happenstance or foresight private homes are well designed for the aging resident, alterations are usually necessary to passage areas, kitchen, bathroom, and bedroom. Taking future needs into consideration when building or acquiring a house can save much effort and huge expenses because the reworking of an existing residence can be substantial and costly.

17.3 Access, walkways, steps, and stairs

In the United States, perhaps the best-known house expressly built for use by a person with restricted mobility is the Top Cottage in Hyde Park, New York, which President Roosevelt designed to accommodate his wheelchair. Fortunately, most persons can walk about freely, but designing a house for a wheelchair user will certainly make it convenient for every inhabitant, even if disabled, including aging persons who are no longer as agile and powerful as in their youth. Thus, a—real or imaginary—wheelchair is a good instrument to assess the suitability of passageways.

Passageways

Passages to and from a dwelling and within it must be safe and comfortable to use even for a frail person. The walking surface should be flat, without barriers such as stairs or thresholds, and best not sloped. Doors and passageways should be wide enough to allow a wheelchair to pass and turn, and flooring should provide enough friction for safe stepping even when wet. Straight hallways are easier to pass through than passages with turns and corners. Passages must be well illuminated, as should be all other rooms of the dwelling.

No steps

Flights of stairs, steps, and thresholds often make it impossible for wheelchair users to roll up or down; they make moving difficult and cause many falls, often resulting in serious injuries.

Even low sills at doorways or shower stalls can cause stumbling, as do the rims of carpets and loose rugs.

Lifts and elevators facilitate moving about if a residence spans several stories. Elevators connecting the ground floor with the next one are fairly simple and relatively inexpensive to set up, especially when the dwelling is planned to accommodate them and installation done early during construction or remodeling°.

Doors

Doors (and windows) must be easy to open and close, even when an additional insect screen or storm door is present. They require clear space in front to provide access. All exterior doors should open to the outside to allow a quick escape in case of a fire. Controls must be handy and require little strength to operate yet provide security. Push bars and lever handles are easier to operate than round knobs. A transparent or translucent curtain can replace some interior doors.

17.4 Kitchen

Work flow in the kitchen

The kitchen is one of the most frequently used and important rooms of the house for many people. This is often a gathering room, a social place, and a communication and message center; although, basically it is the location to prepare, serve, and store food. The idea of an efficient "work triangle" underlies many design concepts°. Its corners are the three areas of primary activity: storing (refrigerator, cabinets, etc.), preparing/cooking food (gas or electric burners, conventional or microwave oven, coffee/tea maker), and cleaning (sink, dishwasher, garbage disposal). Several classic human factors principles, augmented by newer ergonomic findings, apply to kitchen design:

1. Kitchen design and components should facilitate the workflow at the most used activity areas, among these areas, and for serving the prepared food and cleaning used utensils.

2. Items should be stored near the point of use.

3. Cabinets and other storage facilities should be within comfortable reach. The contents should be in sight; therefore, shelves should be at or below eye height and not so deep that items located in front obscure those at the back.

4. Items on storage shelves and in cabinets should be easy to reach. For this, shallow shelves are beneficial.

5. Retrieving items should be easy, not requiring excessive body bending, stretching, or twisting. Rollout shelves are advantageous at low height.

6. The workspace for the hands, at counters and sinks, should be at about elbow height or slightly below. This facilitates manipulation and visual control. Counter and sink may be put lower than usual for wheelchair-bound persons.

7. Stove, oven, refrigerator, and dishwasher openings should be at no-bend heights.

8. All counters should have a work surface that extends into the room, or at least a recessed toe space, so that one can step close in.

9. Traffic of others should not cut through the patterns of workflow of the work triangle.

17.5 Bedroom, bath, and toilet

Bedroom

Most of us stay about one third of the day in the bedroom, and weak or ill persons spend even more time in it. Hence, it is important to pay attention to its ergonomic features. The bed should be at a height that makes lying down and getting up comfortable. In the past, many different mattress properties have been promoted, especially for people with back problems, ranging from hard through pliant to soft; when no objective criteria appear convincing, users select the support that pleases their individual preferences.

The bedroom must be spacious enough to allow maneuvering space. It should supply easily reachable shelving and hanging for clothing, linen, and bedding. It should contain direct-access storage for medical supplies. It should allow emergency access and have an emergency exit. As a rule, the bedroom must provide privacy and be near a bathroom.

Bathroom

Bathrooms deserve particular ergonomic attention because they are essential for healthy living. Basic equipment includes bathtub, shower, toilet, and washbasin. Additionally, bathrooms usually contain storage facilities for toiletries, towels, and other supplies. Unfortunately, many traditional bathroom designs in the United States are difficult to use for aged people and persons with disabilities. Major problems are narrow doors and tight space so that users who need canes, walkers, and especially wheelchairs find it difficult to move about.

Tub and shower   Bathtub and shower, the two main areas for cleansing the whole body, are sites of numerous accidents. Most danger stems from the slipperiness of wet surfaces against skin, including bare feet. The most hazardous is the bathtub with slanted slick surfaces. Arising from it and stepping over its high sides are not easy for most people and particularly difficult for older people who have balance and mobility deficiencies. Proper handrails and grab bars within easy reach can be of great help when getting in and out. In spite of the use difficulties, many people prefer a conventional tub to a shower.

A shower stall is easier to use because its low enclosure rim at the floor poses little hindrance to walking in and out. A rimless design usually takes more floor space but is the best solution for a wheelchair user.

Control handles   Using the control handles for hot and cold water is often difficult for impaired persons, particularly when they have to reach across a tub or shower basin to access them. In some setups, the controls for hot and cold water move in the same direction whereas in other designs faucets turn in opposite directions; standardization of human-engineered solutions would be advantageous. To prevent scalding, thermostatically adjusting the water temperature is a good ergonomic solution, helpful to all users.

Washbasin   Proper height of the washbasin is important. The basin may be difficult to use if it is too far away such as when inserted in a cabinet so that one cannot step close to it but must lean forward over it. Big faucets often reduce the usable opening area of the washbasin.

Toilet   The toilet obviously is of great importance for elimination of alimentary wastes while keeping the body clean. The literature contains information about suitable toilet design, sizing, shaping, height, and location for western-style toilets. Throughout the world, however, different customs and installations prevail for which few ergonomic recommendations are available. Proper handrails and grab bars aid persons with mobility problems, such as caused by back pain, in sitting down and getting up. Personal hygiene systems installed in toilets are often of great help, as are self-cleaning provisions and any other features that facilitate maintenance.

17.6 Light, heating, and cooling

Windows are important to many persons: they are visual and emotional connections with the outside and, furthermore, they provide light to the inside of the house. If such natural light fails, electrical lamps supply illumination, so arranged that they light up what must be seen, yet not producing deep shadows or glare that can blind the eye.

Automatic lights are advisable in passageways, bathrooms, and bedrooms. Manual switches and all other controls as well as electrical outlets should be located at about hip height so that persons both standing and sitting can reach them naturally. They should be easy to operate, at best by a simple push, and not require fine fingering. *Many table and floor lamps, often found in North America, provide impressive examples of how to make it difficult, instead of easy, to turn lights on and off: they have push–pull or turn switches at the socket of the light bulb, hidden high up inside the lampshade so that one cannot see them. Due to their location, they are hard to find by groping and they are tricky to operate due to the required arm bending and the awkward switch design.*

Automatic settings for climate control are desirable because they require no judgment, decision, or action by the person. Many people prefer a floor-heating system with its uniform warmth to the often drafty and loud forced-air arrangement common in North America.

17.7 Home office

Office at home

Changing work practices usually combine with the development of new work technologies. A striking example is the appearance of the so-called home office. Modern computers, stationary or portable, even handheld; fax machines and scanners; and especially mobile phones have enabled many people to work from home, part-time or full-time, instead of commuting daily to an office building. Setting up an office workplace in the home often starts with occasional use of the kitchen table, of a space in the den, or of a spare room in the home, to work there: talking with colleagues and customers, writing letters and memos, producing brochures and articles, drawing and inventing, ordering wares, and sending bills. Use of wireless electronics, even when traveling, has made working away from the company building widespread.

The more frequent and intensive working at home becomes, the more important it is to apply ergonomic thinking to the layout of the home office. Wherever the office workplace, the same human engineering principles apply:

1. Provide furniture that comfortably supports body movements and postures.

2. Provide work tools and equipment that facilitate execution of the tasks without overloading human capabilities, particularly regarding repetitive movements.

3. Provide lighting at the workplace that is suitable for the task, which does not produce direct or indirect glare.

4. Provide a suitable thermal climate and acoustic environment.

5. Provide comfortable and appealing work surroundings.

Chapter 18, next in this book, deals with these issues in detail.

Summary

Laying out and arranging a dwelling must be one of the oldest tasks for humankind. One should think that modern design procedures and building techniques incorporate, at a basic level, knowledge on how to design a home to accommodate the inhabitants in terms of ease of use and safety, especially when they are not agile because of pregnancy, age, or disability—however, such human factors design is not common in the European and American tradition.

Numerous publications provide advice on how to design homes that suit pregnant women and children and accommodate disabled and older persons. This chapter summarizes much of this ergonomic information and applies it primarily to the topics of access, walkways, steps and stairs; lighting, heating, and cooling; and to rooms of special importance, namely the kitchen, bedroom, bath, and toilet.

Notes° and more information

The text contains markers, ° , to indicate specific references and comments, which follow.

17.1 Designing for mother and child: See Kroemer 2006, 2006b; Kroemer et al. 2003; Lueder et al. 2007.

17.2 Designing for impaired and elderly persons: For references to architectural and ergonomic design guidelines see Kroemer 2006.

17.3 Remodeling: Richard Atcheson describes renovation needs of a home to accommodate older residents in vivid detail in his article "Our Old House" in *Modern Maturity* (November–December 2000, **43R**: 6, 62–71. (Reprinted in Kroemer 2006.)

In the United States, the American Association of Retired Persons (AARP, http://www.aarp.org) provides free brochures on age-friendly buildings and modifications of habitats.

17.4 Kitchen: The "work triangle" in the kitchen initially was the result of motion-and-time studies done by Lillian Gilbreth and her husband Frank in the 1920s.

Office design

One of the best-known office buildings is that of the *Uffici* (Offices) in Florence, Italy, finished in 1581. It forms the shape of the letter U around an open court so that every room has a window to the outside, providing natural light and ventilation. Heating was done, if needed, by coal fires and stoves in the rooms, and bad odors abounded when the windows had to be kept closed. Into the 1900s, large office buildings, and the arrangement of the offices within, still followed that example. Then, new technology in lighting and air-conditioning allowed radically different designs.

The layout, equipment, and decor of our office can affect our quality of life; in addition, well-being can also appreciably influence our productivity at work. Consequently, there are both ergonomic and economic reasons to design offices to truly fit their human occupants so that they can work efficiently and effectively.

18.1 Office rooms

Office designs can vary from open plans to closed, walled-in individual rooms. Open plans include the stereotypical "paperwork factory", taking up a vast room filled with straight rows and columns of desks and chairs. Figure 18.1 depicts an example of such a circa-1990 layout with computers on the desks.

Office landscape versus individual offices

Often, low partitions subdivide the large space into smaller semiprivate cubicles. In spite of their popularity as the target of jibes and jokes, if done appropriately, partitioned cubicles can indeed help provide some privacy to employees within vast

FIGURE 18.1   Computerized paperwork factory, circa 1990

open spaces. At the other extreme of office layout are the separate, walled-in offices shared by a few employees or used by just one person, usually of some rank. In reality, many office buildings house various designs; they have some areas that feature open plans, and other departments or floors within the company may have contained spaces.

Office landscape

The idea of an "office landscape" originated in the 1960s. The basic concept was to utilize a large office space, landscaped to create the appearance of an irregular terrain similar to one in a park or large garden by arrangements of furniture, office machines, dividers, flowers, bushes, and small trees. This often resulted in rather pleasant office environments that could easily accommodate changes in interior arrangements. One popular draw was the use of plants, partly for their pleasing appearance and partly for their assumed abilities to prevent noise propagation and to absorb air contaminants; unfortunately, they do next to nothing in either respect. Still, they can be pleasing to the eye: aesthetics are important.

Disadvantages of open designs include lack of privacy, which is an issue when business requires the exchange of confidential information. Many organizations use a mixture of office designs: wide open spaces here, sections with shoulder-high cubicle dividers there, and closed-off individual offices in other parts. Another clear drawback of the wide office landscape is disruptive noise—from people and equipment—which may reduce performance (see Chapter 6) and job satisfaction.

Stepwise office design

The process of office design or remodeling involves several steps. The first step is to evaluate the needs of the people who will inhabit the office, the tasks they will perform, the work tools and equipment they will use, and their preferences and work styles. Specific statements of these functional requirements guide the actual design.

The next step is to identify a range of design options from which practical solutions can be chosen; this usually involves some compromises. The third step involves evaluating the candidate designs (more on this below). The last step is to select the final design, implement it, and put it to use. Employees should be involved in all phases of the process so that their needs are truly met and they do not feel manipulated or ignored.

Evaluating different designs

An example procedure to evaluate several candidates for a new (or updated) office uses four designs, Layout #1 through Layout #4. These candidate designs are to be evaluated by a number of criteria:

- Construction cost (Constrcost)
- Ability to expand when needed (Expand)
- Running costs (Runcost)
- Appeal to employees and clients (Appeal)
- Expected efficacy and effectiveness of use (EfficEffec)
- Time that is expected to pass until the chosen Layout can be occupied (Availability)

A panel of experts that includes managers and employees can do the scoring. Assume the raters gave the scores (with 1 worst and 10 best) listed in Table 18.1.

Rating scores

According to the raw scores, office Layout #2 is the preferred solution and Layout #4 a strong contender. However, applying weights to the rating changes the outcome: with weighted criteria, office Layout #4 becomes the preferred solution, with design #1 not far behind. Definition of criteria for judging the candidate designs, including their weighting, eliminates any irrational decision such as the muddled, "I like this better." The process outlined above is logical and transparent and forces all decision makers to follow the same rules.

Flexibility via new technology

Current architectural, engineering, and construction technology allows designing almost any office environment one might wish. Yesterday's technology tied us to one office, and often to one place therein, because we were bound by wired phones and computers, cabled machines, and stationary equipment; now, most of us can be flexible in location, work schedules, and habits.

Table 18.1 Raw and weighted scores of four candidate office designs

	Raw scores			
	Layout #1	Layout #2	Layout #3	Layout #4
Constcost	3	1	5	10
Expand	4	10	6	1
Runcost	9	10	1	5
Appeal	7	2	1	10
EfficEffec	8	10	5	1
Availability	1	5	8	10
Total	*32*	*38*	*26*	*37*
	Weighted scores			
	Layout #1	Layout #2	Layout #3	Layout #4
Constcost (×8)	24	8	40	80
Expand (×3)	12	30	18	3
Runcost (×10)	90	100	10	50
Appeal (×7)	49	14	7	70
EfficEffec (×5)	40	50	25	5
Availability (1)	1	5	8	10
Total	*216*	*207*	*108*	*218*

Source: Adapted from Kroemer, K.H.E. and Kroemer, A.D. (2001) *Office Ergonomics.* London: Taylor & Francis.

Looks good, feels good, makes for good work

Pleasing office aesthetics convey an appealing image of the organization; they contribute to attracting new employees and help in retaining them. However, to generate true and lasting positive attitudes of the workers requires more than just facility design: it necessitates treating persons as important contributors to the organization's aims; caring about employees' well-being, on and off the job; and understanding people's concerns about their professional future. In many successful organizations, employees feel they are part of a "family" with extensive interactions among all levels of the corporate hierarchy and a strong prevailing team spirit. They appreciate a supervisor's advice to stay home when they feel sick, to come in late or leave early for important personal reasons, and to switch the mobile phone off and leave the portable computer at home when going on vacation. Being personally and professionally appreciated is very important and makes employees feel engaged with their employer. People also enjoy job amenities that provide convenience, which can create an almost home-like atmosphere at the office. These features go well beyond the usual pleasant employee cafeteria and may include free brand-name coffee bars, fitness rooms, on-site laundry and dry cleaning facilities, the ability to shop and get haircuts and—very important to young parents—on-site childcare. These factors work together to increase employees' job satisfaction, to develop emotional ties to the company providing the job, and hence to enhance job performance. This is of importance to the employer, because primarily the office is a place to work and perform.

18.2 Office climate

The physical environment in the office, in terms of climate, lighting, and sound, generates conditions that may affect the worker's health, comfort, and the ability and willingness to perform. It is the task of office design to provide the best possible conditions and, certainly, minimize any unavoidable adverse effects of environmental factors.

Control of the climate is important because it influences how the people in the office feel and execute their tasks. Climate control (especially of temperature and humidity) influences our well-being and performance. In our work environment, the climate should be neither too hot nor too cold and neither too damp nor too dry; we also want fresh rather than stale air but the airflow should not generate a strong draft.

Working when warm or cold

We are not cold-blooded°, neither in the emotional sense nor physically; instead, our body maintains a rather constant temperature of about 37°C in the brain and chest cavity, regardless of the outside climate. Achieving a constant core temperature is a complex task for the human body's temperature control system, as discussed in some detail in Chapter 8 of this book. The body generates heat energy, and at the same time exchanges energy with the environment. When our office surroundings are cold, we want to prevent excessive heat loss, which we mostly do by choosing suitable clothing. When our environment is warm and it warms up our bodies as well, we must get rid of body heat to prevent overheating. This is rather difficult to do, because we cannot reverse the natural heat flow, which is always from the warmer to the colder. The only biological way to cool our body in a hot environment is to evaporate the water contained in sweat on our skin—but the resultant odor is unpleasant for all in the office. So, we resort to technical means to create that needed outward heat flow; the most complete (and also most expensive) solution is climatizing the office environment: controlling its temperature, humidity, and air movement.

Feeling comfortable

For our well-being and comfort, the temperature difference between exposed skin and the environment is very important, but humidity and airflow also play major roles, as do our attitude and level of acclimatization. If our body's core temperature were to deviate about 2 degrees from its 37°C norm, our body's functions and resultant task performance would suffer severely. However, in a sheltered office environment, the human thermoregulatory system has no problem keeping the core close to constant 37 centigrades. At the skin, in contrast, there are often large temperature differences in various regions. For example, at the same point in time, the toes may be at 25°C, legs and upper arms at 31 degrees, and the forehead at 34 degrees: for most of us, this feels comfortable.

Energy exchanges with the environment

In physics terms, heat energy exchange between our body and the environment takes place via four pathways: by convection, conduction, evaporation, and radiation (see Chapter 8 for more details). In each case, heat irreversibly flows from the warmer to the colder medium.

Convection and conduction can cool or warm

The same thermodynamic rules govern heat exchange by convection and conduction. In both cases, skin contact makes for heat transfer: with air (or fluid) in convection, or with a solid in conduction. The amount of transferred heat depends on the area of human skin that participates in the process, and on the temperature difference between skin and the adjacent layer of the outside medium. An air draft, caused by an open window or a fan, can increase convective heat exchange as the air moves quickly along the skin surface, which helps maintain a temperature (and humidity) differential. Wood and some plastics on office furniture feel warm because their heat-conduction coefficients are below the coefficient of human tissue; so, they keep the warmth close. Metal of the same temperature accepts body heat easily and conducts it away; therefore, it feels colder than wood. If our skin is colder than the air and what we touch, then we gain warmth.

Evaporation makes the body cooler

Heat exchange by evaporation is in only one direction: the human body loses heat. (There is never condensation of water on living skin, which would add heat.) Some water evaporates in the respiratory passages but most appears as sweat on the skin. It requires energy to evaporate sweat water: about 2440 J (580 cal) per cm^3. This energy is taken mostly from the body and hence reduces the heat content of the body by that amount. Hence, evaporating sweat is very effective to cool the body and functions even in a hot environment. Dry air accepts water vapor more readily than humid air; therefore, a breeze feels good on skin because it replaces the humid air there by drier air, which facilitates heat loss through evaporation. There is always some perspiration and hence sweat evaporation going on: that is why our clothes smell when worn too long.

Radiation may cool or warm us

Radiation exchanges heat between two opposing surfaces, for example, between a windowpane and a person's skin. Hence, the human body can either lose or gain heat through radiation, depending whether the skin is warmer (*loss*) or colder (*gain*) than the other surface. The amount of radiated heat depends on the temperature difference between the two surfaces, and on their sizes, but not on the temperature of the air between them.

Heat balance

Primarily, the actual amounts of heat exchanged with our surroundings depend, directly or indirectly, on the difference in temperature between participating body surfaces and the environment. Secondarily, the magnitude of exchanged heat depends on the surface areas that participate; clothing determines how much skin is exposed, hence has a great effect. (Clothing also has other heat transfer properties, such as insulation; see Chapter 8.) Humidity plays an appreciable role when it is very high or very low.

Over time, we feel well in a climate when our body can achieve heat balance, allowing the body to keep its temperature constant at a comfortable level. This occurs when the metabolic energy developed in the body is in equilibrium with the heat exchanged with the environment. The healthy body achieves this by (among other actions) responding to a cold environment by making its skin surface colder, and to a hot environment by making the skin warmer.

Acclimation

We can help to achieve this state of balance by deliberately dressing more lightly in the heat or conversely, when our surrounds are frigid, by covering up in insulating layers.

Adjusting how we dress is partly a social action and hence intentional, but it is also a part of unconscious acclimating: getting used to a hotter environment as summer comes, and getting ready for colder conditions as fall arrives. Continuous or repeated exposure to hot and (not as pronounced) to cold conditions brings about a gradual adjustment of body functions, resulting in a better tolerance of climate stresses. Acclimatization to heat shows increased sweat production, lowered skin and core temperature, and reduced heart rate, compared with a person's first reactions to heat exposure. The process is fast: full acclimatization is achieved within about 2 weeks. A person who is healthy and fit acclimates more easily than a person who is in poor physical condition.

There are no striking differences between females and males with respect to their ability to adapt to changing office climates. Any individual tendencies to feel too warm or too cool can be overcome easily by adjusting clothing habits.

Effects of heat on mental performance

It is difficult to evaluate the effects of heat (or cold) on mental or intellectual performance. However, apparently the mental performance of a nonacclimatized person deteriorates when room temperatures rise above 25°C. That threshold increases to 30 or even 35 degrees if the individual has acclimatized to heat. Brain functions are particularly vulnerable to heat; keeping the head cool improves the tolerance to elevated deep body temperature. A high level of motivation may also counteract some of the detrimental effects of heat. These results stem from laboratory tests; in the office, such high temperatures are rare.

Good office climate

There are several technical ways to generate a thermal environment that suits the physiological needs of people as well as their individual preferences. The primary approach is to adjust the physical conditions of the climate (temperatures, humidity, and air movement), usually done by automatic air-conditioning. ASHRAE recommendations in the United States, international standards by ISO, and national regulations and regional recommendations and customs provide guidance°. The WBGT discussed in Chapter 8 is most often used to assess the effects of warm or hot climates on the human.

If there is no way to control the climate in the office automatically, several time-honored means are at hand to counteract heat:

- Move air swiftly through the room. (This may require old-fashioned paperweights.)
- Stay away from warm surfaces, such as windowpanes, that radiate heat.

- Cool the air and hot surfaces by sprinkling water on warm surfaces.
- Rest during the hottest time of the day. (The "siesta" is an example.)
- Dress lightly.
- If you are cold, dress more warmly, get close to warm radiating surfaces, sit in the sun.

18.3 Office lighting

Lighting affects us strongly in our office work. We like an office environment that allows us to see clearly and vividly what we want to see, which pleases us in terms of contrast and colors. Many individuals and companies value natural light that enters the office through windows because it provides a spacious and natural feel to the workplace. However, to the lighting engineer the disadvantages of daylight count as well. The natural illumination changes over time, and although spots near the windows can be lit glaringly, workplaces in the rear may be too dark. Well-planned artificial lighting overcomes the disadvantages of daylight. "Task lights" lamps at the workstation, controlled by the worker, allow fine adjustments of local illumination.

Lighten the office

One specific task encountered when setting up office lighting is to prevent glare and annoying bright spots in our visual field. Accountants usually like the use of direct lighting where rays from the source fall directly on the work area. This is most efficient in terms of illuminance gain per unit of consumed electrical power, but direct light can produce high glare, poor contrast, and deep shadows. The alternative is to use indirect lighting, where a surface, often the ceiling of the office, reflects the rays from the light source in many different directions so that diffused light reaches the work area. This helps to provide an even illumination without shadows or glare, but is less efficient in terms of use of electrical power. A compromise is to use a large translucent bowl that encloses the light source and scatters the light before it is emitted from the bowl's surface. This can cause some glare and shadows, but is usually more efficient in terms of electrical power usage than indirect lighting. Figure 18.2 shows these kinds of room lighting.

Glare-free lighting

Glare is the experience of intense light that enters the eye and overpowers the ability of the cones and rods in the retina to

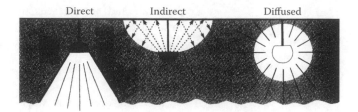

FIGURE 18.2   Direct, indirect, and diffused lighting. (Adapted from Kroemer, K.H.E. and Kroemer, A.D. (2001). *Office Ergonomics*. London: Taylor & Francis.)

FIGURE 18.3   Indirect and direct glare. (Adapted from Kroemer, K.H.E. and Kroemer, A.D. (2005). *Office Ergonomics*, authorized translation into Korean. Seoul: Kukje.)

distinguish shades of gray and colors (see Chapter 5). An example of direct glare is light from a lamp shining straight into your eyes, whereas indirect glare occurs when the light rays are reflected from a shiny surface and from there enter the eye.

Figure 18.3 sketches these conditions. Strong glare temporally disables vision, whereas weak glare acts as a veil in the field of vision, making it difficult to see slight contrasts.

Avoiding glare   Placing the computer in front of a window can make it difficult to discern details on the screen when the strong light from the outside overpowers the ability of the eyes to see subtle images at the workplace. Figure 18.4 shows this example of direct glare and Figure 18.5 illustrates indirect glare coming from a window or lamp. Obviously, repositioning the workplace so that the light sources are on the worker's side remedies the glare situations, and reducing the intensity of the incoming light (turn off the lamp; draw a curtain across the window) would be helpful as well.

The top part of Figure 18.6 shows lamps arranged so that they provide glare-free lighting: locations to the left and right

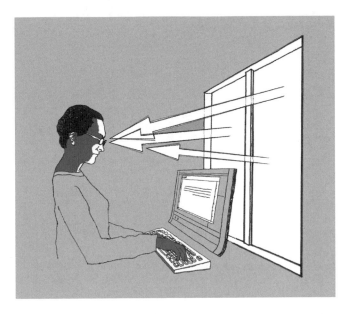

FIGURE 18.4 Direct glare in the office. (Adapted from Kroemer, K.H.E. and Kroemer, A.D. (2005). *Office Ergonomics*, authorized translation into Korean. Seoul: Kukje.)

FIGURE 18.5 Indirect glare in the office. (Adapted from Kroemer, K.H.E. and Kroemer, A.D. (2005). *Office Ergonomics*, authorized translation into Korean. Seoul: Kukje.)

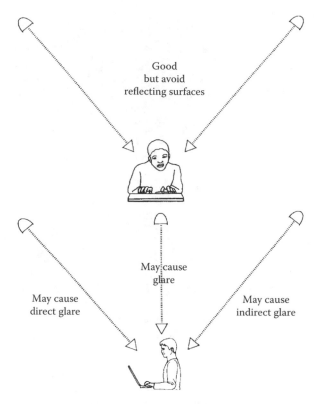

Good
but avoid
reflecting surfaces

May cause
glare

May cause
direct glare

May cause
indirect glare

FIGURE 18.6   Placement of lamps. (Adapted from Kroemer, K.H.E. and Kroemer, A.D. (2001). *Office Ergonomics*. London: Taylor & Francis.)

sides of the operator are not likely to cause indirect glare. However, placing a lamp or bright window in front of a person can cause direct glare. To avoid indirect glare, do not place a light source behind a person because the light could be mirrored on the display surface.

We need light to see

We can see an object only if it sends light that strikes the cones and rods located on the retina at the rear of our eyes. In darkness, no visible light reaches our eyes and we see nothing. When our surroundings become dimly lit, some light reaches the retina and we begin to recognize shapes of objects, which appear in shades of gray. As we receive ever more light from our environment as it gets lit more brightly, cones begin to report colors and details become recognizable. When the surrounds are in full daylight or under equivalent artificial lighting, our cones and rods fully engage and we see all colors, detect fine details, and can read even small text on paper or an electronic screen.

Photometry   We see details well if there is strong luminance contrast between the visual target and its background. *Luminance* is usually understood as the portion of illumination (coming from the sun or an artificial light source, a lamp or luminaire) that is reflected from a surface. However, self-luminous electronic displays, such as computer monitors, present surfaces to the eye that have optical properties similar to the luminance coming from a reflecting surface.

Traditional photometry° commonly uses measurements of incoming light, illumination, to describe the lighting conditions of our visual environment. One of the main reasons for this is the ease of measuring illumination even though luminance is more relevant for our vision.

Recommended office illumination   The optimal lighting conditions for good vision depend on many factors including the task at hand, the objects to be seen, and the conditions of the eyes to be accommodated. General office illumination is usually at about 500 lx but, if there are many dark (light-absorbing) surfaces in the room, it may be up to 1000 lx. If cathode-ray tube (CRT) displays are present, the overall illumination should be between 200 and 500 lx. In rooms with light-emitting displays, illumination of 300 to 750 lx is suitable. The low illuminances may be a bit dim, especially for such tasks as reading text on a paper: it can be helpful to turn on a task light directed at the visual target, which generates more luminance there without appreciably raising the overall lighting level in the office. Care must be taken, however, to avoid glare by shining that light, directly or by reflection, into one's eyes.

Easy things to do   Often, some easy actions can help to alleviate vision-related annoyances, which make us uncomfortable while working at the computer: these simple actions are listed in Table 18.2.

18.4   Office furniture and equipment

Sitting at work   The concept of sitting has different meanings: In many regions on earth, this does imply perching atop a stool or chair, as became a tradition in ancient Egypt and Greece. But in other civilizations, people habitually rest their bodies on a carpet, mat, cushion, or on the bare floor by kneeling, squatting, or sitting with the legs extended, or having a foot or both feet underneath the body.

Table 18.2 Steps to alleviate eye problems resulting from computer use

- *Stop and blink*: When you are working at the computer or reading, pause frequently—say, every 15 minutes — to close your eyes, or gaze away from the screen or page, and blink repeatedly.
- *Take a break*: Every 30 minutes or so, certainly every hour, get up and move about.
- K*eep your distance*: Position your eyes at the same distance from the screen as you would from a book. If you find that uncomfortable, buy a pair of glasses with a prescription designed for computer work, or use "progressive-addition" bifocals, which have gradually changing power from the top to the bottom of the lens.
- *Lower the screen*: Keep the top of the screen below eye level. Gazing upward can strain muscles in the eye and neck.
- *Hold up the document*: Put a support for reading matter next to the screen, at the same distance from your eyes.
- *Clean up*: Clean your screen and your glasses. Dust and grime can blur the images.
- *Lighten up*: Age tends to cloud the lens of the eye and shrink the pupil, sharply increasing the need for luminance. So if reading text strains your eyes, consider installing brighter lights or at least moving the reading lamp closer to the page.
- *Cut the glare*: Position the reading lamp so that light shines from above your shoulder, but make sure it does not reflect into your eyes from the computer monitor. Do not do read or do computer work while facing an unshaded window. Wear sunglasses if you are reading outside, but if you feel that you need sunglasses inside, then the setup of your visual environment is faulty.

If these actions do not reduce the strain from computer work or reading or writing, have an optometrist or ophthalmologist check whether you need to start wearing glasses or to have your prescription changed.

Sitting still

It is tiresome to maintain any body position unchanged over extended periods of time. This includes sitting still for hours on end. We would like to move about, but in the computer office, we are doubly tied to our equipment: by our hands that must operate the keys, and by our eyes that must view screen, text, and keys. These two ties keep our hands and eyes and, through them, our whole body fixed at the workstation. We try to ease the effort of maintaining the rigid posture by sitting on a chair as comfortably as practicable, shifting and slouching as needed, but our body keeps telling us to get up and walk and move around for a while.

Erect standing and sitting

A century or so ago, the office clerks were men who stood at their desks, using ink to write letters and hand-printing entries in ledgers. By the middle of the 20th century, the work posture had changed from standing to sitting, and most office employees in subaltern positions now were females. The idea that erect sitting is healthy sitting had prevailed over standing upright, and office furniture was designed for this body position.

In the late 1800s, body posture had become of great concern° to some physiologists and orthopedists in Europe. In their opinion, the upright (straight, erect) standing posture was balanced and healthy whereas curved and bent backs were unhealthy and therefore had to be avoided, especially in youngsters. Consequently, straight back and neck with the head erect became the recommended posture for both standing and sitting and, logically, seats were designed to bring about such upright body position. But that cliché applied only to lowly employees; managers habitually enjoyed an ample armchair with high back and comfortable upholstery.

Comfortable sitting

The simplistic concept that sitting upright, with thighs horizontal and lower legs vertical, meant healthy sitting lasted, surprisingly, for about a hundred years. Finally, by the end of the 20th century it was generally accepted that people sit any way they like—see Figure 18.7—apparently because freely choosing and changing their posture makes them feel comfortable.

Sitting, as opposed to standing, is suitable when only a small workspace must be covered with the hands: this is typical for much of today's office work. Sitting keeps the upper body stable, which is helpful to execute finely controlled activities. Sitting supports the body at its midchapter and requires less muscular effort than standing, which is important when maintained over long hours. But the seat must support the body, feel comfortable in combination with the other office furniture and equipment, and be suitable for the work tasks. Office furniture should accommodate a wide range of body sizes, varying body postures, and diverse activities; it should further task performance; it should be appealing; and help make people feel well in their work environment.

The human body is made for change

Of course, there is nothing wrong with upright sitting or standing at one's own will, but maintaining an erect back requires tensioning of muscles, as Figure 18.8 demonstrates for sitting

FIGURE 18.7   People sit any way they want to sit

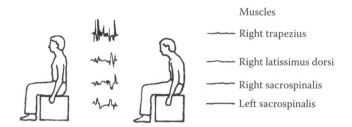

Muscles

——— Right trapezius

——— Right latissimus dorsi

——— Right sacrospinalis

——— Left sacrospinalis

FIGURE 18.8   Activation of back muscles during upright and relaxed sitting

without a backrest. Even when an upright backrest is present, it does not provide much support unless one presses the back against it, which requires muscle effort. Slumping in the seat and moving the body are instinctive actions to take strain and tension away from muscles that otherwise would be working to maintain prolonged postures. Sitting (or standing) still for extended periods of time is uncomfortable: it leads to compression of tissue, it can hinder blood circulation, and lead to accumulation of extracellular fluid in the lower legs. Obviously, the human body is made for change, to move about, as discussed in Chapter 2 earlier in this book.

Comfort and discomfort

Subjective judgments presumably encompass physiological phenomena (such as tissue compression and circulatory events) and incorporate emotional components as well. This becomes obvious when one tries to assess the feeling of comfort, as it relates to sitting. The definition of *comfort* has long been, conveniently and misleadingly, the absence of discomfort°. However, these two aspects are not the opposite extremes on one single judgment scale. Instead, there appear to be two scales: one for the agreeable feelings that relate to comfort, and the other scale for the unpleasant experiences of discomfort, such as not being at ease and the sensations of fatiguing, straining, smarting, and hurting. Using the term *annoyance* (instead of discomfort) avoids the false concept of one scale that has comfort and discomfort as polar opposites. The two scales seem to overlap but they are not parallel to each other. Table 18.3 lists rating categories that distinguish chairs by comfort or annoyance.

Annoying seats

Feelings of annoyance when sitting associate with such descriptors as stiff, strained, cramped, tingling, numbness, not supported, fatiguing, restless, soreness, hurting, and pain. Some of these attributes can be explained in terms of circulatory, metabolic, or mechanical events in the body; others go beyond such physiological and biomechanical phenomena. Users can rather easily describe design features that result in feelings of annoyance such as chairs in wrong sizes, too high or too low, with hard surfaces or sharp edges; but avoiding these mistakes, per se, does not make a chair comfortable.

Comfortable seats

Feelings of comfort when sitting associate with such descriptive words as warm, soft, plush, spacious, supported, safe, pleased, relaxed, and restful. However, exactly what feels comfortable

Table 18.3   Rating of chairs for comfort or annoyance[a]

Comfort statements:

1. I feel relaxed
2. I feel refreshed
3. The chair feels soft
4. The chair is spacious
5. The chair looks nice
6. I like the chair

I feel comfortable

Annoyance (discomfort) statements:

1. I have sore muscles
2. I have heavy legs
3. I feel uneven pressure
4. I feel stiff
5. I feel restless
6. I feel tired

I feel annoyed

Source: Adapted from Helander, M.G. and Zhang, L. (1997). Field studies of comfort and discomfort in sitting. *Ergonomics,* 41, 895–915.

[a] Helander and Zhang used six specific statements about chair comfort or discomfort (annoyance), followed by one general statement. Each ranking employed nine steps from "not at all" to "extremely."

depends very much on the individual and his or her habits, on the environment and task at hand, and on the passage of time.

Aesthetics play a role: if we like the appearance, the color, and the ambience, we are inclined to feel comfortable. Appealing upholstery, for example, can strongly contribute to the feeling of comfort, especially when it is neither too soft nor too stiff but distributes body pressure along the contact area, and if it "breathes" by letting heat and humidity escape as it supports the body.

Dynamic design   In the early 2000s, dynamics was a key word in the design of office chairs, as opposed to the statics of maintained posture. People do move about as they please. Design should encourage and support free flowing motions, as sketched in Figure 18.9, with opportunities for transitory postures at the whim of the person.

FIGURE 18.9   People sit as they want. (Adapted from Kroemer, K.H.E. and Kroemer, A.D. (2005). *Office Ergonomics*, authorized translation into Korean. Seoul: Kukje.)

18.5 Ergonomic design of the office workstation

Successful ergonomic design of the workstation in the office depends on proper consideration of several interrelated aspects: work tasks, work movements, and work activities. All of these must "fit" the person to achieve individual well-being and foster high work output. Of course, job content and demands, control over one's job, and many other social and organizational factors also influence feelings, attitudes, and performance at work, as discussed in Sections III and IV of this book.

Links between person and task

For proper layout of a workstation, it helps to consider three main links between a person and the task.

1. The first link is the visual interface: one must look at the keyboard, the computer screen, or the printed output and source documents.

2. The second link is manipulation: the hands operate keys, a mouse, or other input devices; they manipulate pen, paper, and telephone.

3. The third link is body support: the seat pan supports the body at the undersides of the thighs and buttocks, and the backrest supports the back. Armrests or a wrist rest may serve as support links.

Design for vision

The location of visual targets can greatly affect the body position of the computer operator, as shown in Figures 18.10 and 18.11. Objects upon which we focus our eyesight should be located directly in front, at a convenient distance and height from the eyes. If one is forced to tilt the head up to view the computer screen or to turn it to the side to read a document, eye strain is commonly experienced, often together with pain in neck, shoulders, and back.

FIGURE 18.10 The monitor is set up too far from the eyes and too high, so the operator arches his back and neck in trying to get the image into focus. (Adapted from Kroemer, K.H.E. and Kroemer, A.D. (2001). *Office Ergonomics*. London: Taylor & Francis.)

FIGURE 18.11 A document placed to the side causes a twisted upper body posture. (Adapted from Kroemer, K.H.E. and Kroemer, A.D. (2001). *Office Ergonomics*. London: Taylor & Francis.)

As a rule, the monitor or a source document should be about half a meter from the eyes, the proper viewing distance for the operator. This is the reading distance for which corrective eye lenses are usually ground. A convenient yardstick is to place the screen and source document at slightly less than arm's length. Of course, proper lighting is necessary, as discussed above.

Design for manipulation

In addition to the eyes, our hands are usually very busy doing various office tasks: grasping and moving papers, taking notes, or punching keys. If our hands are engaged in many different activities, the varied manipulation is likely to keep our arms and the upper body moving around in our workspace. Motion is desirable, in contrast to maintaining a fixed posture, such as when tapping on the keyboard over extended time.

When sitting, we do have a large manipulation area available, especially if we move the upper body and, of course, we can cover an even larger area when standing up. However, for finely controlled hand movements, we prefer a space near chest and abdomen, within the areas swept by hand and forearm motions; see Figure 18.12. We can also see acutely there because objects are at a suitable distance from the eyes, and so low as to fit the natural downward direction of gaze (see Chapter 5 in this book).

Design for body motion and support

The body is built to move about, not to remain still. It is uncomfortable and tiresome to maintain a body position without change over extended periods. We experience this often while

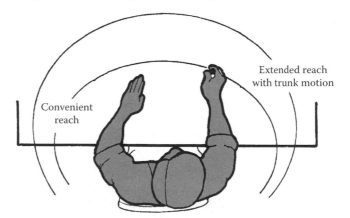

FIGURE 18.12 Convenient and extended reaches. (Adapted from Proctor, R.W. and Van Zandt, T. (1994). *Human Factors in Simple and Complex Systems*. Boston: Allyn and Bacon.)

FIGURE 18.13   Get up and stretch the body. (Adapted from Kroemer, K.H.E. and Kroemer, A.D. (2001). *Office Ergonomics.* London: Taylor & Francis.)

driving a car, and at workstations with immovable desktop computers where the chair is often much less suitable than our car seat. But, unlike when driving, we can get up and move around at will in our office.

A primary aim of ergonomic workstation layout is to facilitate body movement. For the computerized office this means: design it for walking and standing and stretching, not just for sitting. It is a good practice to get up and stretch the body while talking with a colleague and getting some papers or other supplies see Figure 18.13.

"Active sitting"   Design for motion means that the office chair should be comfortable for relaxed and upright sitting, for leaning backward and forward, and for getting in and out. In some chairs, both the seat pan and the backrest follow the motions of the sitting person and provide support throughout the range. Other designs start from the premise that the seat must not be a passive device but an active one: the chair as a whole, or its pan or backrest separately, can automatically change the configuration slightly over time, perhaps in response to certain sitting postures maintained by the person. The change can be in angles, or in stiffness of the material. Seat and back cushions that pulsate were tried in the 1950s to alleviate the strain that military aircrews felt when they had to sit for hours on end to fly extended missions. Should an "intelligent seat" remind us to move, or to get up, after we sit in a static position for too long a time?

Design for variety   For decades, the illusion ruled western office design that one could establish strict norms for dimensions of office furniture that would fit (nearly) everybody. Globalization of trade and the recognition that people are diverse (in sizes, habits, behavior, and preferences) have finished off the one-design-fits-all idea. Instead, diversity of use has led to diversity in design.

The size of office furniture derives essentially from the body dimensions, the work tasks, and the working habits of the people in the office. Main vertical anthropometric inputs to determine the height requirements are lower leg (popliteal and knee) heights, thigh thickness, and the heights of elbow, shoulder, and eye; see Chapter 1 for more details.

At computer workstations, the furniture consists primarily of the seat, the support for the data entry device and the display, and a working surface. It is best, although expensive, to have all of these independently adjustable.

Lumbar spine in relation to pelvis

When one sits down on a hard flat surface, not using a backrest, the lowest protrusions of the pelvic bones (the ischial tuberosities) act as fulcra around which the pelvic girdle rotates under the weight of the upper body. Because connective tissue links the bones of the pelvic girdle to the lower spine, rotation of the pelvis affects the posture of the lower spinal column, particularly in the lumbar region. Leg muscles run from the pelvis area across the hip and the knee joints to the lower legs. Therefore, the angles at hip and knee affect the location of the pelvis and hence the curvature of the lumbar spine.

Seat pan variations

The posture of the spinal column, especially in the lumbar area, has been of major concern for orthopedists and seat designers. Their ideas about proper posture and body support have generated innumerable designs for the seat pan: high and low, hard and cushioned, tilted fore and aft, contoured and flat, saddle-shaped and otherwise curved: all associated with hopes and claims of healthy sitting. No single layout has found universal acceptance.

The surface of the seat pan must support the weight of the upper body comfortably and securely. Hard surfaces generate pressure points, which suitable upholstery, cushions, or other surface materials can avoid by elastically or plastically adjusting to body contours.

The only inherent limitation to the size of the seat for western-style sitting is that its pan should be so short that the front edge does not press into the leg's sensitive tissues behind the knees. A seat pan with a well-rounded front edge, between 38 and 42 cm deep and at least 45 cm wide, fits most western bodies. The height of the seat pan must be widely adjustable, preferably down to about 37 cm and up to 58 cm to accommodate

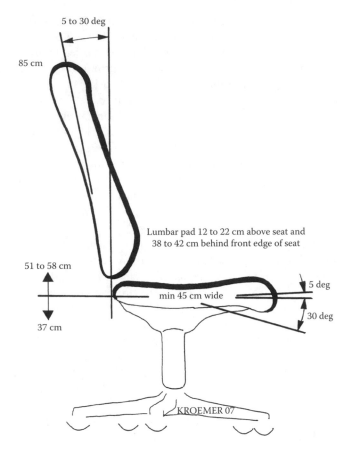

FIGURE 18.14   Major dimensions of seat pan and backrest. (Adapted from Kroemer, K.H.E., Kroemer, H.B., and Kroemer-Elbert, K.E. (2003, amended reprint). *Ergonomics: How to Design for Ease and Efficiency* (2nd Edn). Upper Saddle River, NJ: Prentice-Hall/Pearson Education.)

persons with short and long lower legs. It is very important that the person, while seated on the chair, can easily make all adjustments, especially in height. Figure 18.14 illustrates major dimensions of seat pan and backrest.

Backrest necessary?   Some orthopedists pursue the notion that a backrest is not necessary when sitting because, without it, back muscles act to stabilize the trunk. However, most people think that a backrest is desirable for several reasons. One is that the back support carries some of the weight of the upper body and hence reduces the load that the spinal column must otherwise transmit to the seat

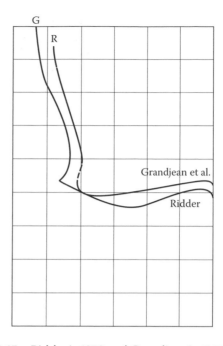

FIGURE 18.15 Ridder's 1959 and Grandjean's 1963 preferred contours of seat pan and backrest. (Adapted from Kroemer, K.H.E. and Kroemer, A.D. (2001). *Office Ergonomics*. London: Taylor & Francis.)

pan. A second reason is that a lumbar pad, protruding slightly in the lumbar area, helps to maintain lumbar lordosis, believed to be beneficial. A third related reason is that leaning against a suitably formed backrest is relaxing.

Backrest shapes   Of course, the backrest should be shaped to support the back comfortably. Around 1960, researchers in the United States and in Switzerland found in experiments that their subjects preferred similar backrest shapes, as depicted in Figure 18.15. In essence, these shapes follow the curvature of the rear side of the human body. At the bottom, the backrest is concave to provide room for the buttocks, and above convex to follow the curve of the lumbar lordosis. Above the lumbar pad, the backrest surface is nearly straight but tilted backward to support the thoracic area. At the top, the backrest is shaped to suit the lordosis of the neck. Shapes such as these have been used successfully for seats in automobiles, aircraft, passenger trains, and for office chairs.

Work surface and keyboard support

The height of the workstation depends largely on the activities to be performed with the hands, and how well and exactly the work must be viewed. Thus, the main reference points for ergonomic workstations are the elbow height of the person and the location of the eyes. Both depend on how one sits or stands, upright or slumped, and how one alternates among postures.

For western users, the table or other work surface of a sit-down workplace should be adjustable in height between about 53 and 70 cm, even a bit higher for very tall persons, to permit proper hand/arm and eye locations. A keyboard tray can be useful if the table is a bit high for a person, but it also may reduce the clearance height for the knees. The tray should be large enough for a trackball or mouse besides the keyboard.

18.6   Designing the home office

While arranging our home office we are inclined to disregard the ballast of old habits and conventions about how to sit at work that governs us in the company office. If you work only occasionally in your home, then the old dining table and the odd kitchen chair probably will not harm you. But as you begin working in your home office for hours, you should become very conscious about the conditions there: in principle, everything said above applies to the home office. So, equip your office with carefully selected furniture, where the components of the workstation fit each other well—and, most important, fit you well.

Sit on an easy chair, a comfortable office chair (even if it is expensive) or semi-sit perched on a kneeling chair: whatever feels good to you, what supports your body well over long periods of time. Perhaps you want to work standing up, at least when you read or make phone calls, for example. This is your own workspace, which you put together for your comfort and ease at work, and it does not have to be similar to anyone else's setup, nor does it have to be expensive, because some simple furniture on the market is well designed.

Get a quality computer with an up-to-date display and with suitable input devices. Select a keyboard that feels comfortable to you (see Chapter 19), but consider voice input that might work for you. If you travel much with your portable computer, consider using it in your office as well.

Select a room with good lighting that is separate, quiet, and well heated and cooled (Chapters 5, 6, and 8). You will probably stay in your home office longer than you expected, and your well-being is worth the effort and money that you spend.

Fitting it all together

The first responsibility for proper office design goes to management, architects, interior designers, and human factors engineers. They determine the overall layout and other important general conditions such as lighting and climatization. However, within the overall boundaries of design, there is much freedom for specific workplace arrangements by section manager, supervisor, and especially by the individual office occupant. Table 18.4 summarizes what the individual can do to make office work agreeable.

Table 18.4 Ergonomic recommendations for individual arrangements at the office workstation

---

This feels good

- Place all the things you must operate with your hands (keyboard, mouse, trackball, pen, paper, telephone)
  —Directly in front of you
  —At elbow height
  —Within easy reach
- Place the display and the keyboard, which you must clearly see, directly in front of you, at your best viewing/reading distance.
- Place the monitor low behind your keyboard. (Do not put it on a tall stand or on the CPU box.)
- Sit on a seat so designed that you can change your posture frequently. If long-time sitting is required, then a tall backrest that can recline helps support back and head.
- Change your body position often. Change helps avoid continued compression of tissues, especially of the spinal column, facilitates blood circulation, counteracts muscular fatigue—and it breaks boredom.
- Support your arms and hands by resting them, as often as feasible, on soft arm rests attached to the seat, and on padded wrist rests at the keyboard, but avoid hard surfaces and, worse, rigid corners and edges that compress the skin tissues.
- Keep the shoulders relaxed, the upper arms hanging down, the forearms horizontal, and the wrists straight.

What to do if you are not comfortable

If your eyes are tired, or hurt, or feel teary or dry:

- Place all the things you must clearly see (display, source document, writing pad, template, keyboard) directly in front of you, to the best viewing/reading distance (which is probably shorter than you have now).

Continued

Table 18.4   Ergonomic recommendations for individual arrangements at the office workstation (Continued)

- Place your monitor low behind your keyboard (Take it off from a tall stand or the CPU unit).
- Make sure you don't have light (from a window or from lamps) reflected in the display, or shine directly into your eyes.
- Talk with your supervisor and, if the condition does not go away, have your eyes checked by an ophthalmologist or optometrist.

If your back hurts:

- Take a break at least every 30 minutes; walk and move your body.
- Make sure that you lean against the backrest of your seat.
- Get a seat that fits your body and accommodates your sitting habits better than your current chair.
- Place all the things you must clearly see (display, source document, writing pad, template, keyboard) directly in front of you, to the best viewing/reading distance (which is probably shorter than you have now).
- Place your monitor low behind your keyboard (Take it off from a tall stand or the CPU unit).
- Place all the things you must operate with your hands (keyboard, mouse, trackball, pen, paper, telephone)
    —Directly in front of you
    —At elbow height
    —Within easy reach
- Talk with your supervisor and get a medical evaluation if the condition does not abate.

If your neck hurts:

- Take a break at least every 30 minutes; walk and move your body
- Place all the things you must clearly see (display, source document, writing pad, template, keyboard)
    —Directly in front of you, at your best viewing/reading distance (which is probably shorter than you have now)
    —Low behind your keyboard (do not use a tall tilt stand, or the CPU unit, under the monitor)
- Place all the things you must operate with your hands (keyboard, mouse, trackball, pen, paper, telephone)
    —Directly in front of you
    —At elbow height
    —Within easy reach

Table 18.4 Ergonomic recommendations for individual arrangements at the office workstation (Continued)

- Talk with your supervisor and get a medical evaluation if the condition does not go away.

If your shoulder hurts:

- Take a break at least every 30 minutes; walk and move your body.
- Put the mouse or trackball next to the keyboard, all at elbow height.
- Operate the mouse or other input device with the other hand. (Yes, you can do so well after just a few minutes.)
- Use armrest and wrist rest often.
- Talk with your supervisor and get a medical evaluation if the condition does not go away.

If your wrist or hand or arm hurts:

- Take a break at least every 30 minutes; walk and move your body.
- Make sure that your wrist remains straight while working the keyboard or other input device.
- Strike keys very lightly.
- Use armrest and wrist rest often.
- Put the mouse or trackball next to the keyboard.
- Talk with your supervisor and get a medical evaluation if the condition does not go away.

If your leg hurts:

- Take a break at least every 30 minutes; walk and move your body.
- Make sure that you have ample room at your workstation to position and move your feet freely.
- If the front portion of your seat presses on the underside of your thighs, lower the seat. (Probably, you also must lower keyboard and monitor, table, or desk or other work surface accordingly.)
- Get another seat that has a soft waterfall shape at its front.
- Use a wide and deep footrest.
- Talk with your supervisor and get a medical evaluation if the condition does not improve.

Source: Adapted from Kroemer, K.H.E. and Kroemer, A.D. (2001). *Office Ergonomics.* London: Taylor & Francis.

Summary

Office designs can vary from open plans to closed, walled-in individual rooms. Disadvantages of open designs include lack of privacy and the propagation of disruptive sound and noise. Many organizations use a mixture of office designs: some wide-open spaces, sections with shoulder-high cubicle dividers, and closed-off individual offices in other parts.

The details of suitable office illumination depend on the tasks at hand, the objects to be seen, and the conditions of the eyes to be accommodated. General recommendations for office illumination are about 500 lx, but, if cathode-ray tube (CRT) displays are present, the overall illumination could be as low as 200 lx whereas in rooms with light-emitting displays, illumination of up 750 lx is suitable. For such tasks as reading text on a paper it can be helpful to use a task light, however, care is necessary to avoid glare.

Neither theories nor practical experiences endorse the idea of one single proper, healthy, comfortable sitting position, such as erect sitting. Instead, many motions and postures may be subjectively comfortable (healthy, preferred, suitable) for short periods of time, depending on one's body, preferences, and work activities. Changing from one posture to another one, moving freely among all the comfortable poses, is important. Consequently, furniture should allow for body movements among various postures by easy or automatic adjustments in its main features, especially in seat height and seat pan angle and backrest position. The entire computer workstation should permit easy variations, for example, in the location (especially height) of the input devices and height and distance of the display.

Within given overall boundaries, there is usually freedom for specific workplace arrangements that fit the task and suit the individual. This is particularly true for the setup of personal home offices. Individual arrangements can be essential to achieve ease of work.

Notes and more information

The text contains markers, °, to indicate specific references and comments, which follow.

18.2 Office climate:

We are not cold-blooded: The physicist Daniel Gabriel Fahrenheit (1686–1736) worked with thermometers and set the temperature of a mix of ice and water to 32°, and the temperature of boiling water 180° higher, at 212°. This is the Fahrenheit scale. In 1742, the astronomer Anders Celsius (1701–1744) suggested a metric scale: zero is the temperature when water freezes and 100° when it boils. This centigrade scale is called the Celsius scale and is in use worldwide except in the United States. For the conversion of temperature degree values from one scale into the other, one must consider the different settings for freezing and boiling temperatures of water as well as the number of degrees between freezing and boiling:

Fahrenheit to Celsius: [degF − 32] [5/9] => degC

Celsius to Fahrenheit: [9/5 degC] + 32 => degF.

(The 5/9 ratio comes from 100/180 and, of course, 9/5 = 180/100)

Good office climate: For listings and discussions of ASHRAE recommendations in the United States, of international standards especially by ISO, and of national regulations and regional recommendations and customs, see Parsons 2005.

18.3 Photometry in office illumination:

In traditional photometry, the apparent reason for measuring illuminance was the ease of doing so. Instruments to measure luminance, which determines how well we can see objects and distinguish their details, are more complex and expensive and harder to use; see Rea 2005.

The traditional relations between lighting and human vision concerned primarily two elements in the visual environment: the objects being viewed, for example, print on paper, and their illumination, such as by daylight or electrical lamps. However, the widespread use of self-luminous sources, such as TVs and computer monitors, has led to some change in emphasis because visual characteristics of these light-emitting objects are not dependent on illuminating light. For more on this topic, check Howarth 2005.

18.4 Office furniture and equipment:

Body posture: In 1884, Staffel had published theories about hygienic sitting postures. He recommended holding trunk, neck, and head erect, with only slight bends in the spinal column in the side view. His recommendation for the desired back posture when sitting was similar to what he advocated for standing in 1889, when he and others reported that farmers and laborers often had back curvatures that diverted from the norm: their spinal columns were either too flat, or overly bent. The experts judged these back postures unhealthy and concluded that they had to be avoided. Staffel and his colleagues were particularly concerned about the postural health of children and therefore they recommended that the children be exhorted to maintain an erect posture of back, neck, and head. Starting in the late 1880s, a great number of "hygienic and healthy" designs for school furniture were proposed: seats, desks, and seat-desk combinations were laid out to promote the upright posture. These ideas were also applied to office furniture. For more information on this topic, see Kroemer et al. 2003.

Comfort and discomfort: The two aspects of comfort and discomfort are two separate scales. These two scales partly overlap but are not parallel. Aesthetics play a role in the perception of comfort. For more information, read Corlett 2005 and Helander 2003.

CHAPTER NINETEEN

Computer design

Within just a few decades, computers have become essential work tools, much-used leisure gadgets and toys of many kinds. Computer technology makes long-distance interaction easy, which is especially important if it is not feasible to travel. Much business communication, formerly done by mail and telephone, now goes by e-mail. Computers opened the world for many persons, including those who are sick or elderly, who would otherwise feel shut in; electronics allow them to communicate directly with others, to shop and bank, and to get the news. The internet provides an abundance of information to anybody with just a few key strokes.

QWERTY keyboards on computers

However, current computer keyboards are not human engineered but, instead, still follow the 1878 QWERTY design. This makes keyboarding unnecessarily difficult and time-consuming, and even causes repetitive injuries to the hands and wrists of keyboarders. Since early in the 20th century, inventors have made many proposals for new designs replacing the keys and keyboards of the old typewriter design; yet, so far no novel solutions have been successful.

This chapter discusses ergonomic aspects of computer design and operation and derives suggestions for better, human-centered designs. This review shows a typical process of solving a difficult technical problem; starting with a hardly workable design, then improving it incrementally, and, ultimately, seeking a fundamentally different, truly good solution.

19.1 Sholes' "type-writing machine" with its QWERTY keyboard

Throughout the 1800s, numerous inventors proposed a great variety of typographical devices° on which manipulation of an input device (usually pressing a key of some sort) generated imprints of letters on paper. Given the technology available at that time, the innards of these "type-writing" machines consisted of complex mechanical lever setups. Apparently, the overriding technical challenge was to find workable mechanisms. So, the inventors of theses machines seem to have paid little attention to the usability of the input side: the design of keys and their arrangements.

Sholes' "type-writing machines"

One of these inventors, Christopher Latham Sholes, from 1868 on obtained eight U.S. patents for various designs of a Type-Writing Machine: two patents in 1868, both with Glidden and Soulé; then one patent in 1876 with Schwalbach; in 1878, five patents: #199,382, followed by #200,351 (with Glidden), then #207,557 and #207,558 and finally, on August 27, #207,559.

Sholes' keyboards

The two first patents, both in 1868, have rectilinear rows of keys: the first patent, #79,265, shows 21 unmarked keys, described (on page 1) as "similar to the key-board of a piano" with 10 shorter keys atop 11 longer keys, akin to the white and black clavier keys. The 36 keys in patent #79,868 are also (as said on page 2) "similar to the keys of a piano or melodeon", but they all lie in one plane, side by side, alternating in length. The keys on the left side, as the operator sees them, show numbers in increasing order whereas the other keys, in the middle and on the right side, carry alphabetic letters. Neither patent contains any explanation for the choice of the respective key layout beyond the just quoted statements.

The following 1876 patent (#182,511) and the first four patents of 1878 all exhibit three straight rows of button-like keys affixed to lever-type bars: 32 keys in the 1876 patent, and 21 keys in the 1878 patents. No labels with letters or numbers are on the buttons, and the texts of the patents provide no explanation or description of any kind.

QWERTY keyboard

Sholes' last patent, #207,559, contains 14 specific technical claims, but none refers to the key layout. One drawing in this patent shows a frontal view of four straight and horizontal

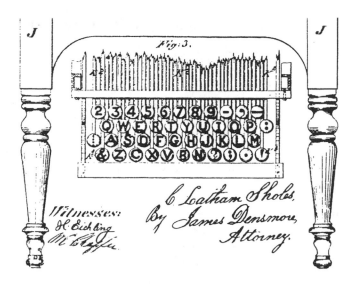

FIGURE 19.1 The QWERTY keyboard in Sholes' 1878 U.S. Patent # 207,559. Sholes and his co-inventors obtained eight U.S. patents for "Type-Writing Machines" with various keyboards: in 1868, patents #79,265 and #79,868 (both with Glidden and Soulé); in 1876, #182,511 (with Schwalbach); in 1878, #199,382 and #200,351 (with Glidden), followed by #207,557, #207,558 and #207,559

rows of key tops. The rows are staggered in height so that the row closest to the operator is the lowest and the farthest row is the highest. Another drawing depicts a top view of the four straight rows, each with 11 round keys, for a total of 44 keys; see Figure 19.1. The key tops carry inscribed numerals, letters, and signs.

Lacking any specific statements by Sholes or his contemporaries either in the patents or written or reported elsewhere, one can only observe that the final 1878 patent #207,559 layout shows some remnants of an alphabetic arrangement. Sholes was a printer by trade, and thus we may surmise similarities to the arrangement of the printer's type case in which, presumably, pieces were assorted according to convenience of use instead of just alphabetically. Another possible reason for the arrangement of the characters and the keys may have been the intent to avoid type bars (in the then-used mechanical lever mechanisms) colliding or sticking together when activated in quick sequence. This may have led to a separation of certain bars and, hence, keys on the keyboard. However, there is no contemporaneous evidence for any of these speculations.

19.2 From typewriter to computer keyboard

Sholes' 1878 invention became a global success, and his puz-zling layout of the keyboard is still in general use, with only the positions of *X* and *C* exchanged and the *M* moved to the first row in English-language versions. The six leftmost keys on the third row, counted from the operator, carry the letters *QWERTY*. Even on today's computers, the term "QWERTY keyboard" serves as a short name for any arrangement in which the letter keys (also called alphabetic or alpha keys) essentially follow Sholes' keyboard layout.

20th century typewriters

None of the numerous patents and proposals by others for alter-nate key and keyboard designs was commercially successful. Changes within the Sholes layout were technically difficult as long as the mechanism relied on mechanical levers. The first significant improvement to typewriters was the introduction of electric motors in the 1950s, which supplied auxiliary energy to strike the type bar—later, a type-ball as on IBM machines—on the inked tape to make the imprint on paper; until then, the typist's hands had provided that energy. In the 1960s, fingertip-operated switches controlling electric and then electronic cir-cuitry began to replace the lever mechanisms of the keys. This drastically reduced both the finger force needed to actuate the key and the key displacement, diminishing the dynamic work (force–displacement integral) required from the operator's fin-gers for every keystroke—but it did not reduce the keying rate.

Typewriters morphing into personal computers

Starting in the 1960s, electronics replaced the mechanical innards: typewriters morphed into computers. The new technol-ogy would have easily allowed relocating keys and redesigning the entire keyboard. However, instead of creating novel solutions, keys were simply added to the original keyboard. These were mostly function keys, placed to the left, behind and especially to the right of the QWERTY set. With an extra numeric keypad and a further cursor control keypad (both commonly on the right-hand side), in the 1980s the total number of keys was customarily just over 100, in some cases about 125, more than double the number of keys on the old typewriter. So many keys took much keyboard space and the big keyboard required large finger and hand movements. Mouse, trackball, touchpad, and other accesso-ries generated new tasks and body motions for the keyboarder.

The 1980s saw a wave of computers entering modern offices, sweeping out the remaining mechanical and even the newer electric typewriters. IBM introduced its first personal computer in 1981. Only ten years later, it stopped producing typewriters. Two decades further on, IBM sold its whole PC business to a Chinese firm.

19.3 Human factor considerations for keyboarding

Obviously, the keyboards on early typographic devices, including Sholes' 1878 invention, were not human engineered. Nevertheless, his QWERTY layout became by default the most commonly manufactured keyboard when his Type-Writing Machine became a global success.

Body posture and effort

Two postural requirements on the user are specific to current computer technology: the computer operator must focus the eyes on the display while keeping the hands on the keyboard. Such prescribed locations of the eyes and the hands fixate the overall positions of the head and upper body, and hence little variation of the body posture is possible. This rigid posture is not likely to be one that computer users would select freely as comfortable.

The spatial arrangement of the keyboard, and the keys on it as per Sholes' design, forced the arms into strong inward twist (pronation) and the hands into a lateral bend (ulnar deviation) at the wrist, and it required complex motions between the ill-located keys. The unfortunate combination of bad posture and hard effort overloaded many typists' hands, wrists, arms, shoulders, and necks.

1920s typists

Klockenberg (1926) provided an illustration of the typical posture of a typist, reproduced as Figure 19.2. In a moving narrative, Klockenberg described how young women, who had chosen typing as their profession, after just a few years on the job found themselves with painful hands, unable to do typing or to perform everyday tasks with their hands, incapable even of lifting their small children. Obviously, the reason for the typists' musculoskeletal injuries was the heavy repetitive work required to operate their typewriters in unsuitable body postures. Today, the posture of a keyboarder usually is much less contorted, but still bound by the needs to keep the fingertips on the keys and the eyes directed at the display.

Heidner's 1915 keyboard designs

Among the early proposals of improved keyboards, one stands out: In 1915, Heidner received U.S. Patent #1,138,474 for his novel layouts. He wrote that his keyboard designs allowed

FIGURE 19.2   Klockenberg's illustration of the posture of a 1920s typist.

"... to write with greater ease, in a less cramped position ... in accordance with the natural form of the hand ... and there being thus much less strain ... writing is rendered considerably less fatiguing." Figure 19.3 illustrates how Heidner divided the keyboard into left and right halves; on them, he arranged the keys in various layouts with straight columns and on straight or curved rows. Heidner's proposals predate many ergonomic recommendations that inventors have put forth for nearly a century; yet, thus far, no typewriter or computer manufacturer has adopted any radically new designs with commercial success.

Repositioning keys

Keying performance has been an issue since the early years of keyboarding. The concerns focused on the total number of keystrokes per time unit and on the ratio between correct and incorrect strokes. Starting in the early 1900s, several patents for new key arrangements appeared that were meant to improve typing performance by overcoming problems of Sholes' QWERTY layout. These early new designs usually relocated keys° within the standard set of keys, but without changing Sholes' original layout of bent columns and straight rows of keys. Dvorak's (1936, 1943) "Simplified Keyboard" is probably the best known and most long-lived of all these plans. However, in 1956 a comparative test by Strong showed

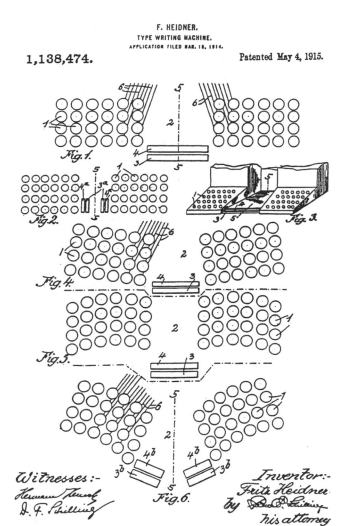

F. HEIDNER.
TYPE WRITING MACHINE.
APPLICATION FILED MAR. 18, 1914.

1,138,474.   Patented May 4, 1915.

FIGURE 19.3 Heidner's 1915 keyboards in his U.S. Patent #1,138,474

that it took a long time to retrain typists to become proficient on the SK, yet they still made more errors on it than on their customary QWERTY keyboard and they gained less improvement in typing rate with ongoing training. These findings apparently discouraged further attempts to reposition certain keys on the regular keypad.

Repetitive work around 1700

In a book first published in 1700, then edited and enlarged in 1713, the Italian physician Bernardino Ramazzini° reported on overuse diseases that appeared in persons who performed "violent and irregular motions" in "unnatural postures of the body". Of course, there were no typewriters or computers in 1713, but Ramazzini reported overuse diseases that occurred among secretaries (*notaries*) and office clerks (*scribes*), saying that their diseases were due to three causes:

> First, constant sitting, secondly, the incessant movement of the hand and always in the same direction, thirdly, the strain on the mind. . . . They must stick to their writing all day long. . . . Incessant driving of the pen over paper causes intense fatigue of the hand and the whole arm because of the continuous and almost toxic strain on the muscles and tendons, which in course of time results in failure of power in the right hand.

Repetitive injuries during the 1800s

The relationships between repetitive work and musculoskeletal strain° were well known in the 1800s. In 1872 and 1887, the British physician Poore referred to other authors who, following Ramazzini, had traced writers' cramp to "muscular impotence" and "spasms" caused by frequently repeated use of the same muscles. Poore stated that this health problem was not limited to writers, but also occurred in tailors, cobblers, fencing-masters, and musicians. He wrote that one might speak of "pianists' cramp" because this overuse disorder occurred so frequently among pianists. In 1892, Osler also associated these spasms "with continuous and excessive use of muscles in performing a certain movement" in writers and musicians and in operators of the single Morse key who suffered from a disabling injury called "telegraphist's wrist". Evidently, during the second half of the 1800s, the associations of repetitive strain disorders with repetitive muscle overuses in certain occupations were well established.

Keying "myalgia"

Until Sholes' 1878 QWERTY key layout, keyboards had been parts of musical instruments, especially of pianos and cembalos. It was known even at that time that serious clavier students and accomplished players are in danger of developing chronic musculoskeletal disorders: the pianist and composer Robert Schumann (1810–1856), for example, lost the use of his right hand. After employment of the new typewriting machine became widespread, the same "myalgia" appeared in typists also. Tenosynovitis and other repetition-related musculoskeletal diseases° of typists concerned physiologists and medical practitioners since the early 1900s.

A slow operator, tapping just 20 words per minute (with five letters per word), performs 12,000 key strokes (20 × 5 × 60 × 2) during a 2-hour work session. A fast keyer, doing 100 words per minute over 6 hours, performs 180,000 keystrokes. Each stroke requires a digit flexion followed by a digit extension. So many motions can create pathomechanical° conditions especially for the flexor tendons° of the digits: see Figures 2.7 and 2.8 in Chapter 2.

Muscles used in typewriting

In 1951, Lundervold published the first report of his groundbreaking studies to attain knowledge about the use of individual muscles while typewriting. He measured electromyographic (EMG) signals on 47 healthy typists and 88 patients, most of them suffering from occupational myalgia in muscles that had been overstrained during repetitive typewriting. Lundervold discovered that muscles became active in typing which were not expected to partake. He found that changes in the EMG records reflected the actual muscle activity and the status of muscle fatigue. His experiments provided support for the long-held opinion that repetitive typing can lead to a cumulative overexertion injury even when doing single keystrokes is not harmful by itself. In Europe, myalgia ailments such as tendinitis, tenosynovitis, and tendovaginitis of the upper extremities became recognized as occupational diseases of typists° in the late 1950s.

Reports first in the early 1970s from Japan, then from Australia, Europe, and the United States indicated that groups of key operators suffered from "local epidemics" of overexertion injuries°. Researching the reasons for cumulative trauma disorders, especially of the carpal tunnel syndrome, associated with repetitive keyboarding activities, turned into topics of engineering and medical concern. In the late 1990s, the (U.S.) National Research Council convened a workshop on work-related musculoskeletal disorders. In the summary of its findings, the NRC (1999, p. 59) stated that musculoskeletal disorders are multifactorial because individual, social, and organizational factors can contribute to their appearance; however, the biomechanical demands of work constitute the most important risk factors.

Overexertion pathomechanics

The biomechanical stresses that generate cumulative trauma disorders, especially the carpal tunnel syndrome (CTS), had become well researched in the 1980s and '90s. More recent experimental findings° support the earlier findings. They explain the anatomy of the human hand and wrist and its kinematics while keyboarding°. They reaffirm the overexertion pathomechanics that especially afflict flexor tendons of the digits in the carpal tunnel; see Figure 2.6 in Chapter 2. Forces in the flexor tendons are often more than three times stronger than the impulses that the fingertips transmit to the key tops. The longitudinal travel of the tendons can be up to 2 cm; wrist position strongly affects the ease of their gliding within the synovial sheaths. Inflammation of the tendons and their sheaths causes swelling of the tissues within the limited space of the carpal tunnel. The resulting pressure can damage the median nerve, causing the carpal tunnel syndrome that affects hand functioning.

19.4　Input-related anthromechanical issues

Sholes' keyboard concept with its mechanical keys required efforts from the typists that exceeded their musculoskeletal capacities. Their hands had to provide all the energy to advance the carriage and the platen and to move the complicated lever mechanisms that made the type bars forcefully strike the inked tape to generate imprints on paper. The hard work led to widespread and well-documented hand/arm overuse disorders. The underlying biomechanical problems arose from several categories of work demands: the physical energy to accelerate the masses of the typewriter mechanisms; the postures of hands, arms, upper body, and neck; the extensive use of mainly the flexor muscles and tendons of the typists' hands; and the rate and repetitiveness of muscle contractions and tendon motions.

Human factors design recommendations

Attempts to lighten the typist's load resulted in many proposals and patents even in the early 1900s; yet, the mechanical nature of the then-available machinery severely hindered new design solutions. In the 1960s, the terms "ergonomics" and "human (factors) engineering" were emerging. Consequently, the importance of fitting task and equipment to the human became widely recognized; reconfiguring the keyboard to suit size and mobility of the hand became an important challenge. In 1969, Remington and Rogers compiled more than 300 publications on keyboard entry devices. In the same year, Kinkead and Gonzalez formulated "human factors design recommendations for touch-operated keyboards". The emerging use of electric and then electronic circuitry would have allowed radically new designs; yet, U.S. standards° in 1968 and 1988 still incorporated the QWERTY layout while adding more keys on its sides.

The convention of binary push keys

Reviews of the issues of keyboard design and novel solutions for operator-centered designs of keys, keyboards, and keyboarding workstations, and for proper ergonomic ways to perform keying work, appeared in the international literature°. Ergonomists voiced concerns in the 1970s that the enlarged QWERTY keyboard of the typewriter, in spite of its well-known human engineering shortcomings, was becoming the de facto standard layout for new communications and computer interfaces. Ergonomic guidelines, first published in 1979 in German, then immediately translated into English, caused great impetus toward human factors considerations for keyboarding. The debate concerned the layout of key sets (the keyboard) in general, and specifically the design of the keys, and their use. Nonetheless, the 1968 and 1988 U.S. standards perpetuated the convention of binary push keys arranged in straight horizontal rows, yet in zigzag columns on the QWERTY part but in straight columns on the rest of the keyboard.

19.5 Possible design solutions

Customary computer keyboards

Neither the original QWERTY keyboard nor its derivatives have been truly "human engineered" in spite of Heidner's 1915 proposals and many other designs that followed. Even on today's customary computer keyboards, all keys are arranged side by side in straight rows (although fingertips are not). Key columns follow two different design rules: on the QWERTY set, the columns are zigzag but overall run slanted to the left (as seen by the operator); on the added key sets, however, the columns are straight. On mobile phones and portable computers, the smallness of keys and keypads seems to intensify the old QWERTY layout problem even if the columns of keys are stratified.

Design issues intermingle with existing technology, user expectations and practices, and with marketing. In spite of these interactions, in the following discussion, improvement ideas are assigned to certain categories, for clarity's sake. Obviously, solutions can combine aspects from two or more groups.

Designing for big changes

A major argument for not changing the basic QWERTY design of the keyboard, brought forth since the 1950s and repeated into the 1990s, is that an alteration of the layout would require retraining the operators and hence would slow keying performance, at least initially. This reasoning is probably true for most minor changes, such as exchanging the letter designations of certain keys. In these cases, apparently the new skills are so close to the old habits that confusion is indeed likely. However, when there is a basic change in design, such as when relocating whole sets of keys, or employing keys that have a clearly different operation mode (ternary keys or joysticks instead of binary keys, for example), then the new procedure is distinctively dissimilar from the old ways and there may be little or no interference by crossover from previous practices. The human is amazingly fast in acquiring novel keyboarding skills as demonstrated by older and recent research° and by the phenomenally fast spread, across the globe, of instant text messaging with its new keyboards and procedures.

Alternative keys and keyboards

In 1981, Litterick called the QWERTY keyboard a "dinosaur in a computer age . . . ideal for people with arms coming out of their chests, and fingers all the same length." In 1983, Noyes published a thorough review of design and use aspects of that keyboard's variations since Sholes' invention.

Redesigning the key

Changing the basic key design can have major consequences for the user's efforts and performance. The traditional mechanical binary key has two activation states: "off" when up, and "on" when tapped down. This principle is in common use even today even though electrical switch and electronics technology would make it easy to replace the old tap-down binary keys by devices that can establish or break connections by various actions. Just a few examples: we can replace the simple binary pat-down key by a ternary device, by a device that responds to movements in many directions other than just down, or by a control that does not need appreciable displacement for activation, or by a multifunction gadget that can react to combinations of positions, displacements, and contact modes, or by a sensor that responds to sound.

"Texter's thumb"

Not all developments are progressive; curious regression to bad old practices may occur. For example, text messaging via the numeric keys of mobile phones commonly requires several key presses, generally done with one thumb, to generate just one letter. The Morse code of the 1800s also required repeated keystrokes to transmit a character. Then, the "telegraphist's wrist" was a rampant repetitive motion injury; this compares to the recent "texter's thumb".

Redesigning the keyboard

Yet, the antiquated Sholes keyboard was not revised but instead kept into the 21st century, even transferred to such new devices as handheld computers and mobile phones. This perpetuated the anthromechanical problems associated with location and the number of keys and their mode of operation.

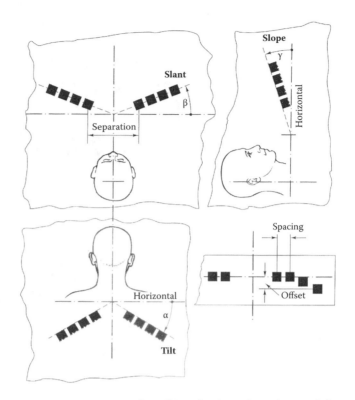

FIGURE 19.4 Terms describing keyboard angles and key positions. (Adapted from Kroemer, K.H.E. (2001). Keyboards and keying. An annotated bibliography of the literature from 1878 to 1999, International Journal Universal Access in the Information Society UAIS 1/2, 99–160. Available on the Internet at www.springerlink.com/index/yp9u5phcqpyg2k4b.pdf.

Heidner's 1915 patent was the first to propose a significantly improved design; numerous later proposals in essence repeated Heidner's ideas to rearrange the angles of slope, tilt, and slant of key sets, and key spacing (see Figure 19.4) to make keying less strenuous and to improve typing performance. In 1962, U.S. Patent #3,022,878, assigned to IBM, placed the keys like a glove around the hands. Several keyboards, commercially available in the later 1900s, had similar features, in some cases with adjustable arrangements.

Ternary instead of binary keys Keyboard design decisively depends on the keys used. The choice of ternary instead of binary keys in Langley's 1988 U.S. Patent #4,775,255 provides a good example. Each toggle key can assume three different positions: "off" when in its center position, "1st on" when pulled from there, and "2nd on" when

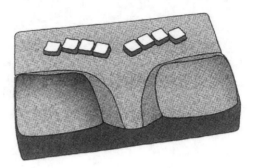

FIGURE 19.5   Langley's 1988 Ternary keyboard with four keys for each hand, U.S. Patent #4,775,255

FIGURE 19.6   Hand positions on Langley's Ternary keyboard

pushed. Langley used chorded activation° (see below), therefore his keyboard needed only eight keys, four for each hand, as illustrated in Figure 19.5. The fingers did not have to move from one key to another but stayed on their one assigned key; this allowed each hand to rest on its support pad, as Figure 19.6 shows. By simultaneous activation of two or more keys, the user can generate any letter or other character.

Repositioning the keyboard

The time-honored piano keyboard, which served as model key set for many early typographic machines including Sholes' early patents, can be broken into several sections to be arranged side by side or elevated in different steps. Examples are traditional organs and modern electronic music keyboards. Although without explaining why he did so, Sholes used in his final 1889 patent a keyboard with arrays of several bent columns of keys with straight and horizontal rows. Tiring arm positions and complex finger movements, and the large energy that the typist had to exert on the keys, caused postural problems in combination with repetitive musculoskeletal trauma. Heidner, in 1915, was apparently the first patent holder who attempted to address the postural problems by breaking the one keyboard into sections and repositioning them into locations and angles that are more suitable for the operator. Many subsequent proposals and patents presented similar ideas, some with adjustment features. Thus far, the more radical designs were not successful on the market, and slight modifications apparently did not stem the flood of repetitive injuries.

19.6 Design alternatives for keyboards

There is no overwhelming reason for using one keystroke for every input bit. The principle of allocating one key each to every character, numeral, and sign follows the tradition of Western writing—it is exact but also inefficient. Using fewer keys, moveable keys, virtual keys, or no keys at all would alleviate, even completely avoid, the past and current overexertion problems and facilitate body movements instead of the fixed posture caused by "hands on keys, eyes on display". Ways to communicate with the computer without manipulating keys are of great interest for everyday use, especially so to persons with physical disabilities. Several avenues are at hand.

Speech and sound recognition

Chords, traditionally used by pianists and other musicians, convey complex sounds composed of single tones, generated by simultaneously triggering keys instead of pressing them sequentially. Chording, the simultaneous activation of two or more keys, was proposed repeatedly for typewriter and computer operation to replace the sequential "one key for one character" procedure. Langley's 1988 U.S. patent, already mentioned, is a recent example for chording design, which can simplify keyboards, cut the number of keys, reduce hand digit movements from key to key, and, altogether, make key input faster while relieving hand strain.

Inputs without keys

There are means to generate inputs to the computer that require no keys of any kind. Already we try to communicate by voice, using speech recognition programs that work reasonably well within confined topic areas, for example, in computer control commands, medical evaluations, or music composition—but in general, speed and especially recognition accuracy leave much to be desired. If the impediments were finally overcome, conversing with the computer would be a natural solution. Of course, many other sounds could transfer information to the computer: whistles, grunts, and groans come to mind readily.

Nonverbal communication

Another set of solutions employs the recognition of body movements°. Examples are

- Hands and fingers for sign language and gestures in addition to activating control devices
- Arms for gestures, making signs, moving or pressing control devices
- The torso for positioning and pressing
- The legs for gestures, moving or pressing devices
- The feet for motions and gestures, for moving and pressing devices;
- The head for positioning
- The mouth for lip movement, use of the tongue or of a blow/suck tube
- The face for grimaces and other expressions
- Eyes for tracking

These solutions encompass sensors that respond to positions, movements, and forces of the operator's body: cameras, force platforms, motion sensors, instrumented gloves, and similar devices are among the state-of-the-art technologies. Other approaches might use surface electromyograms (EMGs) associated with muscle actions°; at the moment, employing neural signals such as electroencephalograms (EEGs) related with brain activities still appears to be wishful thinking. However, certainly there are many thus far unexploited possibilities: in the future, nanoergonomics° may make it possible to access the human central nervous system and to integrate important features of human cognition into computers, thus allowing direct interaction.

19.7 Designing for new syntax and diction

Klemmer (in 1950) and Lockhead and Klemmer (in 1959) recognized how changes in syntax and diction, new abbreviations, words, and phrases, can interact with changes in key and keyboard technology. Stenographers and court reporters have long demonstrated that using shorthand code for words and parts of sentences, chunking, batch processing, and related techniques save time and effort. The telegram style in the middle of the last century created a new short mode of written expression; in recent times, e-mailing and instant text messaging were changing communication manners, word-writing mode, spelling, and punctuation: "thru" largely replaced through, "u r r8" is a sample of text messaging style, and "rgds" is shorthand for closing with best regards, indicating that most text is decipherable without vowels.

Language changes constantly. New technologies fuel shifts in communication; conversely, the morphing of our language can also give designers opportunities for creating novel devices for communicating with computers.

19.8 Designing "smart" software

Cleverly designed software facilitates work with computers by reducing the demands on manipulative skills and short-term memory°. Software can use context to distinguish between mail and male, then and than; it can complete words after the input of only a few first letters; can provide a set of stock responses to given circumstances; it can anticipate the use of certain words; and it can correct sentence structure and insert punctuation. With further advances in artificial intelligence, the auspices seem unlimited.

19.9 Designs combining solutions

In 1878, when Sholes received his patent #207,559, and for about 80 years thereafter, the mechanical nature of the old type-writing machine made changes toward user friendliness difficult. Today's electronics and, perhaps, tomorrow's nano-technology allow fundamentally new solutions. We may sort these into certain categories, as done above; yet, overreaching combinations are most likely to succeed. A current example is the blending of traditional keys with speech recognition and smart software. However, this approach still largely employs the outdated Sholes key and keyboard, making the job hard for persons with disabilities and, in fact, everybody else because it still requires repetitive manipulation and makes the operator's body maintain a constrained posture via "fingers on keys, eyes on monitor".

It appears that abolishing the predominant principle of *one stroke for one input on dedicated binary keys* is an essential step, although keys of some advanced nature may remain in use. Smart software should facilitate the keyboarder's tasks. The general acceptance of innovative mobile phones and game consoles, for example, show that the general public is willing to accept, even embrace, fundamental new techniques in human–computer interaction, together with technology-related neologisms in our language. Further innovations are likely to address both input, currently still mostly via Sholes' 1878 QWERTY keyboard, and feedback, now generally by visual monitor. This would improve the whole loop of feedforward and feedback between computer and human.

Summary

The current challenge for layout of the computer workplace is specific to the existing computer technology. While meeting the requirements of eyes on the screen and hands on the keyboard, it must provide a comfortable body support and an agreeable work environment. The prescribed locations of the eyes and the hands fixate the overall position of head and upper body, which does not allow much variation of the body position over time.

Because the nature of repetitive trauma disorders and, hence, means to avoid them were well documented even in the 1960s, one should have expected that key and keyboard designs for typewriters and then for computers would have been improved accordingly. However, no new input concepts took hold. Instead, even current electronic devices essentially follow the established paths; they created novel uses, but they perpetuated the old key-related overuse problems. New technologies, at hand and emerging, allow the designer to establish innovative solutions for human–computer interfaces that make communication easier and more efficient.

An argument against changing the basic QWERTY design of the keyboard has been that an alteration of the layout would require retraining the operators and hence would slow their keying performance, at least initially. This reasoning is probably true for most minor changes, such as exchanging the letter designations of keys, or repositioning them slightly. In such cases, carryover confusion is indeed likely because the new uses are close to the old habits. However, fundamental changes in keys and keyboards design, or even abandoning these altogether, create distinctively novel use procedures, which, because dissimilar from the old ways, should not suffer from interference by crossover from previous practices.

The human is amazingly fast in acquiring novel keyboarding skills as shown by older and recent research; a striking recent demonstration is the phenomenally fast spread, across the globe, of instant text messaging with its miniature keys and keyboards and new use procedures.

Notes and more information

The text contains markers, °, to indicate specific references and comments, which follow.

19.1 Sholes' "Type-Writing Machine" with Its QWERTY keyboard:

Numerous typographical devices: see the listings in Kroemer 2001; Kroemer et al. 2001, 2005.

19.3 Human factors:

Relocated keys: Dvorak's "Simplified Keyboard" of 1936 and '43 and Strong's 1956 report: see Kroemer 2001.

Ramazzini reported on overuse diseases: see pages 43 and 254 in Wright's 1993 translation of the Latin text.

Repetitive work and musculoskeletal strain, local epidemics, occupational diseases: Osler 1892, Poore 1972, 1887; see the listings by Kroemer 2001; Kroemer et al. 2001, 2005.

Pathomechanisms: Armstrong 2006, Marras et al. 2006.

Affliction of flexor tendons: Goodman et al. 2005; *forces:* Dennerlein 2005, 2006; *travel* Ugbolue et al. 2005; *gliding* Ettema et al. 2007; Zhao et al. 2007.

Recent experimental findings: anatomy of the human hand and wrist: Lee et al. 2005.

Kinematics while keyboarding: Baker et al. 2007.

19.4 Input-related anthromechanical issues:

US Standards: 1968, Proposed USA Standard for a General-purpose Keyboard; 1988, ANSI/HFS 100-1988 Standard.

Reviews in the international literature: see the listings by Kroemer 2001; Kroemer et al. 2001, 2005.

Ergonomic guidelines: Cakir et al. 1979.

19.5 Possible design solutions:

Designing for "big changes": Older and recent research: Kroemer 2001, Marklin et al. 2004.

Designing for communication by speech and sound recognition: Chording: Kroemer 2001; Kroemer et al. 2001, 2005; Noyes 1983a.

Repositioning the keyboard: Proposals and patents: Dennerlein 2006; Kroemer 2001; Kroemer et al. 2001; Noyes 1983b; Rempel 2008.

19.6 Design alternatives for keyboards:

Recognition of body movements: Kroemer 2001, Kroemer et al. 2001, 2003; McMillan 2001, McMillan et al. 2001.

Use of EMGs: Reddy et al. 2006.

Nanoergonomics: Karwowski 2006.

19.8 Designing "smart" software:

Reducing the demands on manipulative skills and short-term memory: Czaja et al. 2001, 2003; Fisk et al. 2004.

CHAPTER TWENTY

Workplace design

The driver's workspace in a motor vehicle provides examples of good and bad design. The good examples generally relate to a supportive and comfortable seat and to suitable designs and arrangements of displays. Most hand control arrangements are good examples; in contrast, the usual arrangement of the foot-operated controls illustrates just about the worst design thinkable. The driver cannot see the pedals but must move the foot quickly between accelerator and brake, both operated in the same forward direction although they cause the opposite effect: making the vehicle move faster or slower. In principle, the driver's workspace is a bad design because it forces the driver to maintain the same body position over long periods of time in order to keep the feet close to the pedals, the hands on the steering wheel, and the eyes focused on the road ahead.

20.1 Sizing the workplace to fit the body

One of the classic examples of equipment designed without paying proper attention to the human operator appears in Figure 20.1. It shows a lathe, which, after a long technological development, is one of the machines that functions well—but needs an operator who should be built much shorter and much wider than normal to easily attend the machine. Of course, it should be the other way around: all the controls to be manipulated, and the cutting tools to be observed, should be arranged in size and location to fit the actual human body.

Obviously, body dimensions are of importance for the design of large pieces of equipment and of workspaces, especially those anthropometric data° that describe overall size (stature,

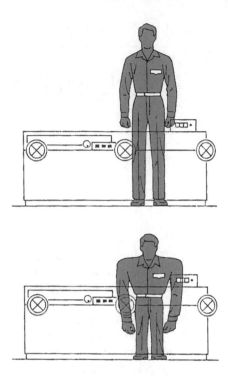

FIGURE 20.1 A lathe with a real and an imagined operator. (Adapted from Eastman Kodak Company (Ed.) (1983). *Ergonomic Design for People at Work*, New York: Van Nostrand Reinhold.)

for example) and which identify eye location as well as hand height and location, because they indicate where objects should be placed that need to be seen and manipulated.

The worker's body size also plays a major role in determining the working height, for example, of a workbench. If it cannot be adjusted for alternating use by small and big persons, then such a simple means as a platform can provide help by raising a short person (see Figure 20.2); however, a raised platform may become a stumbling hazard. The other major determinant of the proper work height is the task: Figure 20.3 illustrates that the same person works best at distinct heights to do different jobs.

Another example of simultaneously considering several human engineering aspects is the design of consoles, where instruments and hand controls are arranged around the upper body of the operator, as shown in Figure 20.4. Such design allows both quick reaches to controls and good viewing of instruments, all about an extended arm's length from the operator.

FIGURE 20.2 Use of a platform to stand on can be of help to a shorter person to operate a machine designed for a taller operator. (Adapted from ILO (International Labour Office) (Ed.) (1986). *Introduction to Work Study* (3rd Edn). Geneva: International Labour Office.)

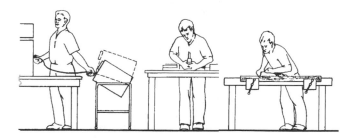

FIGURE 20.3 Different working heights suit the same operator for doing different tasks

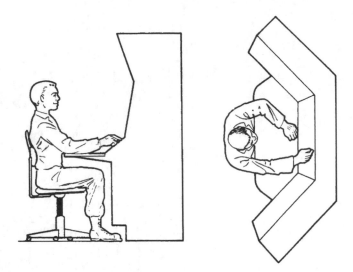

FIGURE 20.4 A console arranged around the operator. (Adapted from Kroemer, K.H.E., Kroemer, H.B., and Kroemer-Elbert, K.E. (2003, amended reprint). *Ergonomics: How to Design for Ease and Efficiency* (2nd Edn). Upper Saddle River, NJ: Prentice-Hall/Pearson Education.)

20.2 On the feet or sitting down?

There are tasks that require walking about, such as to pick up material, take it to the workbench, and, after the work is done, return it to another location; moving around like that can be quite healthy for a fit person. In this case, the work-bench is probably best a bit below elbow height, as shown in Figure 20.5. There also may be periods in which the worker does best while sitting: for this, a tall seat is appropriate, as also shown in Figure 20.5. In both cases, the work surface should extend out some distance beyond the vertical front surface of the workbench so that one can step up close to it, or extend the legs under it when sitting. If such legroom is missing, as shown in Figure 20.6, sitting becomes very awkward.

On the shop floor, elaborate seats are usually not suit-able. Instead, simple designs, robust and easy to clean, are of use. Various kinds of stools and lean-ons, such as shown in Figure 20.7, can take some load off the feet, at least temporarily.

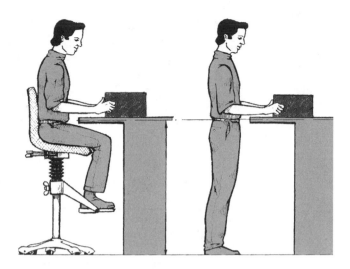

FIGURE 20.5 A workplace suitable for sitting and standing. (Adapted from ILO (International Labour Office) (Ed.) (1986). *Introduction to Work Study* (3rd Edn). Geneva: International Labour Office.)

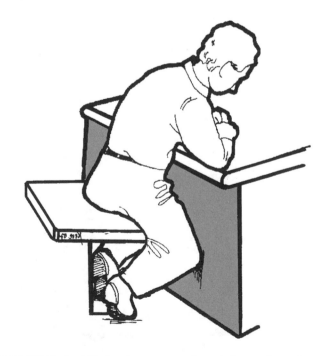

FIGURE 20.6 Missing legroom makes sitting awkward

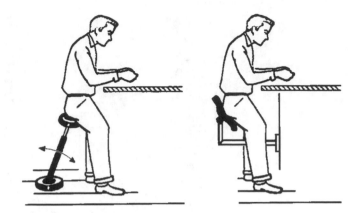

FIGURE 20.7 Examples of stools and lean-ons. (Adapted from Kroemer, K.H.E., Kroemer, H.B., and Kroemer-Elbert, K.E. (2003, amended reprint). *Ergonomics: How to Design for Ease and Efficiency* (2nd Edn). Upper Saddle River, NJ: Prentice-Hall/Pearson Education.)

As mentioned above, requiring a person to maintain the same position over long hours is not a good human factors solution: we humans need to move our bodies, not maintain the same posture. Keeping the same posture is particularly tiring if that involves standing on one foot, as shown in Figure 20.8 where, in order to operate a pedal with the right foot, the operator has to support his whole body weight on his left foot. Birds can stand on one leg for a long time, but people find that very tiring.

Even when we have a chair to sit on, we should not sit still. However, this does not mean that we should do awkward and possibly harmful motions, such as shown in Figure 20.9 where

FIGURE 20.8 Standing on one foot in order to operate a pedal with the other foot

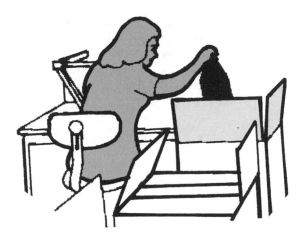

FIGURE 20.9 Packing arrangement that requires bending the body to the right

the workplace is so badly set up that the operator has to bend over to the right side to place items into a container.

Modern technology has generated many tasks and jobs that require sitting, for example, in airplanes and land vehicles, and in many existing offices. Many automobile seats are quite comfortable and supportive—so, some of them are available for use in the office. The times of the hard-surface minimal "task chair" are largely over and now office chairs are on the market that are well-designed yet inexpensive. They provide such features° as sketched in Figure 20.10: a contoured and padded

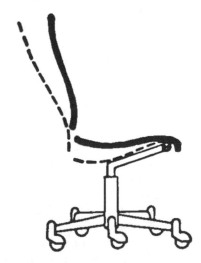

FIGURE 20.10 Essential features of a supportive work seat

seat pan with a smooth "waterfall" contour at the front which avoids pressure behind the knee, and a high backrest that conforms to the shape of the back and even provides a head rest. Both seat pan and backrest can incline and decline, independently or together.

20.3　Manipulating, reaching, grasping

The work task together with the worker's body size determines the necessary size of openings, such as of doors and hatches°, that must allow passing through them, often with equipment worn that makes the body more bulky. Bulkiness may also play a role, together with the need to move the tools, for determining the size of openings in machinery or cabinets to provide access for the hand to do adjustments: Figure 20.11 shows examples.

Quite a few work tasks require finely controlled handling of delicate objects and instruments, such as in repair and assembly work. Exact manipulations are done most easily at about elbow height, in front of the chest and close to the body, which allows steady and secure arm and hand motions; see Figure 20.12. Also, this location facilitates good visual control because it is at a close viewing distance at a well-declined angle of the line of sight°, as Figure 20.13 illustrates.

Repeated reaches, such as in assembly work, are easy if the supply bins are carefully set along the periphery of hand/arm movements, as shown in Figure 20.14. In addition to the bin, tools needed for the job may be hung close by, as Figure 20.15 illustrates, so that the workers can pull them in and simply release them when no longer needed.

20.4　Handling loads

Moving a sizable object by hand from one location to another is often called, in a funny tautology, manual materials handling, as opposed to mechanical "handling". Load handling may remain on the same level as in pushing/pulling, turning, or carrying, or it may involve lifting and lowering. Of course, several of these activities may occur together. Often, load handling is part of other work, such as inspection, assembly, cleaning, polishing, positioning, sorting, and placing.

	Approximate dimensions in cm	
	A	B
For screwdriver use so that hand can trun up to 180 deg.	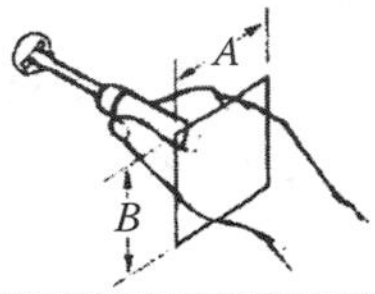 11	12
For use of pliers and similar tools	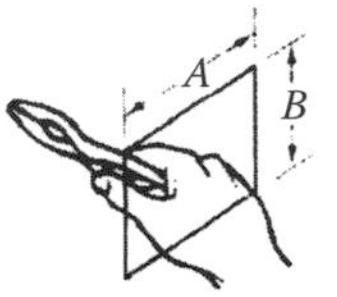 13	12
For turning T-handle wrench up to 180 deg.	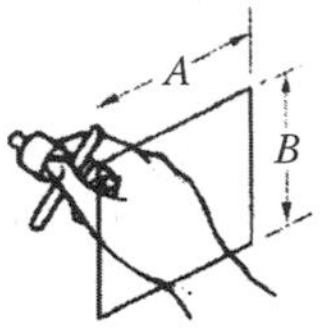 14	16
For turning open-end wrench up to 60 deg.	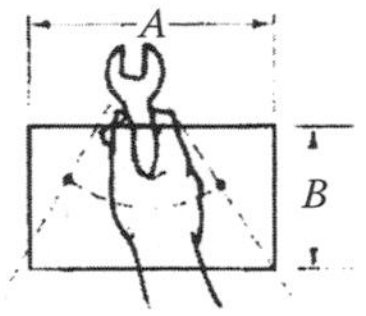 27	20
For turning allen-type wrench up to 60 deg.	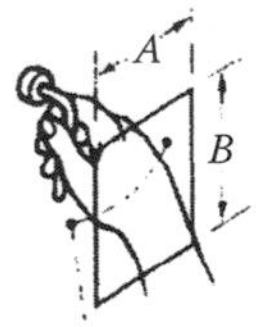 12	16
For using test probe and similar devices	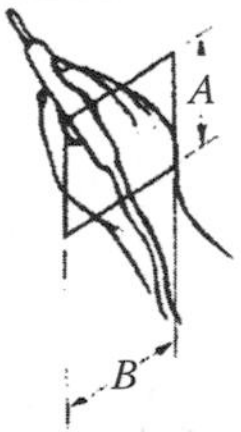 9	9

FIGURE 20.11 Minimal opening sizes (in centimeters) that allow one hand to pass when holding a tool. (Adapted from Kroemer, K.H.E., Kroemer, H.B., and Kroemer-Elbert, K.E. (2003, amended reprint). *Ergonomics: How to Design for Ease and Efficiency* (2nd Edn). Upper Saddle River, NJ: Prentice-Hall/Pearson Education.)

FIGURE 20.12 Work area suitable for finely controlled manipulation. (Adapted from ILO (International Labour Office) (Ed.) (1986). *Introduction to Work Study* (3rd Edn). Geneva: International Labour Office.)

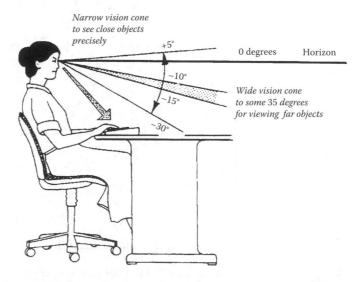

FIGURE 20.13 Vision cones for seeing close and far objects

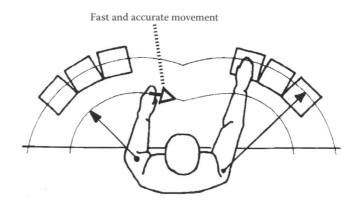

FIGURE 20.14 Supply bins placed along the area of easy reach

FIGURE 20.15 Tools and bins, well arranged at the workplace. (Adapted from ILO (International Labour Office) (Ed.) (1986). *Introduction to Work Study* (3rd Edn). Geneva: International Labour Office.)

FIGURE 20.16 Avoid carrying loads. (Adapted from ILO (International Labour Office) (Ed.) (1986). *Introduction to Work Study* (3rd Edn). Geneva: International Labour Office.)

Moving objects around by hand often leads to injuries of the handler. Therefore, it is advisable to either avoid this altogether by proper design of the task, or by letting some machinery do the job. If people must move objects on the same level, instead of carrying they should pull or push using a dolly, cart, trolley, or conveyor°, as shown in Figure 20.16.

Lifting or lowering loads is one of the most injurious activities. Particularly avoid lifting from the floor: if that absolutely must be done, don't lift in front of your feet and knees, but try to do so between your legs; see Figures 20.17 and 20.18. Figure 20.19 demonstrates that it is easier to lift an object from a workbench or other high location than from the ground.

FIGURE 20.18 If you must lift a load from the floor, try to do it between your legs. (Adapted from Kroemer, K.H.E., Kroemer, H.B., and Kroemer-Elbert, K.E. (2003, amended reprint). *Ergonomics: How to Design for Ease and Efficiency* (2nd Edn). Upper Saddle River, NJ: Prentice-Hall/Pearson Education.)

FIGURE 20.17 Do not try to lift a bulky load in front of your feet and knees. (Adapted from Kroemer, K.H.E., Kroemer, H.B., and Kroemer-Elbert, K.E. (2003, amended reprint). *Ergonomics: How to Design for Ease and Efficiency* (2nd Edn). Upper Saddle River, NJ: Prentice-Hall/Pearson Education.)

FIGURE 20.19 Taking up a heavy load from bench. (Adapted from ILO (International Labour Office) (Ed.) (1986). *Introduction to Work Study* (3rd Edn). Geneva: International Labour Office.)

20.5 Displays and controls

Most of us have learned, through trial and error, to steer a bicycle where we want to go. It is easy to control its direction: we simply turn the handlebar to make the front wheel point to where we want to be; and it is easy to see the path that we are going to take. The relations between control and surroundings are more complicated in most technical systems. In an automobile, for example, there are several controls to determine its movement, and the results of our control actions show on several gauges and through the windows. It takes much longer to learn how to drive a car than to ride a bicycle, and the results of faulty control actions or misreadings of the displays are usually much more severe. Many human–machine systems are even more convoluted: a power plant is run from the control room in an abstract way by pushing buttons and turning knobs, and its functioning is solely displayed via gauges, as Figure 20.20 illustrates. Aircraft cockpits are among the most complex human factors design tasks because they involve many controls and displays. Figure 20.21 sketches the multitude of controls and displays in a large airplane; not surprisingly, the pilots have found ingeniously simple ways to mark objects of specific concern.

Coding° of controls helps to identify them, to point out how to operate them, to indicate the effects of their activation, and to show their status. The major coding practices use

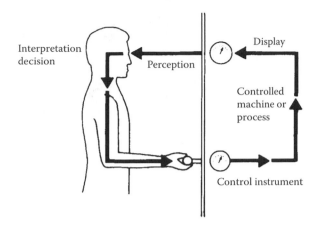

FIGURE 20.20 Operating a system through control action and reading a display

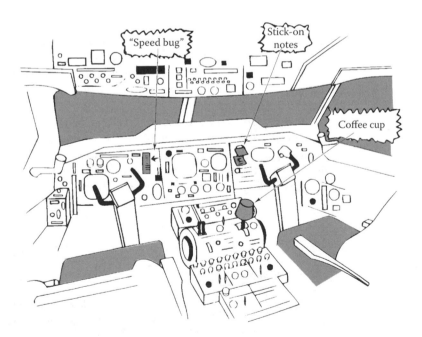

FIGURE 20.21 Aircraft cockpit with improvised markers. (Adapted from Kroemer, K.H.E., Kroemer, H.B., and Kroemer-Elbert, K.E. (2003, amended reprint). *Ergonomics: How to Design for Ease and Efficiency* (2nd Edn). Upper Saddle River, NJ: Prentice-Hall/Pearson Education.)

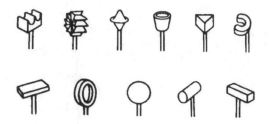

FIGURE 20.22　Easily distinguished control shapes. (Adapted from Kroemer, K.H.E., Kroemer, H.B., and Kroemer-Elbert, K.E. (2003, amended reprint). *Ergonomics: How to Design for Ease and Efficiency* (2nd Edn). Upper Saddle River, NJ: Prentice-Hall/Pearson Education.)

- *Shape:* Appeals to both vision and touch; see Figure 20.22.
- *Location:* To indicate importance and sequence of operation.
- *Size:* To make operation fast and easy.
- *Mode of operation:* Such as pushing, turning, or sliding.
- *Color:* Effective when lit.
- *Labeling:* Effective when read.

Often, several of these coding techniques are used together. In a car familiar to the driver, one is used to finding controls by their location and by their "feel" (produced by shape and size); having to look for them means wasting attention that should be paid to the road and traffic.

A bar knob, shown in Figure 20.23, is easy to grasp and operate (if of proper dimensions) and, by virtue of its shape, indicates its setting on a dial. A toggle switch, Figure 20.24, is also easy to grasp and operate but it has only two or three possible settings. A round knob, Figure 20.25, is handy as well,

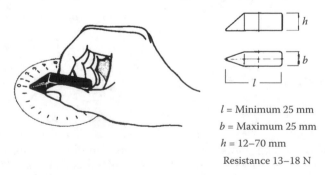

l = Minimum 25 mm
b = Maximum 25 mm
h = 12–70 mm
Resistance 13–18 N

FIGURE 20.23　Bar knob

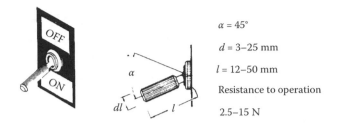

$\alpha = 45°$

$d = 3–25$ mm

$l = 12–50$ mm

Resistance to operation

2.5–15 N

FIGURE 20.24 Toggle switch

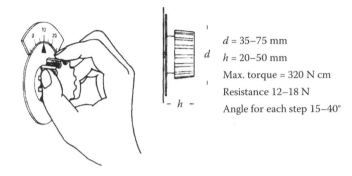

$d = 35–75$ mm

$h = 20–50$ mm

Max. torque = 320 N cm

Resistance 12–18 N

Angle for each step 15–40°

FIGURE 20.25 Rotary knob

but its setting is not obvious unless a special indicator points it out. It can have more settings than a toggle or bar control, and the settings can be continuous or in steps. Push-buttons are fast-operated controls, but they usually have only two settings, which may be difficult to distinguish.

Altimeters, instruments in aircraft to indicate the flying height, have been notorious for being difficult to read, for being easily misread, and for causing numerous emergencies and crashes. Figure 20.26 shows a 1960s model, "improved" from

FIGURE 20.26 Aircraft altimeter, 1960s model

Type of display	 Moving pointer	 Fixed marker, moving Scale	Counter
Ease of reading	Acceptable	Acceptable	Very good
Detection of change	Very good	Acceptable	Poor
Setting to a reading: controlling a process	Very good	Acceptable	Acceptable

FIGURE 20.27   Pointer and counter displays

previous designs: it is still complex, and the pointer can still cover the counter windows.

Figure 20.27 lists use characteristics of three indicator techniques: moving pointer over fixed scale, fixed pointer over moving scale, and digital counter. The main advantage of the moving pointer is that its motion catches the eye and conveys information about both magnitude and direction of change. The advantage of a fixed marker over a moving scale is that a large range can be covered, with only the currently used section appearing in the window. The digital counter is particularly good for exact readings, as Figure 20.28 shows.

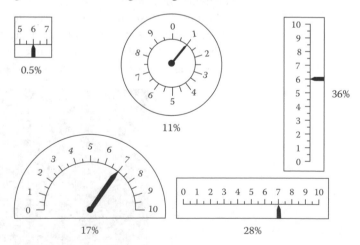

FIGURE 20.28   Reading errors with counter and pointer displays

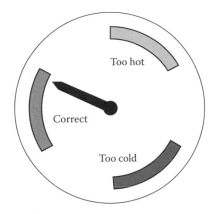

FIGURE 20.29 Two designs of a moving-pointer display

FIGURE 20.30 Status display: too hot, correct, too cold

When comparing moving-pointer displays, such as in Figure 20.29, the location of numbers and scale markers can make a big difference: in a bad design, the pointer obscures numbers and scale indications whereas in a better design, numbers and markers are outside the field over which the pointer moves, and the tip of the pointer just touches the scale markers.

Often, exact information, such as a numerical dial presents, is not needed: all the observer has to know is whether a condition is normal or either below or above. Figure 20.30 presents such a simple but effective gauge. The display may be made even more effective with signal colors, such as green for normal, blue for cold, and red for hot. The designer must keep in mind, however, that the meanings of colors may be different for persons with various cultural roots. Humans do differ: that fact makes ergonomic research exciting and human factors engineering interesting.

Summary

Human-centered design follows these guidelines°.

Consider

Human body size	Establish the dimensions of workspace, equipment, and tools to accommodate the human body.
Human strength	Facilitate exertion of strength (work, power) by object location and orientation.
Human speed	Place items so that they can be reached and manipulated quickly.
Human effort	Arrange work so that it can be performed with least effort.
Human accuracy	Select and position objects so that they can be manipulated and seen with ease.
Importance	Place the most important items in the most accessible locations.
Frequency of use	Place the most frequently used items in the most accessible locations.
Function	Group items with similar functions together.
Sequence of use	Lay out items that are commonly used in sequence in that sequence.
Human safety and comfort	Arrange all details to assure that the human will not be injured, will not suffer health consequences, but instead will be secure and feel comfortable.

Notes and more information

The text contains markers, °, to indicate specific references and comments, which follow.

   20.1 Sizing the workplace to fit the body:

   Anthropometric data: The very first chapter of this book contains a discussion and listing of human body dimensions.

   20.2 On the feet or sitting down:

   Seat features: see Chapter 18 in this book.

   20.3 Manipulating, reaching, grasping: sizes of doors, hatches, passage ways: see Kroemer et al. 2003.

   Line of sight: discussed in Chapters 5, 11.

   20.4 Handling loads:

   Pushcarts, dollys, and other transport means: more in Chapter 4.

   20.5 Displays and controls:

   Coding of controls: see Freivalds 1999, Kroemer et al. 2003.

   Human-centered design, guidelines: more information in many handbooks, such as those by Bridger 2003, Chengular et al. 2003, HFES 2004, Konz et al. 2000, Kroemer et al. 2003, Lueder et al. 2007.

CHAPTER TWENTY-ONE

Making work pleasant and efficient

So far, the text in this book has been concerned mostly with the physical aspects of "fitting work to the human," even though there were repeated references to "fitting the human to work", such as by personnel selection and training. This last section pulls ergonomic information together to demonstrate how work can be both pleasant, which is of primary interest to the individual, and efficient, especially important to management.

21.1 Using our skills and interests; getting along with others at work

Each one of us is different from everybody else; our upbringing, environment, experiences, and personalities all make us special. Every one of us is a unique individual. Yet, polls show consistently—in North America, Europe, and Australia/New Zealand, and probably everywhere else—that, as a general rule, better job skills and education lead to higher job satisfaction and income. For nearly everybody, intrinsic rewards provide us with basic satisfaction with our work. To do a task well often is reward by itself. Doing what we like to do, exercising our talents, learning new skills and techniques and applying them, and accomplishing tasks make us content.

Work worthwhile doing

Work that we perceive as needed, valuable, and useful is worth doing. If our work leads to improvements in procedures, services, and products, that outcome renders us proud and is recompense by itself. Among the most satisfying jobs are those that make

391

other people feel and get better; these jobs usually involve teaching others, helping and protecting others, and caring for others. Persons who feel a calling for what they are doing, generally are deeply content, even if their pay is relatively low and their workload rather high, as a general social survey in the United States showed in 2007°.

Satisfied employees

Seen from the organization's perspective: in many cases, a company's market value is determined about half by its hard assets such as property, plant, and equipment. The remainder is made up of soft attributes: patents, customer base, and especially their human resources. Satisfied and "engaged" employees are a company's most valuable assets.

Engaged to work

To feel engaged to work requires a good match between a person, her or his interests and aspirations, and the job. The job is defined by

- Its tasks including equipment and procedures used
- Its rewards for good performance such as recognition, potential for advancement, and pay
- Its many-faceted working conditions, which include the social relations among people at all levels, often loosely labeled "organizational climate"

Job performance is the product of motivation and ability. Situational factors at work can stymie or enhance performance; examples are the organizational setup and the climate (both psychological and physical) as well as tools and equipment are discussed earlier in this book.

Motivation and job performance

Motivation incites, directs, and maintains our behavior toward goals. Motivation and job performance are evidently related; a motivated person who desires to do well is willing to expend effort to do so. Understanding what motivates people can help us be aware of why we behave on the job as we do and can tell management how to make people effective and satisfied at work. A number of theories exist that try to explain motivation and needs (some of them are outlined in Chapter 13). They fall essentially into two groupings. One focuses on the individual and personal inherent traits; the other places the work environment at the forefront.

Needs-based motivation

Several motivation theories focusing on the individual are needs-based. Personal needs range from basic physiological necessities (food, shelter) to higher-order wants (self-esteem, self-actualization). Individuals strive to first cover their basic needs and then work their way up. A typical hierarchy of needs has several levels: at the base are physiological needs for food and water and shelter, followed by safety needs and wants for economic and physical security. Higher up are social needs, for belonging and love; the next level contains esteem needs, including self-confidence, recognition, appreciation, and respect. The highest-order needs concern self-actualization; at this level, an individual achieves full potential and perceives personal success and satisfaction.

Performance and expectance

Expectancy theories assume that an individual's motivation (and resulting satisfaction) depends on the difference between what the person's environment offers versus what he or she expects. This can be expressed, in its simplest form, as a model that explains motivation as a function of coupled expectations:

{Work will lead to performance} *then*

{Performance will lead to reward}.

The organization provides rewards for its employees: pay, promotions, and formal recognition. (It can be a negative outcome: getting fired, for example.) The value of the reward is highly individual; it depends on how the employee rates the anticipated reward. The expectancy component is crucial; there must be a connection between how hard one tries and how well one performs. To illustrate, a factory worker on an assembly line may have little incentive to increase her rate of production because the overall speed of the assembly line, and hence the performance, remain unchanged regardless of her efforts. In contrast, a truck driver can deliver more the faster he drives, so he may be motivated to exceed the speed limits.

Proud to work

Surveys of the top U.S. companies provide a summary of the various factors underlying individual job satisfaction and performance. The following comments illustrate why organizations make the "100 Best Companies to Work For" list. Employees say: "This is a positive place. We do good things. We succeed. People are friendly. They feel challenged. They feel respected and valued. And they respond with loyalty." And the company maxims: "Provide good tools and equipment. Treat people the way they want to be treated. Strive for mutual respect and for an atmosphere that makes people proud to work here. Provide career opportunities. Say thank you for a job well done."

"Have a life"

We need to balance our work and our personal life. Only a few people live to work while most want to have a life outside of the work. The times of early industrialization are gone when workers labored for ten or more hours a day, six or seven days a week. Nowadays, the five-day workweek with about eight hours of work a day (less in some industrialized countries) is the norm. Even some high-end management-consulting firms, notorious for extraordinarily demanding work schedules that often leave their employees overtaxed, are looking for ways to reduce the workload and let their employees enjoy life outside of their work. Some companies have eliminated weekend work (including business travel) for their employees; they encourage leaving personal phones and computers turned off outside business hours and to take personal days and uninterrupted vacations without being on-call.

Bad bosses

Although insufficient job performance is often a reason why workers lose their jobs, bad bosses are not infrequent either. A 2007 survey in the United States° showed that two of five supervisors did not keep their word or failed to give credit when due; and that one of four bad-mouthed employees or blamed them for mistakes made by the boss. Such managers often create severe dissatisfaction among the employees, many of whom quit their jobs, not leaving their company but rather their boss. Bad managers also create problems for their company because they generate poor morale, reduce production, and increase employee turnover.

Taking charge

Good working conditions, including pleasant personal relations at all levels, are important, often more so than small differences in payment received. Abusive relationships between employers and employees were not rare in the past, when the "little people" where dependent on the "big bosses". These medieval conditions have been replaced, almost uniformly, by the understanding that employees at all levels are responsible for making a business prosper or fail, and that they all should pull in the same direction. Yet, if for whatever reasons a person feels dissatisfied with the conditions and treatment at work, there are a number of constructive ways, many listed in Table 21.1, to determine the causes of discontent so that the conditions can be improved. Both managers and employees alike have similar factors to consider that make their work lives more pleasurable and proficient, although probably seen from different points of view.

Table 21.1 Taking charge of personal relations at work

This will contribute to making you feel good about your work:

- As a potential employee, you should strive to understand the environment of an organization when you consider taking up work there: "environment" includes both the organization itself and the individuals within it.

- Examine the components of a company—such as its structure, strategies, policies, conventions and "culture"—to determine that they agree with your own personality, and your personal goals and beliefs.

- Choose a company that fits your own personal style as much as possible.

- Consider both internal and extrinsic factors that contribute to job satisfaction: spend some time with your prospective colleagues and supervisors and determine how you feel about them because you will be with them many hours at work.

- Examine your own needs and be as realistic as possible about what they are. Does the job meet your physical needs, such as sufficient pay and fringe benefits? What about your emotional needs, such as recognition and a sense of achievement? How are the work conditions, the management setup, the dress code, and the like? Make sure they are appropriate for you and pleasing to you.

- Take advantage of any programs available to you, especially if offered by your company, to increase your skills.

- Technology can make your life easier, but make sure it does not keep you connected to work day and night.

What to do when you are miserable at the job:

- Talk with your supervisor/colleague/supervisee in a nonconfrontational tone about existing problems. It is important that you propose solutions.

- Check your employment contract and the company policies and procedures manual to determine your duties and rights at work.

- Talk with your organization's union steward, social officer, or employee-relations agent.

- Take advantage of any stress-reduction programs your company offers.

- If there is no solution to a problem that bothers you so much that you don't want to work at the organization any more, then look for another position.

Adapted from Kroemer, K.H.E. and Kroemer, A.D. (2001). *Office Ergonomics*. London: Taylor & Francis.

21.2 Setting up our own work, workplace, and work environment

Larger, richer jobs

In the early 1900s, Tayloristic principles favored breaking jobs into small task elements that were then standardized. This usually meant that a worker did the same task repeatedly, resulting in extreme specialization and, unfortunately, often acute tedium. In the mid-1900s, job enlargement—increasing tasks and task variety—and job enrichment—increasing workers' participation and control—grew popular. Several motivation and job satisfaction theories try to explain how job design influences behavior. The so-called job-characteristics model[o] proposes five core dimensions: 1. skill variety, the number of different talents and activities that the job requires; 2. task identity, the specifics of the work that a person does from beginning to end; 3. task significance, the impact of one's work on other persons, procedures, and products. These three factors make the job meaningful. A fourth factor, autonomy, is the degree of independence in planning, controlling, and determining work procedures. This makes the person feel responsible. The final factor is feedback, which provides information about how one's performance is evaluated and perceived. These five job dimensions provide the jobholder the experience of meaningfulness of work, responsibility for work outcomes, and knowledge of results. These experiences generate high motivation and satisfaction for the worker and, of particular importance to management, better-quality work, and lower absenteeism and turnover.

Suggestions for improvements

The design of most workplaces for manual work in production and repair follows well-established traditions, usually determined by the work procedures employed, the objects to be worked on, the machinery used, and the hand tools needed. This commonly leaves little leeway for the individual worker to determine original design details of the workplace or work, although smart management encourages workers' suggestions. In fact, numerous reports and anecdotes show that major enhancements in productivity, safety, and appeal result from suggestions for improvements made by involved personnel. (Evidently, collecting and using such suggestions in the planning and predesign concept stages would have been even more gainful.) When management recognizes and rewards expressions of personal interest in better work, follows up on suggestions, incorporates proposals for novel work and workplace layout, a spirit of teamwork and cooperation develops that provides motivation and impetus to achieve.

Work environment

Chapters 17 through 20 described workplaces, tasks, and work procedures that suit the human body and mind. Special aspects of the environment at work can strongly influence well-being and performance: they prominently concern seeing, hearing, and physical climate, as shown in Chapters 5, 6, 8, and 11.

Vision requires good lighting

The characteristics of human vision discussed in Chapter 5 provide detailed information for designing the environment for proper vision. The most important concept is the provision of suitable workplace lighting that illuminates visual targets as desired. What mostly counts is the luminance of an object (the light reflected or emitted from it) which meets the eye. The acuity of seeing an object is much influenced by strong luminance contrast between the object and its background, including shadows.

Another important point is the avoidance of unwanted or excessive glare. Direct glare meets the eye straight from a light source, such as the sun or a lamp shining into your eyes. Indirect glare is reflected from a surface into your eyes, such as the sun or a lamp mirrored in your computer screen. Use of colors on the visual target, if selected properly, can be helpful, but color vision requires sufficient light, and luminance contrast is usually more important than coloring. Table 21.2 lists recommendations for the lighting at work.

Sounds provide important information

Another essential environmental aspect is that of hearing and sounds at work; see Chapter 6. Sounds convey information about the functioning of machinery and of work progress, they provide verbal communications from co-workers, and, possibly of vital importance, present warning signals about dangers in the work environment. Furthermore, the sound environment may be hazardous to one's hearing if it produces temporary or permanent damage to the ears. Table 21.3 lists recommendations for the acoustic environment at work.

Suitable climate at work

The third important environmental condition at work concerns the physical climate, determined by temperatures, air movement, and humidity, discussed in some detail in Chapter 8. When we work outside, we have to take the climate as is, and our adjustments are in working habits and clothing; inside, however, a variety of low- and high-tech means are available to further well-being and work outcomes, listed in Table 21.4.

Table 21.2   Taking charge of the lighting at work

This will contribute to making you feel good about your work:

- General illumination is best at about 500 lx; if there are many dark (light-absorbing) surfaces in the room, it may be up to 1000 lx.
- If cathode-ray tube (CRT) displays are present, the overall illumination should be between 200 and 500 lx.
- In rooms with light-emitting displays, illumination of 300 to 750 lx is suitable.
- When possible, select indirect lighting, where all light is reflected at a suitable surface (at the ceiling or walls of a room, or within the luminaire) before it reaches the work area. (This helps to avoid direct and indirect glare.)
- Use several low-intensity lights instead of one intense source, placed away from the line of sight. (This avoids direct glare.)
- Place any high-intensity light sources (including windows) outside a cone-shaped range of 60 degrees around the line of sight. (This avoids direct glare.)
- Shine a task light on your visual target if the overall illumination is too dim to generate sufficient luminance.
- Shining a beam of directed light on an object can make stand it out from the background by generating contrasts between illuminated and shaded areas.
- Properly distribute light over the work area, which should have dull, matte, or other nonpolished surfaces. (This avoids indirect glare.)
- The naturally chosen line of sight to a target not more than 1 meter away from the eyes is
    —Straight ahead, neither to the left or right, and
    —Between 25 and 65 degrees below the Ear-Eye line—see Chapter 5. (If the person holds the head erect, the target should be distinctly below the eyes, not higher.)

Notice that these recommendations are not in "hard numbers" because different people prefer various arrangements. What is good for one individual may not suit everybody else.

If you are having difficulties seeing objects clearly, or you feel eye fatigue or eye strain or dry eyes:

- Have a physician (ophthalmologist or optometrist) check your eyes' vision capabilities and advise you about measures that improve and protect your seeing.

Table 21.2 Taking charge of the lighting at work (Continued)

What to do if you are not comfortable:

If you feel eye fatigue or eye strain check and correct, as needed

- The distance from your eyes to visual targets
- The direction of the line of sight to visual targets
- The luminance contrast between the details of the visual target and the background (such as lines on a source document, letters on the screen, objects on the work bench)
- Any direct glare shining into your eyes, or reflected glare from the display or other mirroring surface

If a window is in front of you, with the sun or bright daylight shining into your eyes:

- Draw a curtain or lower blinds, or turn your workstation so that the window is to your side.

If a task light blinds you, turn it off or turn it to the side.

If your white shirt/blouse/sweater or other clothing is mirrored in your computer screen, making it hard to decipher things on the display:

- Change into darker clothes.

If you feel fatigue or strain in neck or back, check and correct, as needed

- The posture of your neck and back, which may be affected by
  —The distance from your eyes to visual targets
  —The direction of the line of sight to visual targets

Table 21.3   Taking charge of the sound environment at work

This will contribute to making you feel good about your work:
- The overall sound level in the office should be between 50 and 65 dB; at other workplaces it should not exceed 75 dB.
- The existing sound level should not change appreciably or dramatically especially if persons do work that requires intense concentration.
- If the surround sound level exceeds about 75 dB or is otherwise disturbing, hearing protection devices should be worn that either cover the whole ear or are inserted into the ear canal. Devices that actively cancel surround sound (see Chapter 6) are best: some of these allow transmission of desired sound, such as voice messages or music.
- Many people find background music pleasant, but the preferences for the kind of music, and its intensity and duration are highly individual. Therefore, music is best presented over individual speakers, such as ear plugs (buds) or caps (muffs), that do not disseminate sound to the environment.

If you are having ringing in your ears, or if your hearing is reduced after exposure to loud sounds:
- Have a physician (otolaryngologist) check your hearing and advise you about measures that improve and protect your hearing.

If it is too loud for you:
- Eliminate the sound at its source:
  —Replace noisy equipment or machines with quieter ones.
  —Place a noisy piece of equipment outside your room.
  —Turn down the sound level of your music.
  —Ask your co-workers for quieter behavior.
- Reduce your exposure to the sound:
  —Wear a hearing protection device.
  —Encapsulate the source of sound.
  —Soften hard surfaces reflecting or transmitting the sound with drapes, carpets, acoustic tiles and the like which dampen or absorb sound.
- If all else fails: move to a different workplace.

If it is too quiet for you:
- Play music.
- Have other people move into your work area.
- If your office is "sound dead" because there is too little reverberation, remove drapery, carpets, acoustic tiles and similar sound-absorbing materials. Bare floors and walls, use large pictures under glass, or windows that have hard surfaces reflecting sounds.

Table 21.4 Taking charge of the climate at work

This will contribute to making you feel good about your work:

- With appropriate clothing and light physical work in the office, comfortable environment temperature ranges are from the low 20s to about 27°C during the summer, but lower during the winter, between about 18°C and the middle 20s.
- The difference between air temperatures at floor and head levels should be less than about 6°C.
- Differences in temperatures between body surfaces and surfaces on the side (such as walls or windows) should not exceed approximately 10°C.
- The preferred range of relative humidity is from 30 to 60 percent, best near 40 to 50 percent.
- Air velocity should not exceed 0.5 m/sec, and preferably remain below 0.1 m/s. Air flow should not generate sound levels above 60 dB.
- If the sun shines onto people, particularly on warm days, they should be able move out of the sun, or to get into the shade behind blinds, curtains, screens, and the like.

These recommendations apply to most conditions in moderate climates, such as northern America and Europe.

- During outside work, where you have no control over the climate, adjust your work intensity, timing and pace (including rest breaks), and your work clothes to achieve a sustainable level of exertion.

If you feel too warm:

- Take off a layer of clothing; bare more skin.
- Lower the room temperature.
- Move away from a heat source such as a radiator, a warm wall or window; get out of the sun.
- Move closer to a cool surface.
- Lower air humidity with a dehumidifier.
- Increase the air movement around you (unless it is very hot air).
- Wet your exposed skin; place a cool/moist cloth on your forehead, neck, or wrists.
- Keep your body at rest, do not exercise it.

Continued

Table 21.4   Taking charge of the climate at work
(Continued)

If you feel too cool:

- Add a layer of clothing; cover more skin.
- Increase the room temperature.
- Move closer to a heat source, such as a radiator, a warm wall, or window; get into the sunshine.
- Move away from a cool surface.
- Move closer to a warm surface.
- Decrease the air movement around (unless it is nice warm air).
- Keep your body moving.

If you feel too dry (dry throat, nose):

- Increase air humidity by evaporating water on a warm surface; use a humidifier; also drink water.

If you draw sparks of static electricity:

- Increase air humidity by using a humidifier or otherwise evaporating water on a warm surface; also check your clothing including your shoes because they may generate an electric charge against the office furniture or carpeting.

Summary

"I like my work."

Too often, work is drudgery, an unpleasant way to earn one's living. However, work can be pleasant if it follows our interests and suits our skills, when it fits our body and mind°.

Notes and more information

The text contains markers, °, to indicate specific references and comments, which follow.

> *21.1 Using our skills and interests, getting along with others at work:*
>
> *Social survey in the United States in 2007:* See T.M. Smith 2007.
>
> *Bad bosses:* See Harvey et al. 2007.
>
> *21.2 Setting up our own work, workplace, and work environment:*
>
> *Job-characteristics model:* See Hackman et al. 1980.
>
> *Work that fits our bodies and minds:* Many handbooks provide guidance for the use of ergonomics/human factors in design. These include, published since 2000, Bridger 2003, Chengular et al. 2003, Konz et al. 2000, Kroemer et al. 2003.

The last page

The motto *Fitting the Human* points out the two approaches in ergonomics. One is mainly in the domain of industrial psychology: matching individuals with work requirements by personnel selection and training.

However, trying to compensate for misguided designs by telling people how to use them properly is a difficult task. The computer keyboard (Chapter 19) and pedals in cars and trucks (Chapter 20) are two examples of long-established designs that originally seemed workable but now are inadequate. It is nearly impossible to teach people how to efficiently and safely use equipment with serious human-engineering deficiencies: consider the clumsy slowness of operating the QWERTY keyboard—instead of using speech recognition—as computer input; and think about the maiming of thousands of people in vehicular mishaps in spite of intensive driver training.

The other ergonomic tactic is to plan carefully all tools and equipment, work tasks and procedures, working hours and shift arrangements, and the physical and social conditions. Accommodating the human is the fundamental and most successful approach, the topic of this book: design the overall work system and its details to fit the human. This makes work safe, efficient, satisfying, and even enjoyable.

References

[The] *Merck Manual of Medical Information,* 1997 Home Edition, Whitehouse Station, NJ, Merck & Co., Inc.

Adams, S.K. (2001). Hand grip strength, 240–246, Hand-grip torque strength, 247–251. In Karwowski, W. (Ed.) *International Encyclopedia of Ergonomics and Human Factors.* London: Taylor & Francis.

Alderfer, C.P. (1972). Existence, Relatedness, and Growth: Human Needs in Organizational Settings. New York: Free Press.

Armstrong, T.J. (2006). The ACGIH TLV for hand activity work. In Marras, W.S. and Karwowski, K. (Eds.) *The Occupational Ergonomics Handbook* (2nd Edn) *Fundamentals and Assessment Tools for Occupational Ergonomics,* Chapter 41. Boca Raton, FL: CRC Press.

Arndt, B. and Putz-Anderson, V. (2001). *Cumulative Trauma Disorders* (2nd Edn). London: Taylor & Francis.

Arswell, C.M. and Stephens, E.C. (2001) Information processing. In Karwowski, W. (Ed.) *International Encyclopedia of Ergonomics and Human Factors,* 256–259. London: Taylor & Francis.

Astrand, P.O., Rodahl, K., Dahl, H.A., and Stromme, S.B. (2004). *Textbook of Work Physiology. Physiological Bases of Exercise* (4th Edn). Champaign, IL: Human Kinetics.

Bailey, R.W. (1996). *Human Performance Engineering* (3rd Edn). Upper Saddle River, NJ: Prentice Hall.

Baker, N.A., Cham, R., Cidboy, E.H., Cook, J., and Redfern, M.S. (2007). Kinematics of the fingers and hands during computer keyboard use, *Clinical Biomechanics,* **22**:1, 34–43.

Basbaum, A.I. and Julius, D. (2006). Toward better pain control, *Scientific American,* **294**:6, 60–67.

Berger, E.H., Royster, L.H., Royster, J.D., Driscoll, D.P., and Layne, M. (2003). *The Noise Manual* (5th Edn). Fairfax, VA: American Industrial Hygiene Association.

Bernard, T.E. (2002). Thermal stress. In Plog, B.A. (Ed.) *Fundamentals of Industrial Hygiene* (5th Edn), Chapter 12. Itasca, IL: National Safety Council.

Bjoerkman, T. (1996). The rationalization movement in perspective and some ergonomic implications, *Applied Ergonomics, 27,* 111–117.

Boff, K.R. and Lincoln, J.E. (Eds.) (1988) *Engineering Data Compendium: Human Perception and Performance.* Wright-Patterson AFB, OH: Armstrong Aerospace Medical Research Laboratory.

Boff, K.R., Kaufman, L., and Thomas, J.P. (Eds.) (1986). *Handbook of Perception and Human Performance.* New York: Wiley.

Booher, H.R. (Ed.) (2003). *Handbook of Human Systems Integration.* New York: Wiley.

Borg, G. (2001). Rating scales for perceived physical effort and exertion. In Karwowski, W. (Ed.) *International Encyclopedia of Ergonomics and Human,* 358–541. London: Taylor & Francis.

Borg, G. (2005) Scaling experiences during work: Perceived exertion and difficulty. In Stanton, N., Hedge, A., Brookhuis, K., Salas, E., and Hendrick, H. (Eds.) *Handbook of Human Factors and Ergonomics Methods,* 11–1 – 11–7. Boca Raton, FL: CRC Press.

Boyce, P.R. (2003). *Human Factors in Lighting.* London: Taylor & Francis.

Bridger, R.S. (2003). *Introduction to Ergonomics* (2nd Edn). New York: McGraw-Hill.

Bullinger, H., Kern, P., and Braun, M. (1997). Controls. In Salvendy, G. (Ed.) *Handbook of Human Factors and Ergonomics* (2nd Edn), Chapter 21. New York: Wiley.

Cakir, A., Hart, D.J., and Stuart, T.F.M. (1979). *Visual Display Terminals* (in German). Darmstadt: Inca-Fiej.

Carayon, P. and Lim S.Y. (2006). Psychosocial work factors. In Marras, W.S. and Karwowski, K. (Eds.) *The Occupational Ergonomics Handbook* (2nd Edn) *Interventions, Controls, and Applications in Occupational Ergonomics,* Chapter 5. Boca Raton, FL: CRC Press.

Casali, J.G. and Gerges, S.N.Y. (2006a). Protection and enhancement of hearing in noise. In Williges, R.C. (Ed.) *Reviews of Human Factors and Ergonomics, Vol. 2,* Chapter 7. Santa Monica, CA: Human Factors and Ergonomics Society.

Casali, J.G. and Robinson, G.S. (2006b) Noise in industry. In Marras, W.S. and Karwowski, K. (Eds.) *The Occupational Ergonomics Handbook* (2nd Edn) *Fundamentals and Assessment Tools for Occupational Ergonomics,* Chapter 31. Boca Raton, FL: CRC Press.

Chaffin, D.B., Andersson, G.B.J., and Martin, B.J. (2006). *Occupational Biomechanics* (4th Edn). New York: Wiley.

Chapanis, A. (1996). *Human Factors in Systems Engineering.* New York: Wiley.

Chengular, S.N., Rodgers, S.H., and Bernard, T.E. (2003) *Kodak's Ergonomic Design for People at Work* (2nd Edn). New York: Wiley.

Ciriello, V.M. (2001). The effects of box size, vertical distance, and height on lowering tasks. *International Journal of Industrial Ergonomics, 28,* 61–67.

Ciriello, V.M. (2007). The effects of container size, frequency, and extended horizontal reach on maximum acceptable weights of lifting for female industrial workers. *Applied Ergonomics,* **38**:1, 1–5.

Corlett, E.N. (2005). The evaluation of industrial seating. In Wilson, J.R. and Corlett, N. (Eds.) *Evaluation of Human Work* (3rd Edn), Chapter 27. London: Taylor & Francis.

Costa, G. (1996). The impact of shift and night work on health. *Applied Ergonomics,* **27**, 9–16.

Cox, T. and Griffiths, A. (2005). The nature and measurement of work-related stress: Theory and practice. In Wilson, J.R. and Corlett, N. (Eds.) *Evaluation of Human Work* (3rd Edn), Chapter 19. London: Taylor & Francis.

Czaja, S.J. and Lee, C.C. (2001). The Internet and older adults: Design challenges and opportunities. In Charness, N., Park, D.C., and Sabel, B.A. (Eds.) *Aging and Communication: Opportunities and Challenges of Technology,* 60–81. New York: Springer.

Czaja, S.J. and Lee, C.C. (2003). Designing computer system for older adults. In Jacko, J. and Sears, A. (Eds.) *Handbook of Human–Computer Interaction*, 413–427. Mahwah, NJ: Erlbaum.

Daams, B.J. (1993). Static force exertion in postures with different degrees of freedom. *Ergonomics* **36**, 397–406.

Daams, B.J. (1994). *Human Force Exertion in User-Product Interaction. Background for Design.* Delft: Delft University Press–IOS Press.

Daams, B.J. (2001). Push and pull data, 299–316, torque data, 334–342. In Karwowski, W. (Ed.) *International Encyclopedia of Ergonomics and Human.* London: Taylor & Francis.

Delleman, N.J., Haslegrave, C.M., and Chaffin, D.B. (Eds.) (2004). *Working Postures and Movements.* Boca Raton, FL: CRC Press.

Dempsey, P.G. (1998). A critical review of biomechanical, epidemiological, physiological and psychophysical criteria for designing manual materials handling tasks. *Ergonomics,* **41**, 73–88.

Dempsey, P.G. (2006). Psychophysical approach to task analysis. In Marras, W.S. and Karwowski, K. (Eds.) 2006a. *The Occupational Ergonomics Handbook* (2nd Edn), *Fundamentals and Assessment Tools for Occupational Ergonomics*, Chapter 47. Boca Raton, CRC.

Dennerlein, J. (2006). The computer keyboard system design. In Marras, W.S. and Karwowski, K. (Eds.) *The Occupational Ergonomics Handbook* (2nd Edn), *Interventions, Controls, and Applications in Occupational Ergonomics*, Chapter 39. Boca Raton, FL: CRC Press.

Dennerlein, J.T. (2005). Finger flexor tendon forces are a complex function of finger joint motions and fingertip forces. *Journal of Hand Therapy* **18**, 2, 120–127.

Deyo, R.A. and Weinstein, J.N. (2001). Low back pain. *New England Journal of Medicine,* **344**, 363–370.

DiDomenico, A. and Nussbaum, M.A. (2003). Measurement and prediction of single and multi-digit finger strength. *Ergonomics,* **45**, 1531–1548.

DiDomenico, A., Nussbaum, M.A., and Kroemer, K.H.E. (1998). Measurement and prediction of finger forces. In Kumar, S. (Ed.) *Advances in Occupational Ergonomics and Safety*, 386–389. Amsterdam, IOS Press.

DiNardi, S.R. (Ed.) (2003). *The Occupational Environment: Its Evaluation, Control, and Management* (2nd Edn). Fairfax, VA: American Industrial Hygiene Association.

Driskell, J.E. and Mullen, B. (2005). The efficacy of naps as a fatigue countermeasure: a meta-analytic integration. *Human Factors,* **47**, 360–377.

Erens, B., Primatesta, P., and Prior, G. (Eds.) (2001). *The Health Survey for England*, London: The Stationary Office. Available on the Internet at www.archive.official-documents.co.uk/document/doh/survey99/hse99-06.htm.

Ettema, A.M., Zhao, C., Amadio, P.C., O'Byrne, M.M., and An, K.N. (2007) Gliding characteristics of flexor tendon and tenosynovium in carpal tunnel syndrome: A pilot study. *Clinical Anatomy*, **20**, 3, 292–299.

Fisk, A.D., Rogers, W.A., Charness, N., Czaja, S.J., and Sharit, J.. (Eds.) (2004). *Designing for Older Adults*. Boca Raton, FL: CRC Press.

Flier, J.F. and Maratos-Flier, E. (2007). What fuels fat, *Scientific American* **297**:3, 72–81.

Folkard, S. and Monk, T.H. (Eds.) (1985). *Hours of Work*. Chichester: Wiley.

Folkard, S. and Tucker, P. (2003). *Shift* work, safety and productivity. *Occupational Medicine,* **53**, 95–101.

Fox, J.G. (1983). Industrial music. In Oborne, D.J. and Gruneberg, M.M. (Eds.) *The Physical Environment at Work*, 221–226. New York: Wiley.

Freivalds, A. (1999). Ergonomics of hand controls. In Karwowski, K. and Marras, W.S. (Eds.) *The Occupational Ergonomics Handbook*, Chapter 27. Boca Raton, FL: CRC Press.

Freivalds, A. (2006). Upper extremity analysis of the wrist. In Marras, W.S. and Karwowski, K. (Eds.) *The Occupational Ergonomics Handbook* (2nd Edn), *Fundamentals and Assessment Tools for Occupational Ergonomics*, Chapter 45. Boca Raton, FL: CRC Press.

Gawande, A. (2008). The itch, *The New Yorker*, 30 June, 58-65.

Gibbons, J.D. (1997). *Nonparametric Methods for Quantitative Analysis* (3rd Edn). Columbus, OH: American Sciences Press.

Goodman, H.J. and Choueka, J. (2005). Biomechanics of the flexor tendons, *Hand Clinics,* **21**, 2, 129–149.

Gordon, C.C., Churchill, T., Clauser, C.E., Bradtmiller, B., McConville, J.T., Tebbetts, I., and Walker, R.A. (1989). *1988 Anthropometric survey of U.S. Army personnel: Summary statistics interim report*. Technical Report NATICK/TR-89-027, Natick, U.S. Army Natick Research, Development and Engineering Center.

Grandjean, E. (1973) *Ergonomics of the Home*. London: Taylor & Francis.

Grandjean, E. (1987). *Ergonomics in Computerized Offices*. London: Taylor & Francis.

Grandjean, E. (1988). *Fitting the Task to the Man* (4th edn). London: Taylor & Francis.

Greenberg, J. and Baron, R.A. (2003) *Behavior in Organizations: Understanding and Managing the Human Side of Work* (8th Edn). Upper Saddle River, NJ: Prentice-Hall/Pearson Education.

Hackman, J.R. and Oldham, G.R. (1980). *Work Redesign.* Reading, MA: Addison-Wesley.

Harvey, P., Stoner, J., Hochwarter, W., and Kacmar, C. (2007). Coping with abusive supervision: The neutralizing effects of ingratiation and positive affect on negative employee outcomes. *Leadership Quarterly,* **18**: 3, 264–280.

Havenith, G. (2005). Thermal conditions measurement. In Stanton, N., Hedge, A., Brookhuis, K., Salas, E., and Hendrick, H. (Eds.) *Handbook of Human Factors and Ergonomics Methods*, Chapter 60. Boca Raton, FL: CRC Press.

Heidner, F. (1915). Type-writing machine, Letter's Patent #1,138,474, United States Patent Office.

Helander, M.G. (2003) Forget about ergonomics in chair design? Focus on aesthetics and comfort! *Ergonomics,* **46**, 1306–1319.

Helander, M.G. and Zhang, L. (1997). Field studies of comfort and discomfort in sitting. *Ergonomics,* **41**, 895–915.

Hendrick, H.W. and Kleiner, B.M. (2001a). *Macroergonomics. An Introduction to Work System Design.* Santa Monica, CA: Human Factors and Ergonomics Society.

Hendrick, H.W. and Kleiner, B.M. (Eds.) (2001b). *Macroergonomics. Theory, Methods, and Applications.* Mahwah, NJ: Erlbaum.

Herzberg, F. (1966). *Work and the Nature of Man.* New York: Thomas.

Herzog, W. (2008). Determinants of muscle strength. In Kumar, S. (Ed.) *Biomechanics in Ergonomics* (2nd Edn), Chapter 7. Boca Raton, FL: CRC Press.

HFES (Human Factors and Ergonomics Society) 300 Committee (Ed.) (2004) *Guidelines for Using Anthropometric Data in Product Design,* Santa Monica, CA: Human Factors and Ergonomics Society.

Hinkelmann, K. and Kempthorne, O. (1994). *Design and Analysis of Experiments, Vol. 1: Introduction to Experimental Design.* New York: Wiley.

Hinkelmann, K. and Kempthorne, O. (2005). *Design and Analysis of Experiments, Vol. 2: Advanced Experimental Design.* New York: Wiley-Interscience.

Hornberger, S., Knauth, P., Costa, G., and Folkard, S. (Eds.) (2000). *Shiftwork in the 21st Century.* Frankfurt: Lang.

Horne, J. (1988). *Why We Sleep – The Functions of Sleep in Humans and Other Mammals.* Oxford, UK: Oxford University Press.

Howarth, P.A. (2005). Assessment of the visual environment. In Wilson, J.R. and Corlett, N. (Eds.) *Evaluation of Human Work* (3rd Edn), Chapter 24. London: Taylor & Francis.

Hsiao, H., Long, D., and Snyder, K. (2002). Anthropometric differences among occupational groups. *Ergonomics,* **45**, 136–152.

Hughes, R.E. and An, K.N. (2008). Biomechanical models of the hand, wrist, and elbow in ergonomics. In Kumar, S. (Ed.) *Biomechanics in Ergonomics* (2nd Edn) Chapter 14. Boca Raton, FL: CRC Press.

ILO (International Labour Office) (Ed.) (1986). *Introduction to Work Study* (3rd Edn). Geneva: International Labour Office.

ILO (International Labour Office) (Ed.) (1988). *Maximum Weights in Load Lifting and Carrying, Occupational Safety and Health Series #59*. Geneva: International Labour Office.

ILO (International Labour Office) (Ed.) (1996). *Ergonomic Checkpoints*. Geneva: International Labour Office.

Jarrett, A. (Ed.) (1973) *The Physiology and Pathology of the Skin.* London: Academic Press.

Juergens, H.W. (2004). Erhebung anthropometrischer Masze zur Aktualisierung der DIN 33 402 – Teil 2 (in German). Dortmund/ Berlin/Dresden: Schriftenreihe der Bundesanstalt fuer Arbeitsschutz und Arbeitsmedizin..

Juergens, H.W., Aune, I.A., and Pieper, U. (1990). *International Data on Anthropometry, Occupational Safety and Health Series #65.* Geneva: International Labour Office.

Kamadjeu, R.M., Edwards, R., Atanga, J.S., Kiawi, E.C., Unwin, N., and Mbanya, J.C. (2006). Anthropometry measures and prevalence of obesity in the urban adult population of Cameroon: An update from the Cameroon Burden of Diabetes Baseline Survey, *BMC Public Health*, 6/228, doc 10 1186/1471-2458. Available on the Internet at http://www.biomedcentral.com/1471-2458/6/228/ prepub*http://www.biomedcentral.com/1471-2458/6/228.*

Kapandji, I.A. (1988). *The Physiology of the Joints*, Edinburgh: Churchill Livingstone.

Karwowski, W. (Ed.) (2001). *International Encyclopedia of Ergonomics and Human Factors.* London: Taylor & Francis.

Karwowski, W. (2006). From past to future. *Human Factors and Ergonomics Society Bulletin*, **49**: 11, 1–3.

Karwowski, W. and Marras, W.S. (Eds.) (1999) *The Occupational Ergonomics Handbook*. Boca Raton, FL: CRC Press.

Kincaid, R.D. and Gonzalez, B.K. (1969). *Human Factors Design Considerations for Touch-Operated Keyboards*, Final Report 12091. Honeywell Inc.

Klemmer, E.T. (1958). *A Ten-Key Typewriter*, Research Memorandum RC-65, Yorktown Heights, NY: IBM Research Center.

Klockenberg, E.A. (1926). *Rationalization of the Typewriter and Its Operation* (in German). Berlin: Springer.

Knauth, P. (2006). Workday length and shiftwork issues. In Marras, W.S. and Karwowski, K. (Eds.) *The Occupational Ergonomics Handbook* (2nd Edn), *Interventions, Controls, and Applications in Occupational Ergonomics*, Chapter 29. Boca Raton, FL: CRC Press.

Knauth, P. (2007a). Extended work periods. *Industrial Health*, **45**, 125–136.

Knauth, P. (2007b). Schicht- und Nachtarbeit, shift work and night work (in German). In Landau, K. (Ed.) *Lexikon Arbeitsgestaltung.* Stuttgart: Gentner.

Konz, S. and Johnson, S. (2000). *Work Design: Industrial Ergonomics* (5th Edn). Scottsdale, AZ: Holcomb Hataway.

Kroemer, K.H.E. (1974). *Designing for muscular strength of various populations*. AMRL-Technical Report 72-46. Wright-Patterson AFB, OH: Aerospace Medical Research Laboratory.

Kroemer, K.H.E. (1986). Coupling the hand with the handle. *Human Factors*, **28**: 3, 337–339.

Kroemer, K.H.E. (1994). Locating the computer screen: How high, how far? *Ergonomics in Design*, January, 40.

Kroemer, K.H.E. (1997). *Ergonomic Design of Material Handling Systems*. Boca Raton, FL: CRC Press.

Kroemer, K.H.E. (1998). Relating muscle strength and its internal transmission to design data. In S. Kumar (Ed.) *Advances in Occupational Ergonomics and Safety,* 349–352. Amsterdam: IOS Press.

Kroemer, K.H.E. (1999). Assessment of human muscle strength for engineering purposes: basics and definitions. *Ergonomics,* **42**, 74–93.

Kroemer, K.H.E. (2001). Keyboards and keying. An annotated bibliography of the literature from 1878 to 1999, *International Journal Universal Access in the Information Society UAIS* 1/2, 99–160. Available on the Internet at *www.springerlink.com/index/yp9u5phcqpyg2k4b.pdf*. An abbreviated and edited treatment of this bibliography appears in Chapter 6 of our book *Office Ergonomics* (Kroemer and Kroemer, 2001, see below).

Kroemer, K.H.E. (2006). *"Extra-Ordinary" Ergonomics: How to Accommodate Small and Big Persons, the Disabled and Elderly, Expectant Mothers and Children.* Boca Raton, FL: CRC Press.

Kroemer, K.H.E. (2006a). Designing children's furniture and computers for school and home. *Ergonomics in Design,* **3**, 8–16.

Kroemer, K.H.E. (2006b). Designing for older people. *Ergonomics in Design* **4**, 25–31.

Kroemer, K.H.E. (2008). Anthropometry and biomechanics: Anthromechanics. In Kumar, S. (Ed.) *Biomechanics in Ergonomics* (2nd Edn) Chapter 2. Boca Raton, FL: CRC.

Kroemer, K.H.E. and Grandjean, E. (1997). *Fitting the Task to the Human* (5th Edn). London: Taylor & Francis.

Kroemer, K.H.E. and Kroemer, A.D. (2001). *Office Ergonomics.* London: Taylor & Francis.

Kroemer, K.H.E. and Kroemer, A.D. (2005). *Office Ergonomics*, authorized translation into Korean. Seoul: Kukje.

Kroemer, K.H.E., Kroemer, H.B., and Kroemer-Elbert, K.E. (2003, amended reprint). *Ergonomics: How to Design for Ease and Efficiency* (2nd Edn). Upper Saddle River, NJ: Prentice-Hall/Pearson Education.

Kroemer, K.H.E., Kroemer, H.J., and Kroemer-Elbert, K.E. (1997). *Engineering Physiology. Bases of Human Factors/Ergonomics* (3rd Edn). New York: VNR-Wiley.

Kroemer, K.H.E. and Robinson, D.E. (1971). *Horizontal Static Forces Exerted by Men Standing in Common Working Positions on Surfaces of Various Tractions*, AMRL-Technical Report 70-114, Wright-Patterson AFB, OH: Aerospace Medical Research Laboratory.

Kuczmarski, R.J., Ogden, C.L., Guo, S.S., Grummer-Strawn, L.M., Flegal, K.M., Mei, Z., Wei, R., Curtin, L.R., Roche, A.F., and Johnson, C.L. (2002). *2000 CDC Growth Charts for the United States: Methods and Development,* DHHS Publication No. PHS 2002-1696, Vital and Health Statistics, Series 11, No. 246. Hyattsville, MD: Department of Health and Human Services,

Kumar, S. (Ed.) (2004). *Muscle Strength*. Boca Raton, FL: CRC Press.

Kumar, S. (Ed.) (2008). *Biomechanics in Ergonomics* (2nd Edn). Boca Raton, FL: CRC Press.

Kumar, S. and Mital, A. (Eds). (1996). *Electromyography in Ergonomics*. London: Francis & Taylor.

Landau, K. (Ed.) (2000). *Ergonomic Software Tools in Product and Workplace Design. A Review of Recent Developments in Human Modeling and Other Design Aids.* Stuttgart: Ergon.

Landy, F.J. and Conte, J.M. (2006). *Work in the 21st Century: An Introduction to Industrial and Organizational Psychology* (2nd Edn). Malden: Blackwell.

Langley, L.W. (1988). *Ternary Chord-Type Keyboard,* Patent #4,775,255, United States Patent Office.

Lavender, S.A. (2006). Training lifting techniques. In Marras, W.S. and Karwowski, K. (Eds.) *The Occupational Ergonomics Handbook* (2nd Edn). *Interventions, Controls, and Applications in Occupational Ergonomics*, Chapter 23. Boca Raton, FL: CRC Press.

Lee, J.C. and Healy, J.C. (2005). Normal sonographic anatomy of the wrist and hand. *Radiographics, 25*: 6, 1577–1590.

Leyk, D., Kuechmeister, G., and Juergens, H.W. (2006). Combined physiological and anthropometrical databases as ergonomic tools. *Journal of Physiological Anthropology, 25*: 6, 363–369.

Liberty Mutual, 2004, Manual materials handling guidelines. Regularly updated on www.libertymutual.com.

Litterick, J. (1981). QWERTYUIOP – Dinosaur in a computer age. *New Scientist, 8 January*, 66–68.

Lockhead, G.R. and Klemmer, E.T. (1959). *An Evaluation of an Eight-Key Word-Writing Typewriter,* Research Report RC-180. Yorktown Heights, NY: IBM Research Center.

Lohman, T.G., Roche, A.F., and Martorel, R. (Eds.) (1988). *Anthropometric Standardization Reference Manual*. Champaign, IL: Human Kinetics.

Lueder, R. and Rice, V.B. (Eds.) (2007). *Ergonomics for Children*. Boca Raton, FL: CRC Press.

Lundervold, A.J.S. (1951). Electromyographic investigations of position and manner of working in typewriting, *Acta Physiology Scandinavia 24, Supplement* 84, 1–171.

Marklin, R.W. and Simoneau, G.G. (2004). Design features of alternative computer keyboards: A review of experimental data. *Journal of Orthopedic Sports Physical Therapy, 34*, 638–649.

Marras, W.S. (2008). *The Working Back: A Systems View*. New York: Wiley.

Marras, W.S. and Karwowski K. (Eds.) (2006a). *The Occupational Ergonomics Handbook* (2nd Edn). *Fundamentals and Assessment Tools for Occupational Ergonomics*. Boca Raton, FL: CRC Press.

Marras, W.S. and Karwowski, K. (Eds.) (2006b). *The Occupational Ergonomics Handbook* (2nd Edn). *Interventions, Controls, and Applications in Occupational Ergonomics*. Boca Raton, FL: CRC Press.

Marras, W.S. and Radwin, R.G. (2006). Biomechanical modeling. In Dickerson, R.S. (Ed.) *Reviews of Human Factors and Ergonomics*, *Vol. 1*, Chapter 1. Santa Monica: Human Factors and Ergonomics Society.

Marras, W.S., McGlothlin, J.D., McIntyre, D.R., Nordin, M., and Kroemer, K.H.E. (1993). *Dynamic* Measures *of Low Back Performance*. Fairfax, VA: American Industrial Hygiene Association.

Martimo, K.P., Verbeek, J., Karppinen, J., Furlan, A.D., Takala, E.P., Kuijer, P.P.F., Jaihiainen, M., and Viikari-Juntura, E. (2008). Effect of training and lifting equipment for preventing back pain in lifting and handling: Systematic review. *British Medical Journal*, **336**:429–431 (published online 31 January 2008).

Maslow, A.H. (1943). A theory of motivation. *Psychological Review*, **50**, 370–396.

Maslow, A.H. (1954). (1970 2nd Edn). *Motivation and Personality*. New York: Harper.

McGill, S.M. (2006). Back belts. In Marras, W.S. and Karwowski, K. (Eds.), The *Occupational Ergonomics Handbook* (2nd Edn) *Interventions, Controls, and Applications in Occupational Ergonomics*, Chapter 30. Boca Raton, FL: CRC Press.

McMillan, G.R. (2001). Brain and muscle signal-based control. In Karwowski, W. (Ed.) *International Encyclopedia of Ergonomics and Human Factors,* 379–381, London: Taylor & Francis.

McMillan, G.R. and Calhoun, G.L. (2001). Gesture-based control. In Karwowski, W. (Ed.) *International Encyclopedia of Ergonomics and Human Factors,* 237–239. London: Taylor & Francis.

Megaw, T. (2005). The definition and measurement of mental workload. In Wilson, J.R. and Corlett, N. (Eds.) *Evaluation of Human Work* (3rd Edn), Chapter 18. London: Taylor & Francis.

Monk, T.H. (2006). Shiftwork. In Marras, W.S. and Karwowski, K. (Eds.). *The Occupational Ergonomics Handbook,* 2nd Edn), *Fundamentals and Assessment Tools for Occupational Ergonomics*, Chapter 32. Boca Raton, FL: CRC Press.

Monk, T.H., Folkard, S., and Wedderburn, A.I. (1996). Maintaining safety and high performance on shiftwork. *Applied Ergonomics,* **27**, 17–23.

Muchinsky P.M. (2007). *Psychology Applied to Work* (8th Edn). Belmont, CA: Wadsworth.

Nachemson, A. and Elfstroem, G. (1970) Intravital dynamic pressure measurements in lumbar discs. *Scandinavian Journal of Rehabilitation Medicine, Supplement* **1**, 1–40.

Nagamachi, M. (2001). Relationships among job design, macroergonomics, and productivity. In Hendrick, H.W. and Kleiner, B.M. (Eds.) *Macroergonomics. Theory, Methods, and Applications*, Chapter 6. Mahwah, NJ: Erlbaum.

NASA, 1989, *Man-Systems Integration Standards, Revision A*, NASA-STD 3000, Houston, TX: LBJ Space Center.

National Research Council (Ed.). (1999). *Work-Related Musculoskeletal Disorders: Report, Workshop Summary, and Workshop Papers.* Washington, DC: National Academy Press.

NIOSH (1981). *Work Practices Guide for Manual Lifting.* DHHS (NIOSH) Publication #81–122, Washington, DC: US Government Printing Office.

Nordin, M. and Frankel, V.H. (1989). *Basic Biomechanics of the Musculoskeletal System.* Philadelphia: Lea & Febiger.

Nordin, M., Andersson, G.B.J., and Pope, M.H. (1997). *Musculoskeletal Disorders in the Workplace: Principles and Practices.* St. Louis, MO: Mosby.

Noyes, J. (1983a). Chord keyboards. *Applied Ergonomics,* **14,** 55–59.

Noyes, J. (1983b). The QWERTY keyboard: A review. *International Journal of Man–Machine Studies,* **18,** 265–281.

Oezkaya, N. and Nordin, M. (1991). *Fundamentals of Biomechanics.* New York: Van Nostrand Reinhold.

Osler, W. (1892). *The Principles and Practice Of Medicine, Ix, Professional Spasms; Occupation Neuroses.* New York: Appleton.

Owen, D. (2006). The soundtrack of your life. Muzak in the realm of retail theatre, *The New Yorker*, April 10, 2006, 66–71.

Parsons, H.M. (1974). What happened at Hawthorne? *Science* **18**, 922–932.

Parsons, K.C. (2003). *Human Thermal Environments* (2nd Edn). London: Taylor & Francis.

Parsons, K.C. (2005). Ergonomic assessments of thermal environments. In Wilson, J.R. and Corlett, N. (Eds.) (2005) *Evaluation of Human Work*, (3rd Edn), Chapter 23. London: Taylor & Francis.

Peebles, L. and Norris, B. (1998). *Adultdata. The Handbook of Adult Anthropometric and Strength Measurements – Data for Design Safety,* DTI/Pub 2917/3k/6/98/NP. London: Department of Trade and Industry.

Peebles, L. and Norris, B. (2000). *Strength Data,* DTI/URN 00/1070. London: Department of Trade and Industry.

Peebles, L. and Norris, B. (2003). Filling "gaps" in strength data for design, *Applied Ergonomics,* **34,** 73–88.

Pew, P.W. and Mavor, A.S. (Eds.). (2007). *Human–System Integration in the System Development Process.* Washington, DC: National Academies Press.

Pheasant, S. and Haslegrave, C.M. (2006). *Anthropometry, Ergonomics and the Design of Work.* London: Taylor & Francis.

Phillips, C.A. (2000). *Human Factors Engineering.* New York: Wiley.

Plog, B.A. (Ed.) (2002). *Fundamentals of Industrial Hygiene* (5th Edn). Itasca, IL, National Safety Council.

Poore, G.V. (1872). "Writer's cramp:" Its pathology and treatment. *Practitioner,* August, 341–350.

Poore, G.V. (1887). Clinical lecture on certain conditions of the hand and arm which interfere with the performance of professional acts, especially piano-playing, *The British Medical Journal, February 26,* 441–444.

Proctor, R.W. and Van Zandt, T. (1994). *Human Factors in Simple and Complex Systems.* Boston: Allyn and Bacon.

Putz-Anderson, V. (1988). *Cumulative Trauma Disorders: A Manual for Musculoskeletal Diseases of the Upper Limbs.* London: Taylor & Francis.

Putz-Anderson, V. and Waters, T. (1991). *Revisions in NIOSH Guide to Manual Lifting.* Paper presented at the Conference "A National Strategy for Occupational Musculoskeletal Injury Prevention." Ann Arbor, MI, April 1991.

Ramazzini 1713 – see Wright.

Rea, M.S. (2005). Photometric characterization of the luminous environment. In Stanton, N., Hedge, A., Brookhuis, K., Salas, E., and Hendrick, H. (Eds.) *Handbook of Human Factors and Ergonomics Methods,* Chapter 68. Boca Raton, FL: CRC Press.

Reddy, N.P. and Gupta, V. (2007). Toward direct biocontrol using surface EMG signals: Control of finger and wrist joint models. *Medical Engineering and Physics,* 29, 3, 398–403.

Remington, R.J. and Rogers, M. (1969). *Keyboard Literature Survey, Phase 1: Bibliography,* TR 29.0042. Research Triangle Park: IBM Systems Development Division.

Rempel, D. (2008). The split keyboard: An ergonomics success story, *Human Factors* **50,** 385-392.

Robinette, K.M. (Ed.) (2009, expected). *Computer Aided Anthropometry for Research and Design.* Boca Raton, FL: CRC Press.

Robinette, K.M. and Daanen, H. (2003). *Lessons Learned from CAESAR: A 3-D Anthropometric Survey.* Paper #00730 in *Proceedings of the 15th Triennial Congress of the International Ergonomics Association, August 24–29.* London: Taylor & Francis.

Rodahl, K. (1989). *The Physiology of Work.* London: Taylor & Francis.

Roebuck, J.A. (1995). *Anthropometric Methods – Designing to Fit the Human Body.* Santa Monica, CA: Human Factors and Ergonomics Society.

Roethlisberger, F.J. and Dickson, W.J. (1943). *Management and the Worker.* Cambridge, MA: Harvard University Press.

Salvendy, G. (1997). *Handbook of Human Factors and Ergonomics* (2nd Edn). New York, Wiley.

Selye, H. (1978). *The Stress of Life* (Rev. Edn). New York: McGraw-Hill.

Sheedy, J. (2006) Vision and work, in Marras, W.S. and Karwowski, K. (Eds.) *The Occupational Ergonomics Handbook* (2nd Edn). *Fundamentals and Assessment Tools for Occupational Ergonomics,* Chapter 18. Boca Raton, FL: CRC Press.

Siegel, J.M. (2003). Why we sleep. *Scientific American*, November, 92–97.

Smith, M.J. and Saintfort-Carayon, P. (1989) A balance theory of job design for stress reduction. *International Journal of Industrial Ergonomics*, **4**, 67–79.

Smith, T.M. (2007). *Job Satisfaction in America: Trends and Socio-Demographic Correlates*. Available on the Internet at http://www-news.uchicago.edu/releases/07/pdf/070827.jobs.pdf.

Snook, S.H. (2000). Back risk factors. An overview. In Violante, F., Armstrong, T., and Kilbom, A. (Eds.) *Occupational Ergonomics. Work Related Musculoskeletal Disorders of the Upper Limb and Back*, Chapter 11. London: Taylor & Francis.

Snook, S.H. (2005). Psychophysical tables: Lifting, lowering, pushing, pulling, and carrying. In Stanton, N., Hedge, A., Brookhuis, K., Salas, E., and Hendrick, H. (Eds.) *Handbook of Human Factors and Ergonomics Methods*, Chapter 13. Boca Raton, FL: CRC Press.

Snook, S.H. and Ciriello, V.M. (1991). The design of manual handling tasks: Revised tables of maximum acceptable weights and forces. *Ergonomics*, **34**:9, 1197–1213.

Sprent, P. (2000). *Applied Nonparametric Statistical Methods*. Boca Raton, FL: Chapman & Hall/CRC Press.

Staff, K.R. (1983). *A Comparison of Range of Joint Mobility in College Females and Males*, Unpublished Master's Thesis. Texas A&M University, College Station.

Stanton, N., Hedge, A., Brookhuis, K., Salas, E., and Hendrick, H. (Eds.) (2005). *Handbook of Human Factors and Ergonomics Methods*. Boca Raton, FL: CRC Press.

Strokina, A.N. and Pakhomova, B.A. (1999) *Anthropo-Ergonomic Atlas* (in Russian). Moscow: Moscow State University Publishing House.

Swink, J.R. (1966). Intersensory comparisons of reaction time using an electro-pulse tactile stimulus. *Human Factors*, **8**: 143–145.

Taylor, F. (1911). *The Principles of Scientific Management*. New York: Norton.

Tepas, D.I. (1999). Work shift usability testing. In Karwowski, W. and Marras, W.S. (Eds.) *The Occupational Ergonomics Handbook*, Chapter 96. Boca Raton, FL: CRC Press.

Ugbolue, U.C., Hsu, W.H., Goitz, R.J., and Li, Z.M. (2005). Tendon and nerve displacement at the wrist during finger movements. *Clinical Biomechanics* **20**: 1, 50–56.

University of Nottingham (2002). *Strength data for design safety, Phase 2*. DTI URN 01/1433. London: Department of Trade and Industry.

Vicente, K.J. (2002). Ecological interface design: Progress and challenges, *Human Factors*, **43**: 62–78.

Violante, F., Armstrong, T., and Kilbom, A. (Eds.) (2000) *Occupational Ergonomics. Work Related Musculoskeletal Disorders of the Upper Limb and Back*. London: Taylor & Francis.

Walji, A.H. (2008). Functional anatomy of the upper limb (extremity). In Kumar S. (Ed.) *Biomechanics in Ergonomics* (2nd Edn), Chapter 8. Boca Raton, FL: CRC Press.

Wang, M.J.J., Wang, E.M.Y., and Lin, Y.C. (2002). *Anthropometric Data Book of the Chinese People in Taiwan.* Hsinchu: The Ergonomics Society of Taiwan.

Wargo, M.J. (1967). Human operator response speed, frequency and flexibility: A review and analysis. *Human Factors,* **9**: 221–238.

Waters, T.R. (2006). Revised NIOSH lifting equation. In Marras, W.S. and Karwowski, K. (Eds.) *The Occupational Ergonomics Handbook,* (2nd Edn) *Fundamentals and Assessment Tools for Occupational Ergonomics,* Chapter 46. Boca Raton, FL: CRC Press.

Waters, T.R. and Putz-Anderson, V. (1999). Revised NIOSH lifting equation. In Karwowski, W. and Marras, W.S. (Eds.) *The Occupational Ergonomics Handbook,* Chapter 57. Boca Raton, FL: CRC Press.

Whitcome, K.K., Shapiro, L.J., and Lieberman, D.E. (2007). How women bend over backwards for baby. *Nature,* Published online 12 December 2007, doi:10.1038/news.2007.374.

Wickens, C.D., Lee, J., Liu, Y., and Gordon-Becker, S. (2004). *An Introduction to Human Factors Engineering (2nd Edn).* Upper Saddle River, NJ: Prentice-Hall/Pearson Education.

Williges, R.C. (2007). *CADRE: Computer-Aided Design Reference for Experiments,* Electronic Book CD-ROM-07-01. Blacksburg: Virginia Polytechnic Institute and State University.

Wilson, J.R. and Corlett, N. (Eds.) (2005). *Evaluation of Human Work* (3rd Edn). London: Taylor & Francis.

Winter, D.A. (2004). *Biomechanics and Motor Control of Human Movement* (3rd Edn). New York: Wiley.

Wright, W.C. (1993). *Diseases of Workers,* Translation of Bernardino Ramazzini's *1713 De Morbis Articum.* Thunder Bay, OH&S Press.

Wu, G., Siegler, S., Allard, P., Kirtley, C., Leardini, A., Rosenbaum, D., Whittle, M., Lima, D.D., Cristofolini, L., Witte, H., Schmid, O., and Stokes, I. (2002). ISB recommendation on definitions of joint coordinate system of various joints for the reporting of human joint motion Part I: Ankle, hip, and spine. *Journal of Biomechanics,* **35**: 543–555.

Zhang, X. and Chaffin, D.B. (2006). Digital human modeling for computer-aided ergonomics. In Marras, W.S. and Karwowski, K. (Eds.) *The Occupational Ergonomics Handbook* (2nd Edn), *Interventions, Controls, and Applications in Occupational Ergonomics,* Chapter 10. Boca Raton, FL: CRC Press.

Zhao, C., Ettema, A.M., Osamura, N., Berglund, L.J., An, K.N., and Amadio, P.C. (2007). Gliding characteristics between flexor tendons and surrounding tissues in the carpal tunnel: A biomechanical cadaver study. *Journal of Orthopedic Research,* **25**: 2, 185–190.

Index

A

Absenteeism, 258, 285, 287, *288*
Absorption
    heat, 155, 157
    foodstuffs, 202
Acclimatization, 164–165, 323–324
Accommodation, visual, 104,
        109–111
Accurate, fast, skillful activities,
        226–232
Actin, 53–55
Action potential, 181, *182*
Active hearing-protection devices
        (HPD), 135–136
"Active sitting", 338
Actual mobility, 45
Acuity, visual, 115–116
Adaptation
    speed of, 188
    syndrome, computer, 238
Adenosine triphosphate (ATP), 61
Adrenalin, 235
Aesthetics, office, 320–321
Afferent signals, 147, 179–180, 194,
        196
Africa, anthropometry data, *4*
Air
    flow, 160
    humidity, 157–160, 323
    temperature, 160, 170, 324–325,
        401–402
Alertness, 219–220
Algeria, *5*
Altimeters, 385–386
Ankle joints, 48
Annoyance, 333
Anthromechanics, 58
Anthropometric data, 3, *4–9, 24*

Anthropometry, 5-15
Anxiety, 236
Apathy, 169
APCFB model, 269–270
Arm
    forces, 63, *65*
    mobility, 33-34
    pain, *345*
Asia, anthropometry data, *4, 26–27*
Assimilation, 202
Astigmatism, 112
Athletics and sports, 59
Audiograms, 131–132
Auditory nerves, 122
Australia, *4*
Autokinetic phenomenon, 113
Autonomic nervous system, 184, 236
"Average person", 16
"Average users", 71
    strength, 93–94
Average value, 16
Axons, 180, *181*

B

Back
    belts, 93
    curvature, 348
    pain, 89, *344*
Backrests, chair, 340–341
Bad bosses, 394
Balance
    heat, 323
    between life and work, 272–273,
        394
    sense of body, 142–143
Bar knobs, 384
Barriers, noise, 133–134

Basal ganglia, 178
Basal metabolism, 203
Bathrooms, 311–312
Bathtubs, 312
Bedrooms, 311–312
Behavior and motivation, 254–260
    apathy and, 169
    APCFB model of, 269–270
Bell-shaped distribution, 71
"Bend the tool, not the wrist", 75
Bicycling, 207, 209
Binary keys, 361–362
Biological clocks, 279–280, 291
Biomechanics, 58, 92–93
Blindness, 112
    color, 113
    night, 113
Blood
    distribution in cold
        environments, 167
    distribution in hot environments,
        165
    thermoregulation and, 154
Blue-collar workers, 238–239
Body rhythms. See Circadian
        rhythms
Body sizes. See also
        Anthropometric data.
    "average person", 16
    common measures and
        applications of, 11–15
    designing objects to fit
        individual, 24–28, 29
    fitting a range of, 28
    height and weight, 3, 5–9
    measurements, 3, 4–9, 9, 16
    methods of measuring, 9, 10
    "normal distribution", 16, 24
    of Russian students, 7, 9, 16,
        17–21
    statics and dynamics, 28, 29
    of Taiwanese, 22–23
    variances in, 3, 24–28
    workplace sizing and, 369–370,
        371–372
Boredom, 219–220, 242
Borg RPE Scale, 218, 244, 245
Bosses, bad, 394
Braille, 141
Brain
    action potential, 181, 182
    anatomy of, 177, 178
    cerebellum, 178
    cerebrum, 178
    dendrites, 180
    electromyograms (EMGs), 182

feedforward/feedback loop,
        182–183, 192
    lobes, 178
    memory and, 190–192
    -nerve network, 177–185
    neurons, 180, 181
    reflexes and, 184
    signal transmission, 181
    during sleep, 292–294
    stem, 178–179
Brazil, anthropometric data, 5
Breaks, rest, 211, 232
Breathing, 123, 179, 184, 208, 293
Britain, anthropometric data, 26–27
Building materials, noise attenuation
        by, 133–134
Byproducts of metabolism,
        202–203

C

Calories (Cal, cal), 201, 203
Cameroon, anthropometric data, 5
Canada, anthropometric data, 9
Cardiac muscles, 56
Carpal tunnel, 35, 37
    syndrome (CTS), 357–358
Carrying, load, 79, 81–85. See also
        Lifting and lowering,
        Material handling.
    on both shoulders, 96
    converting to, 95
    designing for easy, 89–97, 376,
        380, 381–382
    training for safe, 94, 97
    using dollies and carts, 96
    using rollers and conveyors, 97
Cathode-ray tube (CRT) displays,
        329, 330
Cell body soma, 180, 181
Cell phones, 360
Center of mass, pregnant body, 42
Central nervous system (CNS),
        141–143, 184, 185, 188
    actions and reactions, 192–196
Cerebellum, 178
Cerebral cortex, 178, 293
Cerebrum, 178
Chain of strength-transmitting body
        segments, 76, 77
Chairs, 45, 334, 335. See also Sitting
    computer workstation, 339–341
    designs, 372–376
Children, home design for, 307–308

China, anthropometric data, *4, 5,* 9, 16, *26–27*
Choice reaction time, 150, 195
Chromatic aberration, 113
Cilia, 122, 130, 142-143
Circadian rhythms, 153, 277–281
   night work and, 291–292
   shift work and, 298
Circulatory system and thermoregulation, 154, 165
Classification of work demands, 206–207
Climate. *See also* Environment, the; Temperature
   acclimatization, 164–165
   clothing and, 161–165
   control in homes, 313
   designing comfortable, 170, 397, *401–402*
   effects on mental tasks, 169–170
   human thermoregulation and, 153–158
   interacting factors, 160
   measurement and instruments, 160–161
   office, 321–325
   organizational, 266–267
   personal, 161–165
   physical factors in, 158–161
   taking charge of, 397, *401–402*
Clothing, 161–163, 167, 170, 323, 324, *401–402*
Clutter, 188, 191
Cochlea, 122, 135
   implants, 139
Cockpits, aircraft, 382
Co-contraction, muscular, 57–58
Coding of controls, 382
Cold and warm sensations, 145, 147–148, 154. *See also* Temperature
   office climates and, 321–322
   work environment and, 165–170
Color and vision, 105, 113, 116–117
Combustion engines, 202–203
Comfortable seating, 331, 333–334, 348
Common difference tone, 128
Communications, voice, 136
Commuting, 287-288
Complex reactions, 150, 195
Composite population, 24
Compressed work weeks, 286–288
Computer adaptation syndrome, 238

Computer keyboards
   design alternatives for, 363–364
   design history, 349–353
   designing for new syntax and diction with, 364–365
   design solutions for problems associated with, 359–363
   human factors considerations for, 353–358
   input-related anthromechanical issues and, 358–359
   trays, 342
Computer monitors, 329, 335, 336–337, *343–344*
Computer software, 365–366
Computer workstations, 335–342, 366. *See also* Computer keyboards
   inputs without keys, 363
Concentric muscle effort, 58
Concurrent tones, 128–129
Conduction, heat exchange by, 155, 322
Conduits, organizational, 265–266
Contractile microstructure of muscles, 54–55
Contraction, muscular, 58
Contracts, social, 268
Controlled strength tests, 59
Control lobe, 178
Controls, coding of, 382
Convection, heat exchange by, 155, *156,* 322
"Convenient" mobility, 45
Conveyors and rollers, *97*
Cooling systems, home, 312–313
Coping with stress, 236
Cornea, 103–104
Corti organs, 122, 130
Counter displays, 386–387Creatine phosphate (CP), 54, 61, 74
Critical strength values, 94
Crowding, 253
Cubicles, office, 317–318
Culture, organizational, 266–267

D

5-day workweek, 285
Debt, oxygen, 205–206
Decibels (dB), 123–124, 132
Decision making, 189–192
Degrees of freedom, 34

Dehydration, 166
Demand, task, 239
Dendrites, 180
Deprivation, sleep, 280–281, 296–297. *See also* Circadian rhythms
Design
    for "average users", 71
    for body motion and support, 337–338
    chair, 334
    comfortable climates, 170, 397, *401–402*
    computer, 349–367
    computer workstation, 335–342, *343–345*
    for easy load handling, 89–97, 376, 380, *381–382*
    hand tools, 74, 229, *230,* 376, *377*
    for hard labor, 210–212
    home, 307–314
    home office, 342
    human-centered, 369–388
    illumination, 117
    keyboard, 359–366
    kitchen, 310–311
    for manipulation, 337, 376, *378*
    for mobility, 43–45
    for motivation and performance, 268–271
    for new syntax and diction, 364–365
    for "no noise", 133
    of objects to fit individual bodies, 24–28, *29*
    office, 317–348
    of office workstations, 335–342
    for pregnancy, 308
    principles of design, 27–28
    for reaching and grasping, 376, *379*
    for simple movement control, 183
    for tactile perception, 146–150
    for variety, 338–339
    muscle strength data in, 63–65
    of spoken messages, 137
    of walkways and stairs, 309–310
    of warning signals, 137–138
    wheelchairs, 335–342
    workstations, 335–342
Diction, designing for new, 364–365
Digits, hand, 35, *36*
Diopter, 109, *110,* 110–111
Directional hearing, 128

Discontent, measuring, 218–219
Discs, spinal, 39-42, 89–93
    compression of, pressure in, 90-93
    "slipped", 90
    strains in the bent back, 90–92
Disguised pauses in work, 283
Displacements
    in body joints, 43
    hands and forceful activities with, 74
Displays and controls, 382–387
Distance
    hearing, 128
    between persons, 252
Distress, heat, 166
    situational, 235
Diurnal rhythms, *see* Circadian rhythms
Diversity versus monotony in work, 220, 242
"Diving" into containers, 224
Division of labor, 255
Dollies and carts, *96*
Doors, home, 310
    openings, 16
Doppler effect, 128
Dry bulb (DB) temperature, 159
    thermometers, 159
Dynamic anthropometry, 29
Dynamic strength, 58, *61,* 69-70, 79
Dynamics. *See* Statics and dynamics

E

Eardrum, 122
Ear-Eye (EE) line, 108, 227–229, *398*
Ears. *See also* Hearing
    anatomy of, 121–123
    ear-eye (EE) line, 108, 227–229
    noise effect on, 129–130
    plugs and muffs, 135
    sense of body movement and, 141–143
    surgical implants, 138–139
    vestibulum, 142–143
Eccentric muscle effort, 58
Efficiency, energy, 210–211
Eight-hour workdays, 285
Elbow, 11–14, 76–77
Elderly persons, home design for, 308–309
Electricity, sensing, 145–146, 149

Electroencephalogram (EEG), 293, 364
Electromyogram (EMG), 182, 364
Electroreceptors, 143
Elevators, 310
Emission coefficient, 155, 157
Emotions and stress, 235–236
Employees
  absenteeism, 258, 285, 287, *288*
  bad bosses and, 394
  balancing life and work, 272–273, 394
  happy, 254, 260, 391–392
  pride, 393
  taking charge, 394, *395,* 397, *398–402*
Encephalon, 293
Endocrine system, 236
Endurance, 166, 215, 232
Energy
  content of food and drink, 203
  consumption, 200–205
  efficiency at work, 210–211
  exchanges with the environment, 322
  expenditures at sample activities, 216
  metabolism and, 167–168, 201–202
  supply to the body during heavy work, 204
  units, 201
England. *See* Britain
Enlargement and enrichment, job, 258
Envelopes, reach, 45, 47
Environment,*See also* Climate
  heat exchanges with, 155
  modifying input signals and sensation about, 188
  physical factors defining, 158–161
Ergonomic design. *See* Design
Ergonomics
  defined, xix
  industrial psychology in, 405
  macro-, 267
  nano-, 364, 368
  personnel selection and training in, 405
ERG theory, 270
Errors during night work, 295–297
Estimates of body size, 3-4
  of muscle strength, 70, 72, 94
Europe, anthropometry data, *4, 26–27*

Eustachian tube, 123
Evaporation, heat exchange by, 157–158, 165, 322–323
Evening type individuals, 279
Event frequency and vigilance, 220–221
Exact manipulations, 226–227
Excessive motions, 33
Exchanges, heat. *See* Heat, exchanges
Exertion of muscle strength
  maximal versus minimal, 98
  static, 70–71
Expectancy theories, 393
Exteroceptors, 187–188
Extrinsic muscles, 72
"Eye-balling" estimates, 72, 94
Eyes. *See also* Vision
  adaptation to light and dark, 114–115
  anatomy of, 103–105
  ear-eye (EE) line, 108, 227–229, *398*
  fatigue, 107
  fixated, 106
  floaters, 112
  focusing, 109–110
  fovea, 105, 115
  incessant changes, 110
  lens, 104–105
  line of sight (LOS), 107, *108,* 228–229
  movement, 106–107
  muscles of, 104
  optic nerve, 105
  problems with, 110–113
  pupils, 104, 110, 112, 115
  retina, 104–105
  rods and cones, 105, 115
  visual control system, 105
"Eyes on the screen", 366

F

Farsightedness, 104, 112
Fast body movements, 72
Fatigue
  body postures and, 222–223, 233
  eye, 107
  keyboarding and, 353–354
  muscle, 61–63
  performance changes during daylong work and, 281–282

Feedforward/feedback loops,
182–183, 383
Fine manipulation of objects, 72
Fingers, 35, 36, 356–357, 360
"Fitting the human to work", 94, 97,
191-192, 225,229, 374,
391, 405
"Fitting the work to the human",
49-49, 77-78, 97, 183,
358, 369-371, 391, 405
Fit range, 28
Fixated eyes, 106
Fixed organizational structures, 265
Flexible organizations, 265, 266
Flextime, 286
Floaters, 112
Focusing, eye, 109–110
Food and drink, energy content of,
203
Foot
    actions, 48–49
    standing on one, 49, 374
    strength, 76–77
    thrust, 77–78
Forced choice, 218
Forceful activities with
        displacements, 74
Fovea, 105, 115
France, anthropometric data, 5,
26–27
Frankfurt plane, 108, 227–228
Freedom, degrees of, 34
Free nerve endings, 144
Frequencies, sound, 123
Frequent movements between
        targets, 72
Frostbite, 169
Furniture and equipment, office,
329–342

G

Germany, anthropometric data, 5,
26–27
Getting along with others at work,
391–395, 396
Glare, 314, 325-328, 326, 346
Glare-free lighting, 325–328, 397
Glaucoma, 112
Globe temperature (GT), 160, 161
Glucose, 202
Glycogen, 202
Goal setting, 259
Golgi organs, 183, 185

"Good" stress, 237, 271
Goose bumps, 168
Grasping, 376
Grips and grasps, 74
Groups, working in, 253–254
Gustation, 147

H

Handedness, 74
Handles, 73, 74
Hands
    carpal tunnel of, 35, 37, 357–358
    couplings between handle and,
        73
    digits, 35, 36, 356–357
    left-handed persons, 74–75
    manipulations, 226–227, 376,
        378
    mobility, 34–39
    muscles, 35, 36, 37
    office workstation design for
        manipulation, 337
    pain in, 345
    sizes, 24, 25, 26–27
    strength, 72–75
    tasks, types of, 72–74
    tendon sheaths, 37–39, 358
    tools, 74–75, 229, 230, 376, 377
    workspace, 47–48
"Hands on the keyboard", 366
Hard labor. See also Work
    designing for, 210–212
    energy consumption, 200–205
    heart rate during, 205–207
    limits of human capacity for,
        207–209, 210
    physiological principles in,
        199–200
    static work, 209, 210
"Have a life", 394
Hawthorne effect, 258–259
Hearing. See also Ears
    aids, 138
    directional, 128
    distance, 128
    Doppler effect, 128
    improving defective, 138–139
    loss, noise-induced, 131–133
    noise, 129–139
    protection devices (HPD),
        134–136
    sounds, 123–129

Heart
  muscles, 56
  rate during work, 205–207, *208,*
      217, 232
  static work and, 209, *210*
  thermoregulation and, 154, 165
Heat
  absorption, 155, 157
  balance, 323
  conduction 155-156
  convection 155-156
  distress, 166
  effects on mental performance,
      324
  evaporation 157-158
  exchanges, 155–158, 322
  evaporation 157-158
  radiation 155-156
  strain, 165–166
  syncope, 166
Heating systems, home, 312–313
Heaviness of work, measuring, 204
Heidner's keyboards, 353–354, *355,*
      359, 361, 362-363
Height, body, 3, *5–9,* 370, *371*
Herzberg's Two-Factors theory,
      257–258, 270, 272
Hierarchy of needs, Maslow's,
      256–257, 270
Homeostasis, 283
Homes
  access, walkways, steps, and
      stairs in, 309–310
  bathrooms in, 311–312
  bedrooms in, 311
  designed for impaired and
      elderly persons, 308–309
  designed for mothers and
      children, 307–308
  kitchens in, 310–311
  light, heating, and cooling in,
      312–313
  offices in, 286, 313–314, 342
  remodeling of, 309
  working in, 286, 313–314
Horizontal push forces, 63, *64*
Hours, working. *See* Working hours
Human-centered design, 369–388
Human-centered organizations, 264,
      273–274, 349, 388
"Human energy machine", 201-202
Human-system integration (HSI),
      267
Humidity, 157–158, 159–160, 323,
      401–402
Hyperopia, 112

Hypothalamus, 178
Hypothermia, 169

I

Illuminance. *See* Illumination
Illumination, 112-117, *119,* 188, 302-
      303,313, 325, 346-348
Impaired persons, home design for,
      308–309
India, anthropometric data, *6*
Individuals in organizations, 269
Industrial psychology, 405
Information processing
  decision making and, 189–192
  memory and, 190–191
  models, 191–192
  sensory receptors and, 185–188,
      189
Infrasonic vibrations, 123
Injuries, work-related, 86
  repetitive work and, 230–232,
      355–356
Inner ear, 122–123
Instrumentation
  climate measurement, 160–161
  muscle strength testing, 60–61
Insulation, clothing, 162
Intelligibility of speech, 136
Intensity, sound, 123–124, *125*
Interactions, personal, 252–254
International Spelling Alphabet, 137
Interoceptors, 185, 187
"Intimate distance", 253
Intra-abdominal pressure and lift
      belts, 93
Intrinsic muscles, 72
Iran, anthropometric data, *6*
Ireland, anthropometric data, *6*
Iris, 104
Isoinertial testing, 61, 70
Isometric muscle effort, 58, 70–71
Italy, anthropometric data, *6*

J

Japan, anthropometric data, *4, 6,*
      26–27
Jet lag, 280
Joints, body
  designing for mobility of, 43–45
  displacement in, 43, *43*

extensive leg and arm mobility, 33–34
hand, 34–39
knee and ankle, 48
rotations in, 34
Judgments, scaled, 218–219

K

Keyboarding injuries, 356–357
Keyboards, computer
design alternatives for, 363–364
design history, 349–353
designing for new syntax and diction with, 364–365
design solutions for problems associated with, 359–363
human factors considerations for, 353–358
input-related anthromechanical issues and, 358–359
simplified, 354–355
trays, 342
Keys, ternary versus binary, 361–362
Kilocalories)kcal, Cal), 203
Kilojoules (kJ), 203, 211
Kinesthetic proprioceptors, 143
Kitchens, 310–311
Knee
angle, 78, 78
joints, 48
Knobs, 384–385
Kyphosis, 39

L

Lactic acid, 62
Landolt rings, 116
Lapses, performance, 297
"Last and bad resort" design principle, 28
Latin America, anthropometry data, 4
Left-handed persons, 74–75
Legs
and arm mobility, 33–34
pain in, 345
Leisure time
shift work and, 299
stress and, 236

Lens, eye, 104–105
Lifting and lowering, 79, 86, 87–88, 89, 380, 381, 382. See also Load carrying, Material handling
converted to carrying, 95
disc strains and, 90–92
intra-abdominal pressure and, 93
learning safe, 94–97
Light and moderate work. See also Work
accurate, fast, skillful activities, 226–232
physiological and psychological principles in, 216–219
repetitive, 230–232
rest breaks from, 232
suitable postures for, 222–226, 331
tiredness, boredom, and alertness at, 219–222, 232–233
Lighting
glare-free, 325–328, 397
home, 312–313
office, 325–329, 346-348
required for vision, 115, 328, 397
taking charge of, 397, 398–399
Limited trunk flexibility, 41–42
Limits, design, 28
Line of sight (LOS), 107, 108, 228–229, 398
Load carrying, 79, 81–85. See also Lifting and lowering
on both shoulders, 96
converting lifting to, 95
designing for easy, 89–97, 376, 380, 381–382
using dollies and carts, 96
using rollers and conveyors, 97
Loading, spinal, 89–92
Lombard reflex, 130
Long-term memory, 190–191
LOSEE, Line of sight angle against Ear-Eye line, 108
Loudness of sound, 123, 126
Low back pain (LBP), 89
Lowering and lifting, 79, 86, 87–88, 89, 380, 381, 382. See also Load carrying, Material handling
Lumbar spine, 41–42, 90–92, 339
Luminance, 114–115, 119, 329
Lungs, 154, 158, 199–204, 215, 241

M

Macroergonomics, 267
Management-prescribed pauses in
    work, 283
Manipulations
    design for, 337, 376, *378*
    exact, 226–227
Maslow's need hierarchy, 256–257,
    270
Material handling, 86
MAX and MIN strength values,
    71–72
Maximal voluntary contraction
    (MVC), 60
Mean value, 16, 93–94
Measurement
    Borg scales, 218, 244, *245*
    climate, 160–161
    heart rate, 205–207, *208*
    muscle strength, 59, 60–61
    psychophysical, 79, 242–245,
        246
    of satisfaction or discontent,
        218–219
    stress, 237–238
Measurements of body size, 3, *4–9*
    applications of common, *11–15*
    hand size and, 24, *25, 26–27*
    mean and average, 16
    methods, 9, *10*
    "normal distribution", 16, *24,
        25*
    of Russian and Chinese adults, 9,
        16, *17–21*
Mechanization, 241
Mechanoreceptors, 143
Meissner corpuscles, 145, 188
Memory, 190–192
Mental activities
    actions and reactions, 192–196
    brain-nerve network and,
        177–185
    climate effects on, 169–170
    decision making, 189–192
    diversity versus monotony in,
        220
    effects of heat on, 324
    information processing,
        185–188, *189*
    mental workload, 238–241
Metabolism, 167–168, 201–202
    basal, 203
    body posture and, 223
    byproducts, 202–203
        respiratory exchange quotient
            (RQ), 204–205
        resting, 203
        work, 203–204
Microsleeps, 296
Middle ear, 122, 138–139
Mitochondria, 54, 201
Mobile phones, 360
Mobility
    actual, 45
    "convenient", 45
    designing for, 43–45
    hand, 34–39, 47–48
    of impaired and elderly persons,
        308
    leg and arm, 33–34
    during pregnancy, 308
    ramps and stairs for, 217
    rotations in body joints, 34
    spine, 39–43
    during work, 31–33
    workspaces and, 45–49
    of young children, 308
Monitors, computer, 329, 335,
    336–337
Monotony versus diversity in work,
    220, 242
Morning type individuals, 279
Mothers, home design for, 307–308
Motion.
    design for simple control of, 183
    excessive, 33
    eye, 106–107
    hand workspaces, 47–48
    muscular co-contraction and,
        57–58
    muscular dynamic and static
        efforts, 58
    office workstations designed for,
        337–338
    preferred, 45
    ranges, everyday, 45, *46,* 338
    reach envelopes, 45, 47
    sensing, 141–143
    static and dynamic strength
        exertion, 70–71
    time, 149–150, 195, 315
    work in, 28, 31–33
    workspaces and, 45–49
Motivation
    behavior and, 254–260
    design for performance and,
        268–271
    expectance and, 393
    job performance and, 392–393
    needs-based, 392–393

Motor nerve cells, 180
Motor nerves, 181
Motor units, 181
Muffs, ear, 135
Muscle pump, 209, *210*
Muscles
    anthromechanics of, 58
    arm, 63, *65,* 72
    in athletics and sports, 59
    basic physics of, 53
    co-contraction, 57–58
    components, 53
    contraction, 54-58
    contractile microstructure of,
        54–55
    controlled strength tests, 59
    dynamic and static efforts,
        58–61
    eye, 104
    fatigue and recovery, 61–63
    hand, 35, *36, 37*
    hard labor and, 199–200
    horizontal push forces and,
        63, *64*
    innervation, 182–183
    intrinsic and extrinsic, 72
    maximum voluntary contraction
        (MVC), 60
    motor units, 181
    "power factory", 53–54
    repetitive work and, 230–232
    sarcomere, 55, *55,* 56, *56*
    sitting effect on, 331–333
    situational conditions affecting
        strength, 60, *61,* 63
    during sleep, 292–293
    smooth and cardiac, 56
    strength, 58–61
    striated skeletal, 55
    tension, 56
    testing strength of, 59–60
    used in typewriting, 357
    work requirements on, 59
Music, 126–129
Muzak, 127
MVC (maximal voluntary
        contraction), 60
Myopia, 111, 112
    night, 114
Myosin, 54, *54,* 54–55

N

Nanoergonomics, 364, 368

Naps, 297
National Institute of Occupational
        Safety and Health
        (NIOSH), 86, 89
Natural grasp, 74
Near East, anthropometry data, *4*
Nearsightedness, 111
Neck pain, *344–345*
Needs-based motivation, 392–393
Nerves
    afferent signal transmission, 147,
        179–180, 194, 196
    auditory, 122
    autonomic nervous system, 184
    -brain network, 177–185
    carpal tunnel, 35, 37
    central nervous system (CNS),
        141–143, 184, 185, 188
    efferent signal transmission,
        179–180, 194, 196
    feedforward/feedback loop,
        182–183
    free endings, 144
    impingements, 179–180
    muscle, 54
    neurons, 180, *181*
    peripheral nervous system
        (PNS), 184, 188
    sensory receptors, 185
    somatic nervous system, 184
    spinal cord, 179–180
    taction nerve sensors, 144–145
Netherlands, anthropometry data,
        6, 6–7
Neurons, 180, *181*
Neurophysiology, 192
Neurotransmitters, 180
Night vision, 113
Night work. *See also* Shift work;
        Working hours
    need for sleep, 291–294
    performance and health
        considerations in,
        295–297
Nociceptors, 143
Noise. *See also* Sound
    annoyance, 129
    barriers, 133–134
    effect on task performance, 130
    effects on hearing, 129–130
    -induced hearing loss, 131–133
    planning for "no", 133
    protection from, 32-135, 318
    shouting in, 130–131
    warning signals, 137–138

Nonnormal datasets, 94
Non-REM sleep, 293–294, 296
Nonverbal communication and
     computers, 364
Non-Western work postures, 45, *46*
Noradrenalin, 235
Nordic Questionnaire, 218–219
Normal distribution of data, 16,
     24, 25
     non-, 94
North America, anthropometry data,
     4, 26–27
Nose, 106, 148

O

Obesity, 3
Occipital lobe, 178
Offices. *See also* Keyboards,
     computer; Workplaces
     aesthetics, 320–321
     climate, 321–325
     evaluating different designs for,
          319
     flexibility via new technology,
          319
     furniture and equipment,
          329–342
     home, 313–314, 342
     lighting, 325–329, 346-348
     office landscape versus
          individual, 317–318
     rooms in, 317–321
     stepwise design, 318–319, *320*
     workstation design, 335–342,
          343–345
Olfaction, 147, 148
Operator strength, 71
Optic nerve, 105
Organizational behavior, 254
Organizations
     APCFB model of, 269–270
     climate and culture of, 266–267
     conduits in, 265–266
     designed for motivation and
          performance, 268–271
     elements of, 263
     fixed structures of, 265
     flexible, 265, *266*
     human-centered, 267, 273–274
     human-system integration (HSI)
          in, 267
     individuals in, 269

     rules and guidelines in, 254,
          263, 266
     strategies of, 263–264
     structure of, 264–265
Outer ear, 122
Overexertion pathomechanics,
     357–358
Overhead work, *62,* 62–63
Overload versus underload, work,
     239–240, 241–242
"Overshoot" phenomenon, 145
Oxygen consumption at work, 204,
     205–206

P

Pacinian corpuscles, 188
Pain
     back, 89, *344*
     low back, 89
     neck, *344–345*
     sensing, 146, 149
     shoulder, *345*
     stress and, 236
"Paperwork factory", 317, *318*
Partitions, office, 317–318
Passageways in homes, 309
Passive hearing-protection devices
          (HPD), 134–135
Pathomechanics, overexertion,
     357–358
Pedal push force, 78, *78*
Pelvis, 339
Peripheral nervous system (PNS),
     184, 188
Permanent threshold shift (PTS),
     129–130
Permeability, clothing, 162
Personal interactions, 252–254
     getting along with others in,
          391–395, *396*
Personal relationships, distances
     in, 253
Personal space, 252–253
Personnel selection and training,
     209, 255, 391, 405
Phases, sleep, 292–294
Photometry, 329, 347–348
Physical workload, 241
Physiological principles
     of muscular work, 53–58,
          199–200, 216–219
     stress reactions, 235

Pitch, 123
Plugs, ear, 135
Pointer displays, 386–387
Policies and procedures,
        organizational, 266
Population, composite, 24
Postures, 28, *29,* 31, *32*
    conversion to functional, 29
    head and neck, 227-228
    keyboarding, 353
    load lifting and carrying, 90–92
    lumbar spine, 339
    sitting, 32, *33,* 223, 225–226,
        233, 312 *32,* 329,
        331–334, 337-339, 346-
        348, 353
    standing, 45
    suitable for light and moderate
        work, 222–226, 233, 331
Power, muscular, 53–54
Preferred motions, 45
Pregnancy
    design for, 308
    skeletal adjustments in, 42–43
Presbyopia, 104
Prevention of noise-induced hearing
        loss, 132–133
Prickly heat, 166
Principles of design, 27–28
Psychology
    industrial, 405
    needs-based motivation,
        392–393
    principles of light and moderate
        work, 216–219
Psychophysical tests, 79, 242–245,
        246
Psychosocial work factors, 254,
        255, 298
Psychrometers, 159
"Public distance", 253
Pulling and pushing, 78–79, *80*
    converting, *95*
Pupil, 104, 110, 112, 115

Q

Quality of work life, 255–256,
        272–273
QWERTY keyboards, 349–353,
        359. *See also* Computer
        keyboards
    design changes, 359–360

R

Radiant temperature, 160
Radiation, heat exchange by,
        155–157, 323
Rapid eye movement sleep, *see*
        REM
Ramps and stairs, 217
Range
    fit, 28
    motion, 45, *46,* 338
    strength data, 63-65, 71-72
Reach envelopes, 45, 47
Reaching, 376, *379*
Reaction time
    simple 149–150, 194–195
    complex 150, 195
Recovery, muscle, 61–63
Reflexes, 184
Remodeling of homes, 309
REM sleep, 293–294, 296
Repetitive work, 230–232, 355–356
Resetting biological clocks, 280
Respiratory exchange quotient (RQ),
        204–205
Response time, 149–150, 195–196
Rest breaks, 211, 232, 281–283
    microsleeps as, 296
Resting metabolism, 203
Retina, 104–105
Rewards, job, 258–260
Rods and cones, eye, 105, 115
Rollers and conveyors, *97*
Rotations
    in body joints, 34
    of shift work, 300, *301*
Round knobs, 384–385
Ruffini organs, 141, 185
Russia, anthropometric data, *7,* 9,
        16, *17–21, 26–27*

S

Safety
    bathroom design and, 311–312
    lifting, 94–97
Sarcomere, 55, *55,* 56, *56*
Satisfaction, job, 218–219, 222,
        232–233, 257, 270–273,
        391–392
    stress and, 237
    Two-Factors theory and,
        257–258, 270

Saudi Arabia, anthrometric data, 7
Scaled judgments, 218–219
Scale markers, 387
Sciatic nerve, 180
Sclera, 103
Seat pans, 339–340, *341*
Sensation
    adaptation and speed, 188
    body balance, 142–143
    body movement, 141–143
    cold and warm, 153–158
    electricity, 145–146, 149
    exteroceptors, 187–188
    information processing and,
        185–192
    interoceptors, 185, 187
    modifying input signals, 188,
        189
    olfactory, 147, 148
    pain, 146, 149
    reaction and response times,
        149–150
    receptors, 185
    skin receptors, 185, *186*
    taction, 143–145
    temperature, 145, 147–148
    vestibulum in, 142
Shift work, 297–302. *See also* Night
        work
    avoiding complications of, 299
    circadian rhythm and, 298
    free time and, 299
    health concerns regarding, 298
    patterns, 300
    rotations, 300, *301*
    sleep and, 302, 303
    suitability of systems, 302
Shivering, 167–168
Sholes' keyboard, 350–352, 362
Short-term memory, 190
Shoulder pain, *345*
Shouting in noise, 130–131
Showers, 312
Simple reaction times, 149-150,
    194-195
Singapore, anthropometriy data, 7
Sitting
    active, 338
    comfortable, 331, 333–334, 348
    furniture and equipment for,
        329
    posture, 31, *32, 33,* 223,
        225–226, 331–333
    versus standing, 331, 372–376
    too much, 226, 330–331

Situational variables in muscle
        strength, 60, *61,* 63,
        69–70
Size of visual targets, 108–109
Sizing the workplace to fit the body,
        369–370, *371–372*
Skeletal muscle, striated, 55
Skin
    sensory receptors, 143-145,
        187–188
    thermoregulation and, 153–154,
        167–168, 322
Sleep
    circadian rhythms and, 153,
        277–281, 291–292
    deprivation, 280–281, 296–297
    inertia, 297
    micro-, 296
    napping, 297
    night work performance and,
        295–297
    phases, 292–294
    REM and non-REM, 293–294,
        296
    requirements, 291–292, 296
    shift work and, 302, 303
Sliding time, 286
"Slipped" disc, 90
Sloan letters, 116
"Smart" software, computer,
        365–366
Smell, sense of, 147, 148
Smooth muscles, 56
Snellen letters, 116
Social contracts, 268
Social distance, 252–253
Software, computer, 365–366
Somatic nervous system, 184
Sound, 123–129, 397, *400. See also*
        Hearing
    damaging, 132
    noise, 129–139
    recognition by computers, 363
Sound pressure level (SPL), 124,
        125, 126, *127*
Speech communications
    components of, 136–137
    intelligibility of, 136
    sound recognition by computers,
        363
Speed and adaptation, 188
Spinal loading, 89–92
Spine and spinal cord
    adjustments for pregnancy,
        42–43

back problems and load
    handling, 89
brain stem and, 178–179
disc strains in the bent, 90–92
feedforward/feedback loop,
    182–183
limited trunk flexibility and,
    41–42
lumbar, 41–42, 90–92, 339
mobility, 39–41
nerves, 179–180
reflexes and, 184
"slipped" disc, 90
Spontaneous pauses in work,
    282–283
Sri Lanka, anthropometry data, 8
Stairs and ramps, 217, 309–310
Standard deviation, 16, 93, 94
Standing
    erect, 331
    on one foot, 49, 374
    versus sitting, 331, 372–376
Static electricity, 402
Static muscle effort, 58
Static postures, 31, 32
Statics and dynamics, 28, 29,
    97–98
    muscular strength and, 58–61,
        70–71
Static torque exertion, 76, 76
Static work, 209, 210, 212, 222
Steady-state work, 206, 208
Strategy, organizational, 263–264
Strength
    "average user", 93–94
    foot, 76–77
    hand, 72–75
    individual factors, 69
    lifting and lowering, 79, 86
    load carrying, 79, 81–85
    maximal or minimal strength
        exertion, 71–72, 98
    muscular, 58–61, 63, 69–70
    operator, 71
    psychophysical tests, 79
    pulling and pushing, 78–79, 80
    static and dynamic exertions,
        70–71
    values, critical, 94
    whole body, 75–89
Stress
    coping with, 236
    defined, 235
    effects of, 237
    eliminating, 236–237
    emotions and, 235–236

good, 237, 271
happy employees and, 254, 260
at leisure, 236
of long working hours, 282
measurement of, 237–238
mental workload and, 238–241
physiological reactions to, 235
psychophysical assessments of
    workload, 242–245, 246
stressors causing, 235
underload and overload in,
    241–242
at work, 235–238
Striated skeletal muscle, 55
Strong and weak operators, 71
Structures, organizational, 264–265
Sweating, 165, 166, 323
    stress and, 236
Synapses, 180–181
Synchronized biological clocks,
    279–280, 291
Syncope, heat, 166
Syntax, designing for new, 364–365

T

Taction, 143–145
    cold and, 169
    designing for, 146–150
Taiwan, anthropometry data, 22–23
Taking charge, 394, 395, 397,
    398–402
Targets, visual, 108–109
Tasks
    complex, 243
    demands and job rewards,
        258–260
    demands and resources, 239
    links between person and, 33
    performance, 240–241, 256
    performed while sleep-deprived,
        281
    specialization, 396
Taste, sense of, 147
Taylorism, 221–222, 255, 258, 396
Teamwork, 253–254
Technology, office, 319, 359
    designing for new syntax and
        diction using, 364–365
    "smart" software, 365–366
Telecommuting, 286
Temperature. See also Climate
    acclimatization, 164–165, 324
    air, 160, 170, 324–325, 401–402

globe (GT), 160
human thermoregulation and, 153–158
measurement of, 159
metabolic byproducts, 202–203
office, 321–325
radiant, 160
sensations, 145, 147–148
taking charge of, *401–402*
Temporary threshold shift (TTS), 129
Tendon sheaths, 37–39, 358
Tenosynovitis, 356
Tension, muscle, 56
Ternary keys, 361–362
Territoriality, 253
"Texter's thumb", 360
Thailand, anthropometry data, *8*
Thalamus, 178
Thermoreceptors, 143
Thermoregulation, human, 153–158, 167, 322
Thrust, foot, 77–78
Tiredness, 219–220
Toggle switches, 384, *385*
Toilets, 312
Tones
  common difference, 128
  concurrent, 128–129
Tools, hand, 74–75, 229, *230,* 376, *377*
Training people, *94-95*, 100, 183, 192, 211, 238, 355, 359, 405
Transducers, 192–193
Treadmills, 207, 209
Trichromatic vision, 116
Trunk
  flexibility, 41–42
  twisting, 223
Turkey, anthropometry data, *8*
24-hour cycles, 279
Two-Factors theory, Hertzberg, 257–258, 270, 272
Typewriters, 350–352

U

Ultrasonics, 123
Underload versus overload, work, 239–240, 241–242
United Kingdom, anthropometry data, *8*

United States, anthropometry data, *8, 8–9*
Utricle, 142

V

Vacation time, 272–273
Variances in body sizes, 3, 24–28
Variety, design for, 338–339
Vasoconstriction, 169
"Veiling glare", 112
Ventilation, clothing, 163
Vertebrae, spinal, 39–41, 179–180
  biomechanics of, 92–93
  disc strains in, 90–92
Vestibulum, 142–143, 187
Vietnamese Americans, anthropometry data, *9, 26–27*
Vigilance and event frequency, 220–221
Visceral system, 184
Vision. *See also* Eyes
  accommodation, 104, 109–111
  acuity, 115–116
  color perception, 105, 113, 116–117
  in computer use, 329, *330,* 335, 336–337
  cones, *378*
  designing illumination and, 117
  in dim and bright viewing conditions, 113–117
  diopter, 109, *110,* 110–111
  ear-eye (EE) line, 108, 227–229, *398*
  focusing, 109–110
  glare effects on, 325–328
  light required for, 115, 328, 397
  line of sight (LOS), 107, *108,* 228–229, *398*
  looking down on the job in, 108, *378*
  luminance and, 114–115, 329
  problems, 110–113
  size of visual target and, 108–109
  trichromatic, 116
  visual field, 106
Visual acuity, 115–116
Visual control system, 105
Vitreous humor, 104–105, 112
Voice communications, 136

W

Warm and cold sensations. *See* Cold
    and warm sensations
Warning signals, 137–138
Washbasins, 312
Water intake, 166
Weight, body, 3, *5–9,* 370, *371*
Western work postures, 45, 46
Wet Bulb Globe Temperature
    (WBGT), 160-161
Wet bulb (WB) thermometers, 159,
    160–161
White-collar workers, 238–239
Whole body strength, 75–76
Wind chill, 163, *164*
Windows, 310, 312–313
Women and skeletal adjustments for
    pregnancy, 42–43
Work. *See also* Hard labor; Light
    and moderate work;
    Motion; Night work;
    Shift work
  avoiding exhausting, 211
  balanced with personal life,
    272–273, 394
  -caused pauses, 283
  in cold environments, 167–170
  days on/off, 283–284
  demands, classifying, 206–207
  disguised pauses in, 283
  energy efficiency at, 210–211
  engaged to, 392
  getting along with others at,
    391–395, *396*
  heart rate during, 205–207, *208,*
    217
  in hot environments, 165–166
  load effects, 217
  management-prescribed pauses
    in, 283
  measuring heaviness of, 204
  metabolism, 203–204
  monotonous, 220, 221–222, 242
  in motion, 31–33
  muscular requirements, 59
  overhead, *62,* 62–63
  overload versus underload,
    239–240, 241–242
  oxygen consumption at, 204,
    205–206
  personal thermal comfort and,
    161–165
  pride in, 393

  pushes and pulling in, 79
  repetitive, 230–232, 355–356
  rest breaks from, 211, 232,
    281–283
  rewards, 258–260
  satisfaction at, 218–219, 222,
    232–233, 237, 257–258,
    270, 271–273, 391–392
  spontaneous pauses in, 282–283
  static, 209, *210,* 212, 222
  steady-state, 206, *208*
  stress at, 235–238, 271
  suitable postures at, 222–226,
    233
  task demands and resources, 239
  task performance, 240–241
  tiredness, boredom, and
    alertness at, 219–220,
    242
  vigilance and event frequency
    in, 220–221
  worthwhile doing, 391–392
Working hours. *See also* Night
    work; Shift work
  brief, 285
  circadian rhythms and, 277–281
  compressed, 286–288
  daily and weekly, 283–288
  flextime, 286
  history of, 284–285
  long, 285–286, 286–288
  rest pauses and time off, 211,
    232, 281–283
  sleep-deprivation and, 280–281
  stress of long, 282
Workload, mental, 238–241
  psychophysical assessments of,
    242–245, *246*
Workplaces. *See also* Offices
  displays and controls, 382–387
  division of labor in, 255
  getting along with others in,
    252–254
  improving, 396–397
  motivation and behavior in,
    254–260
  personal space in, 252–253
  quality of work life in, 255–256,
    272–273
  relocations of, 251–252
  sitting versus standing in, 331,
    372–376
  sized to fit the body, 369–370,
    371–372
  teamwork in, 253–254

Workspaces, 45–49
Workstations, office, 335–342,
    343–345, 366. *See also*
    Computer keyboards
Wrist pain, *345*

Z

Z-lines, 55